Atomic Structure and Lifetimes
A Conceptual Approach

This book presents a new approach to the teaching of introductory graduate courses on atomic structure.

The author's approach utilizes conceptually powerful semiclassical modeling methods, and demonstrates the surprising degree to which the Maslov-indexed EBK quantization elucidates the quantum mechanical formulation of level energies and lifetimes. It merges this with an update and extension of the semiempirical data systematizations developed by Bengt Edlén to describe complex atoms, and adapts them to include the specification of lifetimes. The text emphasizes the historical basis of the nomenclature and methodologies of spectroscopy; however, the interaction mechanisms are presented deductively, based on modern quantum mechanical and field theoretical models rather than tracing their indirect paths of discovery. Many worked examples provide applications to areas such as astrophysics, hyperfine structure, and coherent anisotropic excitation. The book presents a firm foundation for specialists in atomic physics, as well as a capstone application for specialists in astrophysics, chemistry, condensed matter, and other related fields.

LORENZO J. CURTIS received his Ph.D. from the University of Michigan in 1963. He holds the rank of Distinguished University Professor and the designation Master Teacher at the University of Toledo. His primary research interests involve time-resolved atomic spectroscopy and the structure of highly ionized atoms. He has held visiting scientist positions in Sweden, Denmark, Norway, Germany, and France, as well as at Princeton University, Argonne National Laboratory, and Brookhaven National Laboratory in the US. He has published more than 200 refereed scientific articles and is a Fellow of both the American Physical Society and the Optical Society of America. In 1999 he was awarded the degree Philosophiae Doctorem Honoris Causa by the University of Lund in Sweden.

Atomic Structure and Lifetimes

A Conceptual Approach

Lorenzo J. Curtis
University of Toledo

CAMBRIDGE UNIVERSITY PRESS

PUBLISHED BY THE PRESS SYNDICATE OF THE UNIVERSITY OF CAMBRIDGE
The Pitt Building, Trumpington Street, Cambridge, United Kingdom

CAMBRIDGE UNIVERSITY PRESS
The Edinburgh Building, Cambridge CB2 2RU, UK
40 West 20th Street, New York, NY 10011–4211, USA
477 Williamstown Road, Port Melbourne, VIC 3207, Australia
Ruiz de Alarcón 13, 28014 Madrid, Spain
Dock House, The Waterfront, Cape Town 8001, South Africa

http://www.cambridge.org

First published 2003

Printed in the United Kingdom at the University Press, Cambridge

Typefaces Times 10/13 pt. and Helvetica   *System* LaTeX 2_ε [TB]

A catalog record for this book is available from the British Library

Library of Congress Cataloging in Publication data

Curtis, L. J. (Lorenzo J.)
  Atomic structure and lifetimes: a conceptual approach / Lorenzo J. Curtis.
    p.   cm.
  Includes bibliographical references and index
  ISBN 0 521 82939 9 – ISBN 0 521 53635 9 (pbk.)
  1. Atomic structure. I. Title.
QC173.4.A87C87 2003
539′.14–dc21   2003043962

ISBN 0 521 82939 9 hardback
ISBN 0 521 53635 9 paperback

Dedicated to Maj Rosander Curtis

Contents

Preface

The study of atomic spectroscopy was central to the development of modern quantum mechanical theory. Thus, applications to the field of atomic physics are an important feature of any course in quantum mechanics. However, the converse is not necessarily true – a comprehensive course in atomic physics is not simply a study of quantum mechanics. The aspects of atomic physics that are most useful as illustrative examples for a quantum mechanics course usually involve either hydrogen or helium, and the methods used for these systems are very specialized and not particularly exemplary of the methods used for the study of complex atoms and ions. Graduate atomic physics courses often substitute for increased complexity of the atomic system studied an increased elegance in the theoretical representation of the one-electron system. Thus, a course on the Schrödinger theory of hydrogen is followed by a course on the Dirac theory of hydrogen, and that in turn is followed by a course on the quantum electrodynamic theory of hydrogen.

In the study of complex, many-electron spectra, the precision of the optical measurements greatly exceeds the accuracy that can be obtained with even the most sophisticated of currently available theoretical codes. Therefore, predictions based on these very high precision measurements usually rely on semiempirical methods, often utilizing simple semiclassical or parametrized single-particle models.

The approach adopted here will be to provide conceptual and intuitive insights into quantum mechanical phenomena, drawing on measured data, semiclassical models, and semiempirical parametrizations that reveal unexpected regularities among various atomic systems. While quantum mechanics has delegitimized the hope of *ab initio* quantitative predictability based on conceptual pictures, there is more to physics than mathematics. Physics is **not** a way of thinking about mathematics – mathematics is **one** way of thinking about physics (and sometimes not the best initial way).

Although it can be argued that semiclassical models have become anachronistic to a modern course in quantum mechanics, such a claim should not be extended to include a course in atomic physics. While atomic physics can still occasionally provide a testing ground for quantum mechanics, its primary role is to produce accurate measurements of quantities that provide enabling data for other sciences. Conceptual models and semiempirical systematizations are essential tools for predictive interpolation and extrapolation in the fulfillment of that function.

In order to efficiently serve a diverse student population, the subject matter presented here was developed to serve a broad spectrum of backgrounds. A basic undergraduate physics background is assumed, that includes introductory quantum mechanics, theoretical mechanics, and electrodynamics. However, many of the semiclassical and semiempirical developments could be appreciated by a well-prepared undergraduate physics major. At the same time, the choice of subject matter includes many topics that could be useful to an advanced graduate student in astrophysics or condensed-matter physics, as well as in atomic, molecular and optical physics.

A few notes regarding the organization of the subject matter are in order. For completeness, all concepts regarding the semiclassical modeling are introduced in Chapter 2, making it somewhat longer than the other chapters. The semiclassical concepts are applied to specific applications in later chapters, and it may be useful to return to portions of Chapter 2 in the context of the later applications. The subject matter was developed for presentation in a one-semester course, but some selection among topics may be necessary to fit within that time frame. In this context, Chapters 1–9 provide a general foundation for Chapters 10–13, but these latter chapters are relatively independent of each other.

It is a pleasure to acknowledge many contributors to the writing of this book. I would like to thank my home institution, the University of Toledo, for providing me with a stimulating environment, and the freedom to travel extensively and to grow professionally in the secure comfort of a tenured position. I am grateful to the U.S. Department of Energy for their long and generous support of my research, and for allowing me great latitude in pursuing my own curiosity. I am indebted to Professor Indrek Martinson of the University of Lund for providing me with a second research home, where the phenomenological beauty and pedagogical virtue of the semiempirical exposition of spectroscopic data were revealed to me. Many aspects of the book are a direct consequence of my long and illuminating association with the late Professor Bengt Edlén at Lund.

Many colleagues and collaborators have made significant contributions to my understanding of the field and to the writing of the book. I want to thank Gordon Berry, William Bickel, David Ellis, Alan Hibbert, Richard Irving, Philip James, Richard Schectman, Richard Silbar, Constantine Theodosiou, and Elmar Träbert, as well as my many other collaborators and students. I am grateful to Simon Capelin, Lucille Murby, and Brian Watts of Cambridge University Press for their help and encouragement and for the opportunity to present my own personalized pedagogic approach. Finally, I express my deep appreciation to my wife Maj, whose support and patience made the writing of the book possible.

Physical constants and useful interrelations

Quantity	Symbol	Value(uncertainty)	Units
speed of light in vacuum	c	2.997 924 58(exact)	10^8 m/s
Planck constant	h	6.626 068 76(52)	10^{-34} J s
	$\hbar = h/2\pi$	1.054 571 596(82)	10^{-34} J s
	$\hbar c$	1973.2705(13)	eV-Å
electromagnetic coupling constant	α	1/137.035 999 76(50)	
Bohr radius	a_0	0.529 177 2083(19)	Å
Rydberg constant	Ry	13.605 691 72(53)	eV
	Ry/hc	109 737.315 685 49(83)	cm^{-1}
electron mass	m	9.109 381 88(72)	10^{-31} kg
	mc^2	510 998.902(21)	eV
proton : electron mass ratio	M/m	1836.152 6675(39)	
electron g-factor	g_e	$-2.002\ 319\ 304\ 3737(82)$	
proton g-factor	g_p	5.595 794 674(29)	
Bohr magneton	μ_B	9.274 008 99(37)	10^{-24} J/T
		5.788 381 749(43)	10^{-5} eV/T
Coulomb constant	Ke^2	14.399 652	eV-Å
Boltzmann constant	k_B	1.380 6503(24)	10^{-23} J/K
			10^{-5} eV/K

Interrelations:

$$\alpha = Ke^2/\hbar c; \qquad a_0 = \hbar^2/mKe^2; \qquad Ry = mc^2\alpha^2/2,$$

thus

$$\hbar^2/m = 2Ry\, a_0^2; \qquad Ke^2 = 2Ry\, a_0$$

1

Introduction

If, in some cataclysm, all of scientific knowledge were to be destroyed, and only one sentence passed on to the next generations of creatures, what statement would convey the most information in the fewest words? I believe it is the atomic hypothesis (or the atomic fact) that all things are made of atoms – little particles that move around in perpetual motion, attracting each other when they are a little distance apart, but repelling upon being squeezed into one another.

– Richard P. Feynman [97]

1.1 Atomic physics is more than quantum mechanics

With the stirring testimonial above [97] from one of the foremost scientific minds of our time, why is it that the subject of atomic structure is relegated to a chapter near the end of most elementary physics textbooks? Introductory physics texts tend to discuss gravitational interactions extensively, yet most of the examples treated are atomic in nature. Since "weightlessness" occurs when there is no floor to provide atomic charge polarizations to oppose a gravitational attraction, weight must be considered an atomic phenomenon. Barring the remote possibility of experiencing the huge gravitational gradients predicted near a black hole, no one is ever directly injured by a gravitational force, but rather by the atomic polarization that ultimately opposes it. Why is so important a topic as atomic physics not given an early and thorough conceptual presentation?

Part of the answer to this question lies in discovery-oriented pedagogic tendencies. Scientific facts are deemed inextricable from scientific inquiry. The facts are taught in the order that they were discovered, in the context of those experiments that sorted out the valid concepts from among the misconceptions (which, unfortunately, requires programmatic obfuscation to make the misconceptions seem initially plausible). Thus, the first course in physics deals with 18th-century mechanics, and the second course deals with 19th-century electromagnetics. If time permits there is an addendum describing how the gross mistakes that were made in the 18th and 19th centuries were corrected at the beginning of the 20th century. Unfortunately, the accidents of history have trapped subjects such as the relativistic origin of the magnetic field, the nature of continuum thermal radiation, photovoltaics, and atomic physics firmly in the back of the book.

However, an even larger part of the problem lies in the widely held perception that these "modern" topics require a quantitative knowledge of quantum mechanics, and this is thought to exceed the mathematical prerequisites for an elementary course. These pedagogic practices are now being questioned, and many physics educators are asking their colleagues "Is physics just an application of mathematics, or is there more to it?" Since Newton studied physics first, and this later motivated him to invent calculus, perhaps an early detailed conceptual study of atomic physics could provide the motivation for a subsequent rigorous mathematical study of quantum theory.

Because the study of atomic spectroscopy provided much of the impetus for the development of quantum mechanics, most textbooks on quantum mechanics include extensive examples drawn from the field of atomic physics. However, this does not imply that a textbook on atomic structure should contain within it a course in quantum mechanics. Many of the examples drawn from atomic physics that are most suitable for a quantum mechanics course involve the hydrogen atom, which is a special case not particularly well-suited for illustrating the structure of complex atoms. While quantum mechanical theory is an essential part of the study of atomic structure, there are many other important aspects of this subject that can be concealed by an overemphasis on the details of the quantum mechanical formulation.

Historically, one of the most appealing models for the formulation of mechanics was the motion of the planets as observed through their illumination by light from the Sun. Since the energy of optical photons is very small compared to the mass energy of a planet, these observations are very nearly passive. Thus the positions, speeds, and accelerations of the planets can be followed instantaneously, without being altered by the act of observation. The convenience of this characterization is in sharp contrast to examples (such as an electron illuminated by an x-ray photon) in which the energy of the probe is much greater than the mass energy of the object observed, and the act of observation removes that particular object from further consideration.

Thus, one of the strongest motivations for embedding the study of atomic physics inside a rigorous quantum mechanical presentation has little to do with quantization, but has everything to do with its formulation in terms of position probability densities rather than forces. Since planets can be observed passively with photons and electrons cannot, we pedagogically isolate the electrons in the 20th century instead of updating the archaic 17th-century formulation of the planetary Kepler problem to one of position probability densities. Overcoming this historical bias is one of the goals of this book.

1.2 Trajectories versus probabilities

The various pedagogic formulations of physics are often characterized as either "classical" or "quantum mechanical." Most of the differences between these presentations arise not from quantum mechanical or correspondence limit requirements, but rather from a nonessential heuristic tendency to treat macroscopic systems by instantaneous quantities and microscopic systems by time-averaged expectation values. In many cases modern theoretical developments now indicate that the historical assumptions that led to these characterizations may have been ill-founded, and they sometimes unnecessarily fragment physical concepts. A mathematically simple, pedagogically transparent approach will be presented here that uses position probability densities to describe both macroscopic and microscopic systems. This approach will be applied to a number of familiar examples, sometimes with surprising results.

This observational bias led to many misconceptions that required centuries to correct. For example, classical probabilistic formulations were inhibited by the doctrine of Laplacian determinism. In 1776 Pierre Simon Laplace asserted [105] that "The present state of the system of nature is evidently a consequence of the preceding moment, and if we conceive an intelligence that at a given instant comprehends all the relations of the entities of this universe, it could state the respective positions, motions and general effects at any time in the past or future. . . . So it is that we owe to the weakness of the human mind one of the most delicate and ingenious of mathematical theories, the science of chance or probability." The inherent fallacy of this view was emphasized in 1903 by Henri Poincaré in his statement [165] that "It may happen that small differences in the initial conditions produce very great ones in the final phenomena – prediction then becomes impossible." In 1887 Poincaré had entered a contest sponsored by the King of Sweden that contained a challenge to show rigorously that the solar system is dynamically stable. It at first appeared that Poincaré had succeeded, but an error was found. Poincaré's correction of that error is generally regarded as the birth of chaos theory. This indicates the limitations of the linearized approximations that were considered by Laplace. The development of quantum mechanics in 1924 with its inherent Heisenberg uncertainty principle showed clearly that there is a fundamental limitation on the accuracy to which position and velocity can be measured simultaneously. Even when applied to macroscopic systems, modern considerations of quantum gravity indicate that space and time themselves break down for very short distances.

Laplace himself did not seem completely comfortable with Laplacian determinism. A recurring theme of Laplace's work was his lifelong tendency to couple the sciences of probability and astronomy. Consistent with the spirit of Laplacian determinism, probability was viewed by him as a means of repairing the defects in knowledge. However, there are tantalizing passages scattered throughout his writings [105] that suggest that he may have had an inkling (or perhaps a repressed belief) that there are inherently random processes in nature that are not merely the result of our ignorance.

One aspect of his work in which Laplace may have "pried open the first chink in the armor of deterministic physics" was his application of probability to demography and actuarial determination. This inspired his Belgian pupil Adolphe Quetelet to formulate the study of "Staatswissenschaft," which was the forerunner of the modern statistical social sciences. Quetelet's work [166] was heralded as a cure for societal ills, and was championed by the social reformer Florence Nightingale. This subsequently led James Clerk Maxwell, through his reading of an 1850 essay on Quetelet's work written [120] by John Herschel, to adopt a strategy using Laplace's law of errors as a basis for his kinetic theory of gases. Maxwell's formulation of statistical mechanics marked a turning point in physics, since it presupposed the operation of chance in nature [105]. Thus, contrary to popular belief, the "exact sciences" here borrowed from the methods of the "social sciences" and not vice versa.

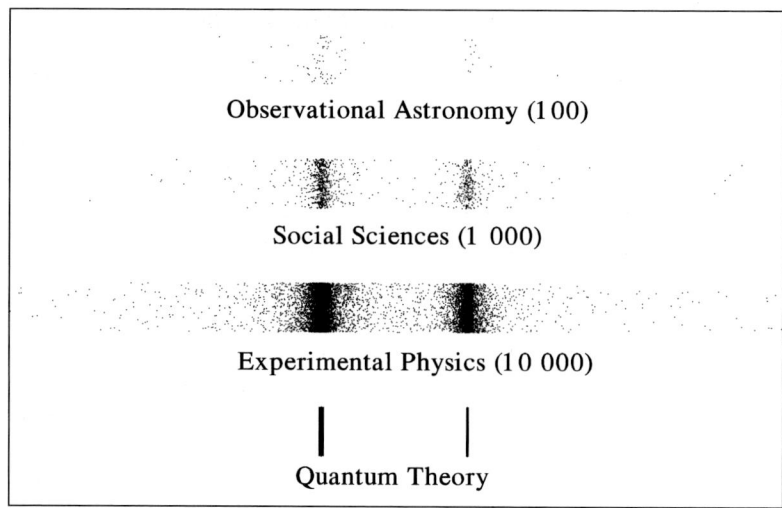

Fig. 1.1. Lorentizan scatter for a hypothetical doublet transition. The line spacing is twenty times the natural line width, and statistical sample sizes are chosen to be characteristic of various fields of inquiry.

Stripped of the mantle of Laplacian determinism, the differences between the "hard" physical sciences and the "softer" social sciences are largely reduced to a question of available statistical sample sizes. This is illustrated metaphorically in Fig. 1.1. Here a simulation is made of a hypothetical doublet transition, in which the line spacing is set at twenty times the natural line width, and the number of photon counts is chosen to match the sample size available in various fields of scientific inquiry. The spread of points was generated using an inverse Lorentzian function of a probability obtained from a random number generator. The plot labeled "Experimental Physics" includes 10 000 counts which is typical of many experiments at the forefront of a field. The plot labeled "Social Sciences" includes 1000 counts, which is the standard sample size used to obtain 3% statistical accuracy in survey research studies. The plot labeled "Observational Astronomy" includes 100 points, and is modeled on a field such as stellar atmospheres. Since the number of known stars is only about 1% of the number of atoms in a gram molecular weight, when a sample of stars is selected that exhibits a desired feature and otherwise has more likenesses than differences, the sample size is often quite small. In the plot labeled "Quantum Theory," only the positions of the two line centers are indicated. The Schrödinger equation yields the energies of the time-independent stationary states, but their radiative decay and their Lorentzian spread require the invocation of the Weisskopf–Wigner approximation. It seems clear that the probabilistic formulation provides a universally applicable technique, which a conceptual reliance on instantaneous motions only tends to fragment.

Even during the time of Laplace, position probability densities were actually (albeit unconsciously) favored over instantaneous positions and velocities in the specification of planetary interactions. Laplace, Gauss, and others calculated the perturbations of the planets by considering the time-averaged loci of their orbits smeared as rings around the Sun, as one automatically assumes when representing them in a Legendre polynomial expansion.

1.3 Semiempirical parametrization

Atomic physics has many different facets. It can be used to test fundamental theory to levels of accuracy that exceed those attainable in virtually any other field. It is also an enabling science, that provides measured structural constants that are essential to, e.g., the interpretation of observations, the design of new types of devices, and the modeling of physical processes. Without dismissing the importance of fundamental quantum mechanical theory, it is not the optimal starting point for all processes that it ultimately governs. The construction of a building is also governed by the laws of quantum mechanics, but the architect must be more concerned with the measured values for Young's modulus than with any theoretical predictions for that quantity that can be obtained from *ab initio* solution of the Schrödinger equation. Similarly, in many applications involving complex atoms, either direct measurements or semiempirical determinations are essential to obtaining the required precision.

While the development of quantum mechanics provided a thorough understanding of the underlying basis of atomic physics, with a few exceptions the accuracy of *ab initio* quantum mechanical methods lag far behind experimental capabilities for atoms more complex than hydrogen and helium. Spectroscopic accuracies are often of the order of parts in 10^8 or better, and theoretical calculations can at best provide a planning guide to definitive experimental measurements. Needed values for energy levels, transition wavelengths, ionization potentials, polarizabilities, fine and hyperfine structure splittings, transition probabilities, level lifetimes, etc., can be determined experimentally for complex atoms more precisely than they can be specified using the best currently available theoretical methods.

Thus, as the experimental methods have continued to improve, many of the semiempirical techniques used prior to the development of quantum mechanics are still in active use. Methods such as the quantum defect formulation of Rydberg series, the fine structure screening parametrization of Sommerfeld, etc., have been greatly refined, and their application can be understood in terms of simple conceptual models. While it is sometimes asserted that quantum mechanics has made conceptual models obsolete, the rejection of a simpler model because a more fundamental approach exists can be extended to a *reductio ad absurdum*. One can reject conceptual models and adopt the Schrödinger approach, but the nonrelativistic scalar nature of this formulation separates spin from space. This leads to a radial wave function that is independent of the total angular momentum, which is of course physically wrong. One could reject the Schrödinger model and adopt the Dirac approach, but this is a single-electron theory that includes the electron's own spin, but relegates spin–spin and spin–other-orbit to perturbative inclusion, and does not include second quantization. Nonetheless, just as the Dirac equation offers some conceptual insights over a "sea" of Feynman diagrams and the Schrödinger picture provides advantages over the Dirac equation in the inclusion of configuration interaction and correlation, the Einstein–Brillouin–Keller semiclassical quantization can provide some very useful insights into various aspects of the quantum mechanical structure of the atom.

2

Semiclassical conceptual models

Why didn't Isaac Newton think about the probability of getting hit on the head when he sat under the apple tree?

2.1 Classical position probability densities for periodic systems

Even for macroscopic objects, a formulation based on position probability densities offers many advantages over the specification of individual positions, velocities, and accelerations for an ensemble of many particles. Since the standard formulation of mechanics as specified from forces, masses, and accelerations does not transcend elementary physics courses, but is replaced by energy and momentum considerations at the quantum mechanical level, why not simply adopt this perspective at the outset?

2.1.1 Probabilistic formalism

If one characterizes a system of particles in some type of periodic motion in terms of kinetic, potential and total energies, their position probability density can easily be specified. If each member of a group of particles moves periodically in one dimension with a period T, and spends a time dt in a length of path dx (through which each passes twice during each cycle), and has a speed $v = dx/dt$, the probability $P(x)dx$ of finding a particular particle between x and $x + dx$ is given by

$$P(x)dx = 2dt/T = 2dx/vT. \qquad (2.1)$$

Using conservation of energy

$$E = \frac{1}{2}mv^2 + V(x) \qquad (2.2)$$

the probability is given by

$$P(x)dx = \frac{2}{T} \frac{dx}{\sqrt{\frac{2}{m}[E - V(x)]}}. \qquad (2.3)$$

The average value of any power k of x is given by

$$\langle x^k \rangle = \frac{2}{T} \int \frac{x^k dx}{\sqrt{\frac{2}{m}[E - V(x)]}} \qquad (2.4)$$

and the average value of any power k of v by

$$\langle v^k \rangle = \frac{2}{T} \int dx \left[\frac{2}{m}[E - V(x)] \right]^{\frac{k-1}{2}}. \qquad (2.5)$$

2.1.2 Example: molecule in a room (classically)

A convenient example is given by the motion of an oxygen or nitrogen molecule in a room. In a one-dimensional case, using a coordinate system centered in a room of width L with impenetrable walls, the period for a particle to make a round trip between the walls is given by $vT = 2L$. Thus the position probability density of Eq. 2.1 is given by

$$P(x) = dx/L \qquad (-L/2 \leq x \leq L/2)$$
$$= 0 \qquad (-L/2 > x > L/2). \qquad (2.6)$$

This model assumes that there are no temperature or density gradients, or other nonuniformities in the room, which could be subjected to experimental testing. One way to perform such tests would be to examine the average position ($\langle x \rangle = 0$) and the root mean square deviation therefrom ($\langle x^2 \rangle = L^2/12$).

2.2 Quantum mechanical oscillation of the localizability

To someone not misled by the false steps in the historical development of physics, the concept of an object possessing both particle and wave properties is less than perplexing. Since one seldom encounters point singularities in daily life, most objects possess structure, and thus have internal degrees of freedom that exhibit characteristic frequencies. While a billiard ball struck by a cue has a well-defined position and speed, it is also ringing with acoustic oscillations, and thereby its shape properties trace out wavelengths as it moves. By the simple act of walking, the periodic processes of heartbeat and respiration cause each of us to simultaneously exhibit particle and wave properties. Rather than instilling conceptual impediments for the sake of subsequently demolishing them, new analogies can be developed that reinforce current knowledge.

2.2.1 Conceptual inclusion of periodic processes

Since (as discussed above) all objects encountered in daily life inherently possess both particle and wave properties, the periodic localization and delocalization of electrons in quantum mechanics can certainly be made plausible. Moreover, the macroscopic concealment of these periodicities can be explained by quantum statistics, since the Pauli exclusion principle prevents the oscillating localizability of a group of fermions from occurring in cadence. This is perhaps less mysterious than the fact that bosons do this coherently, thus causing the ensemble of the whole to mimic the behavior of the individual.

2.2.2 Example: molecule in a room (oscillating localizability)

Thus, if we add to this classical picture a periodic nonlocalization of the molecules, and require that minimum localizability occurs at the walls to avoid energy losses there, the standard quantum mechanical result is obtained

$$P(x)\mathrm{d}x = \mathrm{d}x\, 2\sin^2\left(\frac{N\pi x}{L}\right)\Big/ L \qquad (-L/2 \le x \le L/2)$$
$$= 0 \qquad\qquad (-L/2 > x > L/2) \qquad (2.7)$$

for N even, with a similar expression with the sine function replaced by a cosine function for N odd. Thus, while $\langle x \rangle = 0$ is still valid, now

$$\langle x^2 \rangle = L^2[1 - 6/(N\pi)^2]/12. \qquad (2.8)$$

The position probability densities of both the classical (dashed line) and the oscillating (solid curve) localizability are shown in the top inset of Fig. 2.1. In subsequent sections the position probability densities for the simple harmonic oscillator and the hydrogen atom (Kepler problem) systems will be treated. To illustrate the similarity of the results for these systems, all three are plotted in Fig. 2.1. From this picture it is clear that, in the oscillating case, the macroscopic uniformity of the detection probability as a function of position within the box breaks down if the width of the measuring apparatus becomes too narrow to include a sufficient number of oscillations. This is an illustration of the uncertainty principle.

2.3 Einstein–Brillouin–Keller quantization

This oscillating localizability can be made quantitative by use of the Einstein–Brillouin–Keller (EBK) quantization. Unfortunately, modern textbooks tend to perpetuate an earlier flawed version of this formulation, thus depriving students of the many reliable and conceptually transparent predictions that are obtained by its correct application.

2.3.1 Historical context

Many textbooks present the quantum Hamilton–Jacobi formulation in terms of the Bohr–Sommerfeld–Wilson (BSW) quantization hypothesis. Simple flaws in the details of this hypothesis were pointed out [91] by Einstein already in 1917, and subsequently corrected [19] by Brillouin in 1926 and [129] by Keller in 1958 (hence the modern form of the semiempirical hypothesis is known as the EBK quantization). Important contributions were also made by Langer [140] in 1937 and by Maslov [149] in 1972, and this

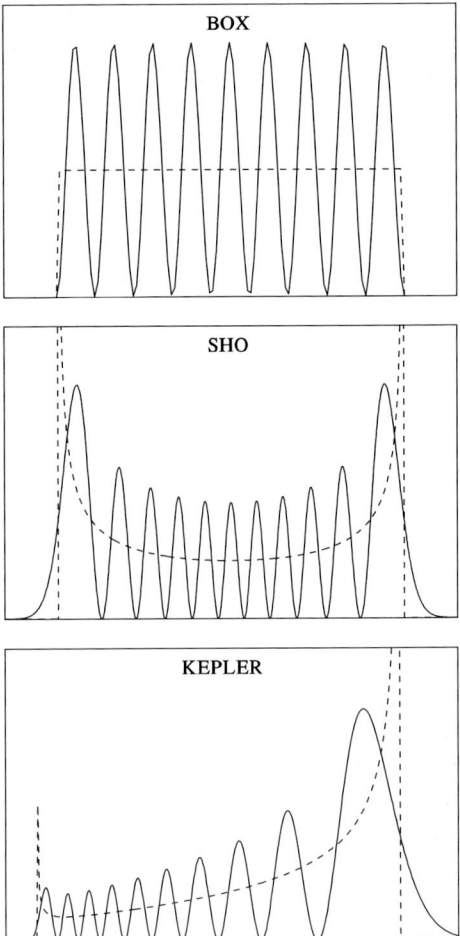

Fig. 2.1. Position probability densities for the particle in a box, the simple harmonic oscillator (SHO), and the Kepler problem.

also forms the basis for the quantum mechanical WKB approximation. The continued exposition of the flawed BSW form in modern textbooks is misleading, and conceals the usefulness of the EBK form in appropriate situations. The EBK quantization involves a contour integration over the space of a coordinate q, and its conjugate momentum p, of the form

$$\frac{1}{2\pi} \oint p\,dq = \left(n + \frac{\mu}{4}\right)\hbar \tag{2.9}$$

where $\hbar$ is the reduced Planck constant, $n = 0, 1, 2, \ldots$, and μ is the Maslov index, which is a detailed accounting of the total phase loss during one period in units of $\pi/2$. For example, each classical turning point (or caustic) and each reflection contributes one unit to μ.

2.3.2 Example: molecule in a room (with EBK quantization)

For the case of the particle in a box discussed above, each of the two walls adds both a classical turning point and a hard reflection, so $\mu = 4$, and the constant momentum p is obtained from this integral

$$p(2L) = (n + 1)2\pi\hbar. \tag{2.10}$$

Notice that, by virtue of the Maslov index, $N = n + 1$. Thus the null solution is formally avoided, and the uncertainty principle is automatically obeyed. In many common cases, such as the simple harmonic oscillator and the hydrogen atom, the Maslov index is determined from the number of classical turning points. In these cases, the Maslov index for a rotation is zero and for a libration (an oscillation between two zeros of the kinetic energy) is two.

2.3.3 Example: simple harmonic oscillator

For a simple harmonic oscillator (SHO) potential

$$V(x) = \frac{1}{2}Kx^2 \tag{2.11}$$

where $\sqrt{K/m} \equiv 2\pi/T$. Denoting the classical turning points as $\pm x_m$, the position probability density of Eq. 2.3 is given by

$$\begin{aligned}
P(x)\mathrm{d}x &= \mathrm{d}x\Big/\left[\pi\sqrt{x_m^2 - x^2}\right] && (-x_m \le x \le x_m) \\
&= 0 && (-x_m > x > x_m).
\end{aligned} \tag{2.12}$$

This probability distribution, together with a corresponding quantum mechanical case, is shown in the middle panel of Fig. 2.1. In the classical case the expectation values can be transformed to the form of Wallis' formula through the substitution $\sin\theta \equiv x/x_m$

$$\langle x^k \rangle = \frac{1}{\pi}\int_{-x_m}^{x_m}\mathrm{d}x\,\frac{x^k}{\sqrt{x_m^2 - x^2}} = x_m^k\,\frac{2}{\pi}\int_0^{\pi/2}\mathrm{d}\theta\,\sin^k\theta \tag{2.13}$$

$$\langle v^k \rangle = \frac{2}{T}\int_{-x_m}^{x_m}\mathrm{d}x\left[\frac{2\pi}{T}\sqrt{x_m^2 - x^2}\right]^{k-1} = \left[\frac{2\pi}{T}x_m\right]^k\frac{2}{\pi}\int_0^{\pi/2}\mathrm{d}\theta\,\cos^k\theta \tag{2.14}$$

which yields the results

$$\langle x^k \rangle = \langle v^k \rangle = 0 \qquad (k\text{ odd}) \tag{2.15}$$

and

$$\langle x^k \rangle = \left(\frac{m}{K}\right)^{k/2}\langle v^k \rangle = \frac{1\cdot 3\cdot 5\cdots(k-1)}{2\cdot 4\cdot 6\cdots(k)}x_m^k \qquad (k\text{ even}). \tag{2.16}$$

For $k = 2$, $\langle x^2 \rangle = x_m^2/2$. This leads to the virial theorem

$$\frac{1}{2}K\langle x^2 \rangle = \frac{1}{2}m\langle v^2 \rangle = \frac{1}{4}Kx_m^2. \tag{2.17}$$

Since the motion is between two turning points the Maslov index $\mu = 2$, and the EBK solution is given directly (in terms of the frequency $\omega \equiv 2\pi/T$) by

$$\left(n + \frac{1}{2}\right)\hbar = \frac{1}{2\pi}\oint p\,dq = 2\frac{m\omega}{2\pi}\int_{-x_m}^{+x_m} dx\sqrt{x_m^2 - x^2}$$

$$= m\omega x_m^2/2, \tag{2.18}$$

corresponding to the correct expression for the quantum mechanical energy

$$E = \frac{1}{2}m\omega^2 x_m^2 = \left(n + \frac{1}{2}\right)\hbar\omega. \tag{2.19}$$

2.3.4 Example: anharmonic oscillator

These expressions can be applied to an anharmonic oscillator of the form

$$V(x) = \frac{1}{2}Kx^2 + \lambda_1 x + \lambda_2 x^2 + \lambda_3 x^3 + \lambda_4 x^4 + \lambda_5 x^5. \tag{2.20}$$

If the anharmonic terms (those involving λ_i) are small, they can be included in first-order perturbation theory by use of their average values in the unperturbed problem. Since the odd powers vanish and the quadratic term can be incorporated into the unperturbed potential, only the quartic term enters. The averages $\langle x^k \rangle$ will be specified by the harmonic potential

$$V_0(x) = \frac{1}{2}(K + 2\lambda_2)x^2. \tag{2.21}$$

Using the virial theorem (Eq. 2.17) to specify the kinetic energy, the total unperturbed energy is

$$E_0 = \frac{1}{2}(K + 2\lambda_2)x_m^2, \tag{2.22}$$

and the total energy E' of the anharmonic oscillator is approximately

$$E' \approx \frac{1}{2}(K + 2\lambda_2)x_m^2 + \frac{3\lambda_4}{8}x_m^4 = E_0 + \frac{3\lambda_4}{2(K + 2\lambda_2)^2}E_0^2 \tag{2.23}$$

which, using the EBK quantization, yields the correct quantum mechanical result (approximating $K + 2\lambda_2 \approx m\omega^2$)

$$E' \approx \left(n + \frac{1}{2}\right)\hbar\omega + \frac{3\lambda_4\hbar^2}{2m^2\omega^2}\left(n + \frac{1}{2}\right)^2. \tag{2.24}$$

2.4 The Kepler problem

The use of position probability densities can be applied to both planetary motion and to atomic spectra by reconsideration of the Kepler problem [107] in this context.

2.4.1 Kepler's laws

An object moving under the influence of a potential $V(r) = -\kappa/r$ obeys Kepler's three laws:

First law: the bound state orbits are ellipses given by

$$\frac{1}{r} = \frac{1 + \varepsilon \cos \varphi}{a(1 - \varepsilon^2)} \tag{2.25}$$

where a and b are the semimajor and semiminor axes and $\varepsilon = \sqrt{1 - b^2/a^2}$ is the eccentricity.

Second law: equal areas are swept out in equal times

$$\frac{\pi ab}{T} = \frac{1}{2} r^2 \frac{d\varphi}{dt} \tag{2.26}$$

where T is the period of the orbit.

Third law: the square of the period is proportional to the third power of the semimajor axis

$$\left(\frac{T}{2\pi}\right)^2 = \frac{m}{\kappa} a^3. \tag{2.27}$$

Here, $\kappa = GMm$ for gravitation and $\kappa = -KQq$ for electrostatics (G is the Newtonian gravitational constant and K is the Coulombic electrostatic constant).

2.4.2 Position probability densities

If we represent this system by a position probability density $P(r)$ in the radial coordinate r, this becomes

$$P(r)dr = \frac{2dt}{T} = \frac{2dr}{T \, dr/dt}. \tag{2.28}$$

Because of the two-dimensional motion in the plane of the orbits, the instantaneous speed is given by

$$v^2 = \left(\frac{dr}{dt}\right)^2 + \left(r \frac{d\varphi}{dt}\right)^2. \tag{2.29}$$

From Kepler's second law, the angular momentum L is constant over the orbit and given by

$$L = m \mid \mathbf{r} \times \mathbf{v} \mid = mr^2 \frac{d\varphi}{dt}. \tag{2.30}$$

Combining these relationships, we obtain

$$\frac{dr}{dt} = \sqrt{v^2 - \left(\frac{L}{mr}\right)^2}. \tag{2.31}$$

Through the application of the virial theorem to this system, it will be shown in subsequent sections that (nonrelativistically) the total energy depends only on the semimajor axis, and is given by $E = -\kappa/2a$. To express the quantities v and L in radial coordinates, we first make use of conservation of energy

$$-\frac{\kappa}{2a} = \frac{1}{2}mv^2 - \frac{\kappa}{r} \tag{2.32}$$

to obtain

$$v^2 = \frac{\kappa}{m}\left(\frac{2}{r} - \frac{1}{a}\right). \tag{2.33}$$

(Notice that for a circular orbit $r = a$ and the expression reduces to the force law $\kappa/r^2 = mv^2/r$.) Next we use conservation of angular momentum

$$L = 2\pi ma^2\sqrt{1 - \varepsilon^2}/T \tag{2.34}$$

where $T = 2\pi\sqrt{ma^3/\kappa}$. From this we obtain

$$\left(\frac{L}{mr}\right)^2 = \frac{\kappa a(1 - \varepsilon^2)}{mr^2}. \tag{2.35}$$

Inserting Eqs. 2.33 and 2.35 into Eq. 2.31, the desired expression becomes

$$\frac{dr}{dt} = \frac{2\pi a}{Tr}\sqrt{2ar - r^2 - a^2(1 - \varepsilon^2)}. \tag{2.36}$$

There are two interesting algebraic manipulations of this square root, given by

$$\sqrt{2ar - r^2 - a^2(1 - \varepsilon^2)} = \sqrt{(a + \varepsilon a - r)(r - a + \varepsilon a)} \tag{2.37}$$

$$= \sqrt{(\varepsilon a)^2 - (r - a)^2}. \tag{2.38}$$

The first alternative provides an exposition in terms of the poles at the apoapsis A_+ and periapsis A_- (aphelion and perihelion for the solar system) defined by

$$A_\pm = a(1 \pm \varepsilon). \tag{2.39}$$

The position probability density along the radial coordinate is given from Eqs. 2.28 and 2.36, with the algebraic manipulation of Eq. 2.37 and the substitution of Eq. 2.39, by

$$P(r)dr = \frac{rdr}{\pi a\sqrt{(A_+ - r)(r - A_-)}}. \tag{2.40}$$

This probability distribution, together with a corresponding quantum mechanical case, is shown in the bottom panel of Fig. 2.1. The average values of the various moments of this coordinate are given by

$$\langle r^k \rangle = \frac{1}{\pi a}\int_{A_-}^{A_+} \frac{r^{k+1}dr}{\sqrt{(A_+ - r)(r - A_-)}}. \tag{2.41}$$

The second alternative again uses Eqs. 2.28 and 2.36, but this time uses the algebraic manipulation of Eq. 2.38. This can be used to convert these integrals to the forms that occur in the simple harmonic oscillator through substitution $x = r - a$. This yields a position probability density

$$P(x)dx = \frac{(x+a)dx}{\pi a\sqrt{(\varepsilon a)^2 - x^2}},$$

(2.42)

and average values can be computed as sums of SHO integrals using

$$\langle r^k \rangle = \frac{1}{\pi a} \int_{-\varepsilon a}^{+\varepsilon a} \frac{(x+a)^{k+1}dx}{\sqrt{(\varepsilon a)^2 - x^2}}.$$

(2.43)

However, there is an even more elegant way of obtaining these quantities in closed form [41].

2.4.3 Calculation of expectation values of radial moments

If one forms the time-averaged integral in terms of the angular coordinate φ directly

$$\langle r^k \rangle = \frac{1}{T} \int_0^T dt\, r^k = \frac{1}{T} \int_0^T \frac{d\varphi}{d\varphi/dt}\, r^k,$$

(2.44)

the radial dependences can be written as functions of φ using Kepler's first and second laws (Eqs. 2.25 and 2.26) as

$$\langle r^k \rangle = \frac{1}{T} \int_0^{2\pi} d\varphi \left[\frac{T}{2\pi ab}\right]\left[\frac{a}{b^2}(1 + \varepsilon \cos\varphi)\right]^{-k-2}$$

(2.45)

which simplifies to

$$\langle r^k \rangle = \frac{b^{2k+3}}{a^{k+3}} \frac{1}{2\pi} \int_0^{2\pi} d\varphi(1 + \varepsilon \cos\varphi)^{-k-2}.$$

(2.46)

This is a well-known integral of the "Laplace type" [108] with the value

$$\frac{1}{2\pi} \int_0^{2\pi} d\varphi(1 + \varepsilon \cos\varphi)^n = (1 - \varepsilon^2)^{n/2} P_n(1/\sqrt{1 - \varepsilon^2}),$$

(2.47)

where $P_n(x)$ is the Legendre polynomial. This yields

$$\langle r^k \rangle = b^k \left(\frac{b}{a}\right) P_{-k-1}(a/b).$$

(2.48)

Notice that this is an unusual application of the Legendre polynomial in which its argument is greater than unity. The full range of positive and negative powers of $\langle r^k \rangle$ can be described in terms of Legendre polynomials of positive index n using the relationship $P_{-n}(x) = P_{n-1}(x)$. The expectation values can thus be written in this very compact form [41]

$$\langle r^k \rangle = b^k \left(\frac{b}{a}\right) P_{|k+3/2|-1/2}(a/b).$$

(2.49)

which will be shown to be very useful in subsequent sections.

Notice that the relationship between Legendre polynomials of positive and negative index relates the expectation values for $k = -q - 2$ and $k = q - 1$

$$\langle r^{-q-1} \rangle = b^{-2q-1} \langle r^{q-1} \rangle, \tag{2.50}$$

which corresponds to a well-known quantum mechanical identity [160].

2.4.4 The EBK quantization

Up to here in this section, r and φ have denoted the two-dimensional cylindrical coordinates defined in the plane of the orbit. In order to investigate the EBK quantization with the full inclusion of the spatial degrees of freedom, the coordinates will now be generalized to the standard three-dimensional spherical polar coordinates r, ϑ and φ, where ϑ is an azimuthal angle specifying the tilt of the orbital plane relative to an arbitrarily chosen z axis. If the normal to the plane of the orbit makes an angle Θ with the z axis, then the ranges of the coordinates over the orbital motion are given by

$$A_- \leq r \leq A_+; \qquad -\Theta \leq \vartheta \leq \Theta; \qquad 0 \leq \varphi \leq 2\pi. \tag{2.51}$$

Thus r and ϑ undergo librations (oscillate between two endpoints) and φ undergoes a rotation.

The generalized momenta can be obtained from the standard Hamilton–Jacobi approach [107]. This involves construction of the Lagrangian in terms of the coordinates q_i and their time derivatives $\dot{q}_i$. The generalized momentum canonical to the coordinate q_i is obtained by differentiating the Lagrangian with respect to $\dot{q}_i$. The Hamiltonian is then formed by re-expressing the $\dot{q}_i$ quantities in terms of the generalized momenta. For this case, the Hamiltonian is given by

$$H = \frac{1}{2m} \left(p_r^2 + \frac{p_\vartheta^2}{r^2} + \frac{p_\varphi^2}{r^2 \sin^2 \vartheta} \right) - \frac{\kappa}{r}. \tag{2.52}$$

By separation of variable technique it can be seen that

$$p_\varphi^2 = \text{constant}; \qquad p_\vartheta^2 + \frac{p_\varphi^2}{\sin^2 \vartheta} = \text{constant}'. \tag{2.53}$$

Denoting the first constant as L_z^2, the second constant as L^2, and the Hamiltonian energy as E, this yields

$$p_\varphi = L_z; \qquad p_\vartheta = L^2 - \frac{L_z^2}{\sin^2 \vartheta}; \qquad p_r = 2m \left(E + \frac{\kappa}{r} \right) - \frac{L^2}{r^2}. \tag{2.54}$$

The phase integrals

$$\left(n_i + \frac{\mu_i}{4} \right) \hbar = \frac{1}{2\pi} \oint dq_i \, p_i \tag{2.55}$$

for the Kepler problem consist of two librations (one in r and one in ϑ) and one rotation (in φ). All three can be integrated by suitably chosen contours that have been discussed in detail in the literature [107]. The phase integrals are:

$$\left(n_r + \frac{1}{2}\right)\hbar = \frac{1}{2\pi}\oint dr \sqrt{2m\left(E + \frac{\kappa}{r}\right) - \frac{L^2}{r}} = -L + \kappa\sqrt{-\frac{m}{2E}} \qquad (2.56)$$

$$\left(n_\vartheta + \frac{1}{2}\right)\hbar = \frac{1}{2\pi}\oint d\vartheta \sqrt{L^2 - \frac{L_z^2}{\sin^2\vartheta}} = L - L_z \qquad (2.57)$$

$$(n_\varphi)\hbar = \frac{1}{2\pi}\oint d\varphi \, L_z = L_z. \qquad (2.58)$$

If the second two are added this yields

$$\left(n_\vartheta + n_\varphi + \frac{1}{2}\right)\hbar = L \qquad (2.59)$$

and if all three are added it yields

$$(n_r + n_\vartheta + n_\varphi + 1)\hbar = \kappa\sqrt{-\frac{m}{2E}}. \qquad (2.60)$$

To obtain a notation consistent with that utilized in the corresponding quantum mechanical formulation, denote

$$\begin{aligned} n &\equiv n_r + n_\vartheta + n_\varphi + 1 \\ \ell &\equiv n_\vartheta + n_\varphi \\ m_\ell &\equiv \pm n_\varphi \end{aligned} \qquad (2.61)$$

where the $\pm$ indicates a counterclockwise or clockwise azimuthal integration, respectively. In this general development, both gravitational and electrostatic potentials were considered. Since this quantization narrows the consideration to an atomic situation, the electrostatic constants $\kappa = \zeta K e^2$ will be used here, where $K = 1/4\pi\epsilon_0$, e is the magnitude of the charge of the electron, and ζe is the central charge (either the nuclear charge Ze or the effective charge of the core). The equations assume the form

$$E = -Ry\,\zeta^2/n^2 \qquad (2.62)$$

$$L = \left(\ell + \frac{1}{2}\right)\hbar \qquad (2.63)$$

$$L_z = m_\ell\hbar \qquad (2.64)$$

where $Ry = (mc^2/2)(Ke^2/\hbar c)^2$ is the Rydberg constant, 13.6 eV. The values for E and L_z agree exactly with the predictions of nonrelativistic Schrödinger theory, and the value for $L^2 = (\ell + \frac{1}{2})^2\hbar^2 = [\ell(\ell + 1) + \frac{1}{4}]\hbar^2$ agrees in the limit of high ℓ.

Noting that $\langle E \rangle = -\kappa/2r$ and $\langle L^2 \rangle = \kappa m b^2/a$, the quantized values of the semimajor and semiminor axes of the elliptic orbit are given by

$$a = a_0 n^2/\zeta \qquad (2.65)$$

$$b = a_0 n\left(\ell + \frac{1}{2}\right)\Big/\zeta \qquad (2.66)$$

where $a_0 = \hbar^2/mKe^2$ is the Bohr radius, 0.529 Å. (The substitutions $Ke^2 = 2Ry\, a_0$ and $\hbar^2/m = 2Ry\, a_0^2$ are often convenient in converting units). Correspondingly, the eccentricity of the orbit is given by

$$\varepsilon = \sqrt{1 - (\ell + 1/2)^2/n^2}. \tag{2.67}$$

The radial turning points can be approximated by

$$A_- = \frac{a_0 n^2}{\zeta}\left[1 - 1 + \frac{1}{2}\left(\frac{\ell + \frac{1}{2}}{n}\right)^2 + \cdots\right] \simeq \frac{a_0\left(\ell + \frac{1}{2}\right)^2}{2\zeta}, \tag{2.68}$$

$$A_+ = \frac{a_0 n^2}{\zeta}\left[1 + 1 - \frac{1}{2}\left(\frac{\ell + \frac{1}{2}}{n}\right)^2 + \cdots\right] \simeq \frac{2a_0 n^2}{\zeta}, \tag{2.69}$$

so to first order the distance of closest approach for a given ℓ is independent of n, and the distance of greatest retrogression is independent of ℓ. Notice that (unlike the BSW result) the periapsis of an $\ell = 0$ s-orbital does not vanish, but has the value $A_- = a_0/8\zeta$. This feature of the Maslov index will be shown to have an interesting subtlety when the corresponding relativistic case is considered.

2.4.5 Position probability densities for these orbits

As was shown earlier in the bottom panel of Fig. 2.1, the position probability densities for these individual orbits have maxima at the classical turning points, where the radial speed passes through zero and changes sign. To convert these radial position probability densities to a corresponding spatial position probability distribution, one must superimpose the envelope of all of the angular values that correspond to the ℓ and m_ℓ quantum numbers.

For an $\ell = 0$ s-state the orbit possesses no magnetic moment, hence it is not possible to select specific values of ϑ or φ by the impression of an external magnetic field, and a measurement will yield a superposition of all angular orientations. Thus the position probability density has a spherical distribution, with maxima on two concentric spherical shells corresponding to the radii of the periapsis and the apoapsis. This fact often confuses students when they first encounter molecular orbitals in a chemistry class, since they incorrectly associate the spherical distribution of the s-state with a circular orbit, rather than its correct identification as the orbit of maximum eccentricity (Eq. 2.67 with $\ell = 0$).

For the case of the $\ell = 1$ p-orbitals, the $m_\ell = \pm 1$ and $m_\ell = 0$ systems can be separated according to the value of ϑ selected through the impression of a magnetic field (along a z-axis in an arbitrary direction), and the position probability specified by superimposing orbits of all values of φ. For the $m_\ell = \pm 1$ p-states, the locus of the orbits lies within $\vartheta = \pm \cos^{-1}(3/2n)$ and forms a ring (smeared by the difference in radii of the periapsis and apoapsis) of radius $\zeta a_0 n^2$.

For the $m_\ell = 0$ p-states, $\vartheta = 90°$, so the planes of all the orbits pass through the z-axis, and the distribution is the envelope of all values of φ. The maximum position probability

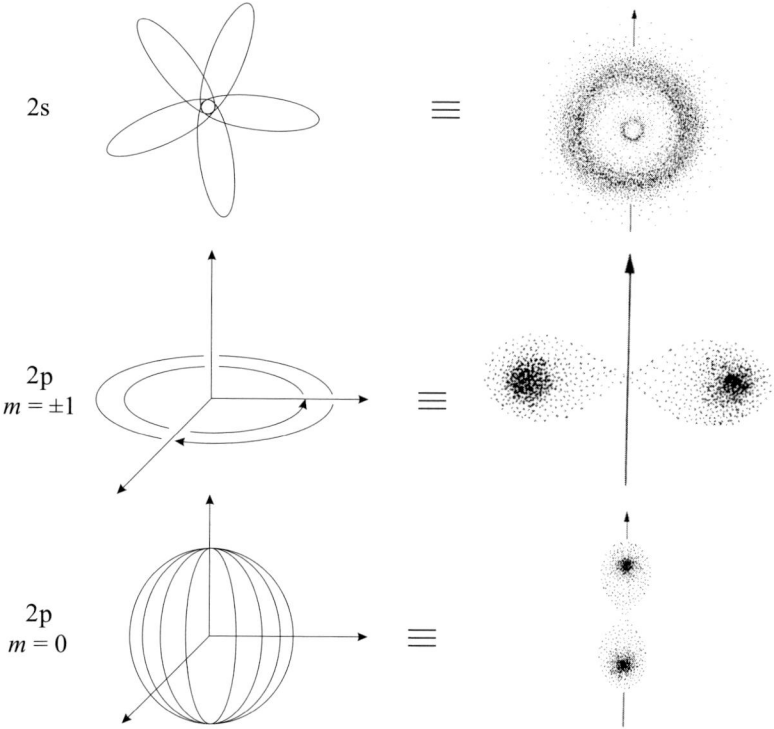

Fig. 2.2. Classical (*left*) and quantum mechanical (*right*) position probability densities for the $n = 2$ orbits. (Quantum mechanical densities after Ref. [184].)

density occurs where all of the orbits intersect at the "north and south poles" of the distribution, forming two lobes.

While this qualitative classical picture is very simplistic, it does make plausible the quantum mechanical distributions, which are shown to the right of the related orbits in Fig. 2.2. The lobed pattern of the p-orbitals here is associated with the carbon bonding of organic molecules, as well as with the macromolecules formed in silicon semiconductors and germanium transistors. Thus the picture of these intersection points as hubs of interaction for atomic orbital electrons provides a useful conceptual model.

2.4.6 Expectation values of r^k

Using these quantized values for a and b, the expressions for $\langle r^k \rangle$ for $-6 \leq k \leq +6$ are given in Table 2.1. These semiclassical expressions differ from their quantum mechanical equivalents only in the eigenvalues that are generated by successive applications of the angular momentum L. In these semiclassical expressions $L^n \to [(\ell + \frac{1}{2})\hbar]^n$, whereas in the quantum mechanical case $L^n \to (\ell + a_1)(\ell + a_2) \cdots (\ell + a_n)\hbar^n$. Simple formulae have been developed [41, 50] to generate values for a_n which are also listed in Table 2.1. The rational fractions $C_{\lambda qi}$ (λ is the order of the Legendre polynomial) can be generated from a

Table 2.1. Expectation values of powers of the radial coordinate.

$$\langle r^4 \rangle = \left(\frac{a_0}{\zeta}\right)^4 \frac{n^8}{8} \left[63 - 70\left(\frac{\ell_{op}}{n}\right)^2 + 15\left(\frac{\ell_{op}}{n}\right)^4\right]$$

$$\langle r^3 \rangle = \left(\frac{a_0}{\zeta}\right)^3 \frac{n^6}{8} \left[35 - 30\left(\frac{\ell_{op}}{n}\right)^2 + 3\left(\frac{\ell_{op}}{n}\right)^4\right]$$

$$\langle r^2 \rangle = \left(\frac{a_0}{\zeta}\right)^2 \frac{n^4}{2} \left[5 - 3\left(\frac{\ell_{op}}{n}\right)^2\right]$$

$$\langle r^1 \rangle = \left(\frac{a_0}{\zeta}\right) \frac{n^2}{2} \left[3 - \left(\frac{\ell_{op}}{n}\right)^2\right]$$

$$\langle r^0 \rangle = 1$$

$$\langle r^{-1} \rangle = \left(\frac{\zeta}{a_0}\right) \frac{1}{n^2}$$

$$\langle r^{-2} \rangle = \left(\frac{\zeta}{a_0}\right)^2 \frac{1}{n^3 \ell_{op}}$$

$$\langle r^{-3} \rangle = \left(\frac{\zeta}{a_0}\right)^3 \frac{1}{n^3 \ell_{op}^3}$$

$$\langle r^{-4} \rangle = \left(\frac{\zeta}{a_0}\right)^4 \frac{1}{2n^3 \ell_{op}^5} \left[3 - \left(\frac{\ell_{op}}{n}\right)^2\right]$$

$$\langle r^{-5} \rangle = \left(\frac{\zeta}{a_0}\right)^5 \frac{1}{2n^3 \ell_{op}^7} \left[5 - 3\left(\frac{\ell_{op}}{n}\right)^2\right]$$

$$\langle r^{-6} \rangle = \left(\frac{\zeta}{a_0}\right)^6 \frac{1}{8n^3 \ell_{op}^9} \left[35 - 30\left(\frac{\ell_{op}}{n}\right)^2 + 3\left(\frac{\ell_{op}}{n}\right)^4\right]$$

$$\langle r^{-7} \rangle = \left(\frac{\zeta}{a_0}\right)^7 \frac{1}{8n^3 \ell_{op}^{11}} \left[63 - 70\left(\frac{\ell_{op}}{n}\right)^2 + \left(\frac{\ell_{op}}{n}\right)^4\right]$$

For the semiclassical model $(\ell_{op})^q = (\ell + \frac{1}{2})^k$

For the quantum mechanical case, the odd and even powers are

$$(\ell_{op})^{2q+1} = \frac{(2\ell + q + 1)!}{(2\ell - q)! \, 2^{2q+1}}; \qquad (\ell_{op})^{2q} = \sum_{i=0}^{q} (-1)^{q+1} \frac{C_{\lambda q i} \, (\ell + i)!}{(\ell - i)!}$$

simple algorithm, and the results are given in Table 2.2. Thus the correct quantum mechanical expressions can be routinely generated from these semiclassical formulae, provided that the simple quantum number ℓ is replaced by an operational definition ℓ_{op} prescribed by these algorithms.

Table 2.2. The coefficients $C_{\lambda q i}$ that connect the semiclassical and quantum mechanical expressions for the $\langle r^k \rangle$ through the relationship given in Table 2.1. (From Ref. [50].)

		$C_{\lambda q i}$					
λ	q	$i = 0$	$i = 1$	$i = 2$	$i = 3$	$i = 4$	$i = 5$
2	1	0	1				
3	1	1/3	1				
4	1	5/6	1				
4	2	0	0	1			
5	1	3/2	1				
5	2	4/5	4/3	1			
6	1	7/3	1				
6	2	14/5	3	1			
6	3	0	0	0	1		
7	1	10/3	1				
7	2	101/15	5	1			
7	3	36/7	36/5	3	1		
8	1	9/2	1				
8	2	27/2	22/3	1			
8	3	761/35	332/15	13/2	1		
8	4	0	0	0	0	1	
9	1	35/6	1				
9	2	145/6	10	1			
9	3	1315/21	48	21	1		
9	4	64	576/7	144/5	16/3	1	
10	1	22/3	1				
10	2	119/3	13	1			
10	3	3124/21	177/2	15	1		
10	4	10736/35	2124/7	84	34/3	1	
10	5	0	0	0	0	0	1

Several interesting insights can be gained by inspection of Table 2.1. It can be seen that $\langle r^k \rangle$ is independent of ℓ in two cases, $k = 0$ and -1, corresponding to the normalization integral and the expectation of the $1/r$ potential. The latter provides an exposition of the "accidental degeneracy" peculiar to the Coulomb potential (the total energy for a given n is independent of ℓ in the nonrelativistic approximation). The $k = -1$ and -2 cases mark a symmetry axis in Table 2.1 (arising from the relationship between values for $k = -q - 2$ and $k = q - 1$ noted in Eq. 2.50) that breaks $\langle r^k \rangle$ into two families of solutions with respect to its asymptotic behavior along a Rydberg series. For fixed ℓ and increasing n the quantity $x = n/(\ell + \frac{1}{2})$ becomes large, in which case $P_q(x) \propto x^q$, resulting in the asymptotic dependence for $n \gg (\ell + \frac{1}{2})$

$$\langle r^k \rangle \propto n^{2k} \qquad (k \geq -1)$$

$$\propto n^{-3} \left(\ell + \frac{1}{2} \right)^{2k+3} \qquad (k \leq -2). \tag{2.70}$$

This also occurs (with the appropriate expression for the ℓ dependence) in the quantum mechanical formulation. Thus, for large n, the average values of positive powers of r do not depend on ℓ (an asymptotic analogue of the accidental degeneracy), and the average values of quadratic and higher powers of the reciprocal of r all have the same functional dependence on n. The fact that perturbations which decrease faster with r than the Coulomb potential all contribute to the same $1/n^3$ term to the energy corrections is the underlying reason for the success of the Rydberg formula in describing complex atoms. This is because, by adding a quantum defect δ to the Balmer formula (Eq. 2.62 for a complex of core charge ζ), one obtains

$$Ry \frac{\zeta^2}{(n - \delta)^2} = Ry \frac{\zeta^2}{n^2} \left(1 - \frac{\delta}{n} \right)^{-2} \simeq Ry \frac{\zeta^2}{n^2} + Ry \frac{2\zeta^2 \delta}{n^3} + \cdots, \tag{2.71}$$

so a perturbation that adds an incremental energy ΔE to the Coulomb binding energy of the atom can be described by a quantum defect

$$\delta = -\frac{n^3 \, \Delta E}{2 Ry \, \zeta^2}. \tag{2.72}$$

Since the leading term in all $\langle r^k \rangle$ with $k \leq -2$ is $1/n^3$, that contribution will be accurately accounted for by a constant empirical value for δ.

2.4.7 Perturbations

In treating perturbations to the simple Coulomb potential, we shall rewrite all dynamical aspects as functions of powers of the radial coordinate, and thereby specified in terms of $\langle r^k \rangle$. In order to make this formulation for the kinetic energy, we make use of the virial theorem. For an unperturbed potential

$$V_0 = -\kappa/r \tag{2.73}$$

(to treat the two cases simultaneously, for gravitation $\kappa \equiv GMm$ and for atomic systems $\kappa \equiv KZe^2$), the nonrelativistic kinetic energy is

$$T_0 = p^2/2m, \tag{2.74}$$

so the virial theorem for a total energy $E_0 = T_0 + V_0$ yields

$$E_0 = -\langle T_0 \rangle = \frac{1}{2} \langle V_0 \rangle = -\frac{\kappa}{2a}. \tag{2.75}$$

If we now add to this a small perturbation

$$\Delta E = \kappa' r^k \tag{2.76}$$

the energy of the perturbed system can be approximated by

$$E = E_0 + \langle \Delta E \rangle = -\kappa/2a + \kappa' \langle r^k \rangle. \tag{2.77}$$

This will now be applied to several specific examples.

2.4.8 Relativistic corrections to the kinetic energy

The relativistic energy can be written as

$$T = \sqrt{(mc^2)^2 + (pc)^2} - mc^2 \tag{2.78}$$

which can be binomial expanded to yield

$$T \cong mc^2 \left[1 + \frac{1}{2} \left(\frac{p}{mc} \right)^2 - \frac{1}{8} \left(\frac{p}{mc} \right)^4 + \cdots \right] - mc^2. \tag{2.79}$$

This can be simplified using the expression for the nonrelativistic kinetic energy ($T_0 = p^2/2m$), to obtain

$$T \cong T_0 - T_0^2/2mc^2 + \cdots. \tag{2.80}$$

The perturbation then becomes

$$\langle \Delta E \rangle = \langle T \rangle - \langle T_0 \rangle = \langle T_0^2 \rangle/2mc^2. \tag{2.81}$$

The average of the square of the nonrelativistic kinetic energy can be written in terms of quantities related to powers of r using

$$\langle T_0^2 \rangle = \langle (E_0 - V_0)^2 \rangle = E_0^2 - 2E_0\langle V_0 \rangle + \langle V_0^2 \rangle. \tag{2.82}$$

Since $E_0 = \langle V_0 \rangle/2$, this yields

$$\langle \Delta E \rangle = -\frac{1}{2mc^2} \left[\langle V_0^2 \rangle - \frac{3}{4}\langle V_0 \rangle^2 \right] + \cdots \tag{2.83}$$

which involves averages of two powers of r

$$\langle \Delta E \rangle = -\frac{\kappa^2}{2mc^2} \left[\langle r^{-2} \rangle - \frac{3}{4}\langle r^{-1} \rangle^2 \right] + \cdots. \tag{2.84}$$

In terms of the semiaxes of the elliptic orbit this is

$$\langle \Delta E \rangle = -\frac{\kappa^2}{2mc^2} \left[\frac{1}{ab} - \frac{3}{4a^2} \right] + \cdots. \tag{2.85}$$

This correction has two interesting applications in the history of physics. It occurs in Einstein's Special Relativity calculation for the precession of the perihelion of the planet Mercury, which yielded a 7.2 arcsec/century correction. Einstein's recomputation using General Relativity yielded a value six times as large (as will be discussed in a later section).

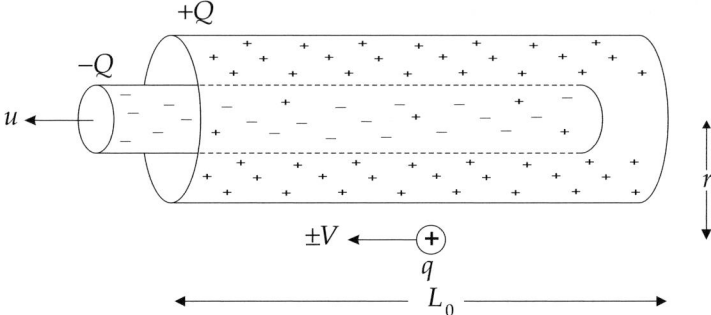

Fig. 2.3. Model for a current-carrying wire.

It also occurs in Sommerfeld's calculation of the fine structure of the hydrogen atom. If we include the results of the EBK quantization for the semimajor and semiminor axes of the elliptic orbit, we obtain Sommerfeld's result

$$\langle \Delta E \rangle = -\frac{Ry\,\alpha^2 Z^4}{n^3}\left[\frac{1}{\left(\ell+\frac{1}{2}\right)} - \frac{3}{4n}\right] \tag{2.86}$$

where we have used the electromagnetic constants $\kappa \to KZe^2 = 2Ry\,a_0 Z$ and incorporated the fine structure constant $\alpha \equiv Ke^2/\hbar c$ into the formulation.

2.4.9 Relativistic corrections to the potential energy

In addition to the relativistic modifications to the kinetic energy, the magnetic corrections to the electrostatic potential also comprise a relativistic correction. The deduction from Coulomb's law and Special Relativity of the magnetic force law for a current-carrying wire can be so concisely and transparently done that it can be argued that it should appear in all physics books at all levels. However, it is seldom presented or even mentioned. The development will therefore be included below.

From Gauss's law $EA = 4\pi KQ$, it follows that the electric field E due to a cylinder of long length L_0 and charge Q, evaluated at a distance r from the center of the cylinder ($A = \pi r^2 L_0$, the area of the imaginary Gaussian cylinder), is given by $E = 2KQ/Lr$. As a model for a current-carrying wire, consider two concentric superimposed cylinders of charge, as shown in Fig. 2.3.

One cylinder is positively charged and stationary relative to the laboratory, and the other is negatively charged and drifts with a small speed u ($\approx$ mm/minute). A particle of charge q moves in the same (or opposite) direction and parallel to the negative (electron) drift with a significantly larger speed $\pm V$ ($\approx$ m/s) at a distance r from the center of the cylinders. The Coulomb force on a moving test charge due to the electric field set up by the two cylinders is given by

$$F = \frac{2Kq}{r}\left[\frac{Q}{L_0\sqrt{1 - V^2/c^2}} - \frac{Q}{L_0\sqrt{1 - (\pm V - u)^2/c^2}}\right]. \tag{2.87}$$

Here we have included the apparent relativistic contraction of the two line charges as seen by the moving test charge (as was demonstrated [95, 187] by Voigt in 1887 and further developed [90] by Einstein in 1905). This can be binomial expanded to yield

$$F \approx \frac{2Kq}{r}\frac{Q}{L_0}\left[\left(1 + \frac{1}{2}\frac{V^2}{c^2} + \cdots\right) - \left(1 + \frac{1}{2}\frac{V^2 \mp 2Vu + u^2}{c^2} + \cdots\right)\right]. \quad (2.88)$$

To first order in u/c this becomes

$$F \approx \pm\frac{2KqV}{rc^2}\left(-\frac{Qu}{L_0}\right). \quad (2.89)$$

Although u is small, Q/L_0 is large, and their product can be identified as the current in the wire (for 1 ampere through a 1-mm-diameter copper wire, $u \approx 10^{-4}$ m/s and $Q/L_0 = 10^4$ C/m). Thus we can define

$$I \equiv -Qu/L_0; \qquad B \equiv 2(K/c^2)I/r; \qquad F = \pm qVB \quad (2.90)$$

and the Biot–Savart law neatly drops out. Contrary to a common student misconception, the interaction is just the familiar attractive or repulsive Coulomb force, modified only by a relativistic correction for the retarded value of the apparent charge per unit length because of the relative motions of the individual charge distributions. The right-hand rules are merely a mnemonic designed to determine the sign of the force. Since these hand rules are used twice in a two-step process, the handedness is arbitrary ("two rights don't make a wrong, but neither do two lefts"). Adding another dimension to describe an interaction that proceeds within a plane can create confusion, and the right-handed algorithm can be replaced by a simple conceptual picture. If the positive test charge moves in the *same* direction as the electron drift, it will see the positive charge as the more severely contracted, and feel a net repulsion. If the positive test charge moves in the *opposite* direction from the electron drift, it will see the negative charge as the more severely contracted, and feel a net attraction. It is unfortunate that Woldemar Voigt's 1887 formulation of the apparent length contraction was not available in 1820, when Hans Christian Ørsted's lecture demonstration revealed the relationship between currents and magnetic fields. The study of electricity and magnetism today still bears the divisive artifacts of the time-ordering of these discoveries.

The model treated above simplifies the development by use of two-dimensional cylindrical geometry and a field source with no net charge. The magnetic interactions that affect an orbiting electron can be formulated in terms of the three-dimensional, point-charge form of the Biot–Savart law (in standard SI symbols)

$$\mathbf{B} = \frac{K}{c^2}\frac{Ze(\mathbf{r} \times \mathbf{v})}{r^3}. \quad (2.91)$$

If this field is produced by an orbiting charge, it can be written in terms of the orbital angular momentum

$$\mathbf{L} = m(\mathbf{r} \times \mathbf{v}). \quad (2.92)$$

Introducing the anomalous magnetic moment of the electron

$$\boldsymbol{\mu}_s = -g_e\frac{e}{2m}\mathbf{S} \quad (2.93)$$

The energy of the interaction can be computed by adopting a seemingly pre-Copernican viewpoint, computing the interaction energy in a frame in which the electron is at rest and is orbited by the nucleus of charge Ze. Thus, combining these three equations with the expression for the energy of a magnetic dipole in an external magnetic field, we obtain

$$\Delta E = -\boldsymbol{\mu}_s \cdot \mathbf{B} = g_e \frac{ZKe^2}{2(mc)^2} \frac{\mathbf{L} \cdot \mathbf{S}}{r^3}. \tag{2.94}$$

Having made this calculation, it is now necessary to transform the result into the normal frame of reference in which the nucleus, and not the electron, is at rest. This transformation introduces an additional relativistic correction called the "Thomas precession" [113], which occurs because time dilation causes an observer on the nucleus and an observer on the electron to disagree about the time that it takes for the one particle to make a complete revolution about the other. The transformation requires that the Larmor frequency due to the electron spin be corrected by the addition of the transformational Thomas frequency. The Thomas precession is in a direction opposite to the electron Larmor precession, and involves all of the same quantities except for the electron g-factor. The interaction energy after transformation to the rest system of the nucleus is therefore

$$\langle \Delta E \rangle = (g_e - 1) \frac{ZKe^2}{2(mc)^2} \left\langle \frac{\mathbf{L} \cdot \mathbf{S}}{r^3} \right\rangle. \tag{2.95}$$

To a first approximation, $g_e \cong 2 + \alpha/\pi$, so the factor in front is approximately $(g_e - 1) \cong 1 + \alpha/\pi$ (and not $g_e/2 \cong 1 + \alpha/2\pi$ as appears in some textbooks [113]).

Using the EBK quantization

$$\langle r^{-3} \rangle = b^{-3} = \frac{Z^3}{a_0^3 n^3 \left(\ell + \frac{1}{2}\right)^3}. \tag{2.96}$$

Since there is no semiclassical approximation to the electron spin in this model, it is necessary at this point to move away from the EBK semiclassical model and use quantum mechanical expressions. However, before doing so it is worth noting that there are some advantages to the EBK model over the Pauli–Schrödinger quantum mechanical treatment.

Notice that, because of the presence of the Maslov index in the factor $(\ell + \frac{1}{2})$ in the denominator of the semiclassical expression for $\langle r^{-3} \rangle$, the EBK value for the spin–orbit interaction is not indeterminate for an s-state. This is because, owing to the Maslov index, an s-state has a small but finite periapsis in the nonrelativistic semiclassical formulation (it will be shown later that this periapsis collapses for high Z in the relativistic case). Compare this to the corresponding quantum mechanical expression for this quantity

$$\langle r^{-3} \rangle_{\text{QM}} = \frac{Z^3}{a_0^3 n^3 \ell \left(\ell + \frac{1}{2}\right)(\ell + 1)}. \tag{2.97}$$

The appearance of the factor ℓ in the denominator leads to problems for the specification of s-states in hydrogen, because the wave function does not vanish at the origin. This leads to the introduction of the so-called "Darwin term" [78]. The origin of this term lies in a phenomenon in the Dirac theory called "Zitterbewegung" (described in more detail in Chapter 4). Rather than moving smoothly, the electron undergoes rapid fluctuations in its position due to the virtual photon field. These fluctuations are on the order of the Compton wavelength of the electron, and cause the electron to see a smeared-out Coulomb potential at the nucleus. When first-order perturbation theory is applied, the Darwin term contributes only to s-states, since only they have a nonzero probability of being found at the origin in the Schrödinger model.

It is often stated that the Darwin term "has no classical analogue," but that statement clearly ignores the EBK semiclassical model. Since the Maslov index automatically provides s-states with a small but finite periapsis, they do indeed have zero probability of being found at the origin, which is a clear semiclassical analogue of the Zitterbewegung.

In order to complete the specification of the spin–orbit term, it is necessary to compute the angular momentum quantities. As will be described in more detail in Chapter 4, this can be done using the vector model. In terms of the total angular momentum $\mathbf{J} = \mathbf{L} + \mathbf{S}$

$$2\langle \mathbf{L} \cdot \mathbf{S} \rangle = \langle \mathbf{J} \cdot \mathbf{J} - \mathbf{L} \cdot \mathbf{L} - \mathbf{S} \cdot \mathbf{S} \rangle$$
$$= [j(j+1) - \ell(\ell+1) - s(s+1)]\,\hbar^2 \tag{2.98}$$

and the result is (substituting in the fine structure constant)

$$\langle \Delta E \rangle = (g_{\mathrm{e}} - 1)\, Ry\, \alpha^2 Z^4 \left[\frac{j(j+1) - \ell(\ell+1) - s(s+1)}{2n^3 \ell\left(\ell + \frac{1}{2}\right)(\ell+1)} \right]. \tag{2.99}$$

Notice that the quantum mechanical expression $\ell(\ell + \frac{1}{2})(\ell + 1) = (\ell + \frac{1}{2})^3 - \frac{1}{4}(\ell + \frac{1}{2})$ reduces to the EBK result in the correspondence limit. One of the most fascinating connections between nonrelativistic and relativistic quantum mechanics is obtained by combining the relativistic corrections to the potential and kinetic energies. If one sets the g-factor of the electron equal to its non-QED "Dirac" value $g_{\mathrm{e}} = 2$ in Eq. 2.99 and combines it with Eq. 2.86, the result for both the relativistic kinetic and potential energy corrections is

$$\langle \Delta E \rangle_{\mathrm{Total}} = -\frac{Ry\, \alpha^2}{n^3} \left\{ \left[\frac{1}{\left(\ell + \frac{1}{2}\right)} - \frac{3}{4n} \right] \right.$$
$$\left. - \left[\frac{j(j+1) - \ell(\ell+1) - s(s+1)}{2\ell\left(\ell + \frac{1}{2}\right)(\ell+1)} \right] \right\} \tag{2.100}$$

which reduces (with a small amount of algebra) for either $j = \ell + \frac{1}{2}$ or $j = \ell - \frac{1}{2}$ to

$$\langle \Delta E \rangle_{\mathrm{Total}} = -\frac{Ry\, \alpha^2}{n^3} \left[\frac{1}{\left(j + \frac{1}{2}\right)} - \frac{3}{4n} \right]. \tag{2.101}$$

This is the exact form obtained from the solution of the relativistic Dirac equation, and differs only from the Sommerfeld formula for the kinetic energy corrections by the substitution $\ell \to j$. Thus the ℓ dependence present in both the spin–orbit and relativistic mass contributions vanishes when they are combined. Therefore, this result automatically contains the Darwin correction to the s-state that was needed to repair the flaw in the Schrödinger calculation (that does not occur in the EBK result).

This correction also has a connection to Einstein's General Relativistic formulation of the advance of the perihelion of Mercury. When General Relativity is applied to the gravitational problem, the Schwarzschild solution of the Einstein field equations assumes the form [107]

$$\langle \Delta E \rangle = \frac{\kappa L^2}{(mc)^2} \langle r^{-3} \rangle. \tag{2.102}$$

So, just as in the case of the atomic spin–orbit interaction, there is here another correction beyond the relativistic kinetic energy of Special Relativity that depends on the inverse cube of the radius vector. Also as in the case of the gravitational problem, this extension provides a correction that connects the kinetic and total energy results. In the case of fine structure, the Sommerfeld formula could be corrected by the substitution $\ell \rightarrow j$. In the case of the advance of the perihelion of Mercury, the result of the Special Relativity calculation (7.2 arcsec/century) can be corrected by including this term, thus obtaining an overall multiplicative factor of six. This yields the observed result 43 arcsec/century.

2.4.10 The core polarization model

In order to precisely characterize highly excited states in complex atoms, a core polarization model is often adopted [87, 190]. In this model, the outer active electron is considered to orbit a core (consisting of a nucleus and a deformable cloud of inner passive electrons) of effective charge ζ and electrostatic polarizabilities α_d and α_Q that lead, respectively, to field-induced dipole and quadrupole moments d and Q.

The induced dipole moment of the core is proportional to the electrostatic field of the electron (which varies as $1/r^2$). Since the energy of the interaction is given by the dot product of the dipole moment and the strength of the electric field, the dipole interaction energy is proportional to $1/r^4$. By similar arguments the interaction energy of the induced quadrupole polarization is proportional to $1/r^6$. A more detailed development of these quantities is presented in Section 3.5.

Thus, in the context of this model, the presence of a many-electron core can be included by a perturbation of the form

$$\langle \Delta E \rangle = -(Ke^2/2)[\alpha_d \langle r^{-4} \rangle + \alpha_Q \langle r^{-6} \rangle]. \tag{2.103}$$

In terms of the expectation values (and noting that $Ke^2/2 = Ry\, a_0$) this becomes

$$\langle \Delta E \rangle = -Ry\, a_0 \left[\alpha_d \left(\frac{\zeta}{a_0} \right)^4 \left(\frac{1}{2n^3 \ell_{\text{op}}^5} \right) \left(3 - \frac{\ell_{\text{op}}^2}{n^2} \right) \right.$$
$$\left. + \alpha_Q \left(\frac{\zeta}{a_0} \right)^6 \left(\frac{1}{8n^3 \ell_{\text{op}}^9} \right) \left(35 - 30 \frac{\ell_{\text{op}}^2}{n^2} + 3 \frac{\ell_{\text{op}}^4}{n^4} \right) \right]. \tag{2.104}$$

Note that this perturbation splits the "accidental ℓ degeneracy" of the pure Coulomb potential. This expression will be applied to the study of high n and ℓ states in Chapter 3. It can also be utilized to gain insight into the quantum defect method formulation of a Rydberg series. If the perturbation is expressed as a quantum defect, the leading term will be given by Eq. 2.72 as

$$\delta = -\frac{n^3 \Delta E}{2Ry\,\zeta^2} \approx \frac{\alpha_d \zeta^2}{4a_0^3 \ell_{op}^5}\left(3 - \frac{\ell_{op}^2}{n^2}\right).$$

(2.105)

We shall return to this in the next chapter in connection with the Ritz expansion.

2.4.11 Gauss's formulation of the perturbations of the planets

Perturbations on a planetary orbit due to other planets can also be described by this formalism. Since these effects are usually measured over hundreds of years, the planets involved have completed many orbits, and the time-averaged approach is appropriate. The situation is very similar to the quantum mechanical case, in which instantaneous positions can be replaced by time-averaged position probability densities. As a simple pedagogic model, a procedure first utilized by Gauss is followed here, in which each planet is replaced by a uniform circular ring having the same mass as the planet and a radius corresponding to the appropriate moments of the planet's orbits about the Sun. For precessing elliptical orbits with planes tilted relative to the ecliptic plane the time-averaged mass distribution becomes toroidal.

The potential energy at any point in space due to a ring of mass M and a test mass m is given by

$$V(r, \vartheta) = GMm \sum_{j=0}^{\infty} \frac{r_<^j}{r_>^{j+1}} P_j(\cos\alpha) P_j(\cos\vartheta),$$

(2.106)

where G is the gravitational constant, α and ϑ are the angles between the axis of the ring and the source and field points. The quantities $r_<$ and $r_>$ denote, respectively, the lesser and greater of the two radial coordinates that specify the positions of the ring and the test mass. If both the source and field masses are concentric rings then $\alpha = \vartheta = \pi/2$.

We denote the mass and radial coordinate of the ith perturbing planet by M_i and R_i, the mass and radial coordinate of the planet under consideration by M_p and r, and the mass of the Sun by $M_\odot$. The unperturbed energy is thus given by

$$E = GM_\odot M_p \langle r^{-1} \rangle.$$

(2.107)

For perturbing planets interior to the orbit considered, the potential energy is given by the infinite series

$$\langle \Delta E \rangle = -GM_p M_i \left[\langle r^{-1} \rangle + \left(\frac{1}{2}\right)^2 \langle R_i^2 \rangle \langle r^{-3} \rangle + \left(\frac{1 \cdot 3}{2 \cdot 4}\right)^2 \langle R_i^4 \rangle \langle r^{-5} \rangle + \cdots \right].$$

(2.108)

For perturbing planets outside the orbit considered, the potential is given by the infinite series

$$\langle \Delta E \rangle = -GM_p M_i \left[\langle R_i^{-1} \rangle + \left(\frac{1}{2}\right)^2 \langle R_i^{-3} \rangle \langle r^2 \rangle + \left(\frac{1 \cdot 3}{2 \cdot 4}\right)^2 \langle R_i^{-5} \rangle \langle r^4 \rangle + \cdots \right].$$

(2.109)

Table 2.3. Precession of the perihelion of Mercury (in arcsec/century) due to perturbations from the other planets. (After Ref. [67].)

Perturbing planet	Three-term ring model	Clemence
Venus	267.359	277.856
Earth and Moon	94.696	90.038
Mars	2.434	2.536
Jupiter	157.646	153.584
Saturn	7.611	7.302
Uranus	0.143	0.141
Neptune	0.044	0.042
Total	529.933	531.499

This approach contains both a multipole expansion and a perturbation expansion. The inclusion of additional higher-order terms in the multipole expansion above would not be justified unless the perturbation expansion were extended beyond first order.

Through the use of these simple three-term expansions, together with data for the planetary masses, orbit radii, and orbital eccentricities, calculations for the advance of the perihelion of Mercury arising from planetary perturbation have been made [67]. These results are summarized in Table 2.3. This yields a value 529.933 arcsec/century, in very close agreement with the more sophisticated calculations of Clemence, [27] which yielded 531.499 arcsec/century.

It was the discrepancy between calculations of this type and the precise measurements of the orbit of Mercury that were available that led Einstein to search for the relativistic corrections described earlier. It is interesting to note that the difference between the results of this simple model and the sophisticated calculations of Clemence is only 1.6 arcsec/century. This is much smaller than the observed 43 arcsec/century discrepancy (described at the end of Section 2.4.9) that required the formulation of General Relativity for its resolution.

2.4.12 Semiclassical self-consistent field (SCSCF)

The quantum mechanical formulation of complex many-electron atoms is often described using the Hartree method [117], which iteratively forms a self-consistent central charge distribution by the superposition of single-electron position probability densities. A semiclassical analogue of the Hartree method has been developed [62] that not only provides conceptual insights into this procedure, but also yields surprisingly accurate results.

The use of time-averaged position probability densities permits the composite of a system of many electrons to be easily superimposed. The radial EBK equation can be written, in terms of the standard quantum numbers $n_r = n - \ell - 1$ and the EBK quantized orbital angular momentum $L = (\ell + \frac{1}{2})\hbar$, as twice the integral from the periapsis to the apoapsis

$$\left(n - \ell - \frac{1}{2}\right)\hbar = \frac{1}{\pi}\int_{A_-}^{A_+} dr\, p(r) \tag{2.110}$$

where the radial momentum is given by

$$p(r) = \sqrt{2m\left(E + \frac{KZe^2}{r}\right) - \frac{\left(\ell + \frac{1}{2}\right)^2\hbar^2}{r^2}}. \tag{2.111}$$

Here we have set $\zeta \to Z$, since we are considering only a single electron in the field of an unscreened point nucleus. Let us now extend the central field to consist of a point nucleus and a second electron that can be described as a spherically symmetric distribution of charge. To describe this we can define an average charge per unit thickness $e\sigma(r)$ on a spherical shell of radius r and thickness dr, due to transits of the electron, given by

$$\sigma(r)dr = \frac{2dt}{T} = \frac{2m}{T}\frac{dr}{p(r)}, \tag{2.112}$$

where we have used $p(r) = m\,dr/dt$. Thus, the charge per unit thickness can be obtained from the momentum and the normalization condition

$$1 = \int dr'\sigma(r'). \tag{2.113}$$

In this case, the EBK equation will still be valid, provided the radial momentum is modified so that

$$p(r) = \sqrt{2m\left(E + \frac{KZe^2}{r} - V_{\text{eff}}(r)\right) - \frac{\left(\ell + \frac{1}{2}\right)^2}{r^2}\hbar^2} \tag{2.114}$$

where $V_{\text{eff}}(r)$ is the effective central potential energy due to the second electron. Using Gauss's law of electrostatics, the electric field E is given by the charge enclosed within a sphere of radius r to be written as

$$\mathsf{E}(r) = \frac{Ke^2}{r^2}\int_0^r dr'\sigma(r'). \tag{2.115}$$

The corresponding potential is obtained by performing a line integral along the field

$$V_{\text{eff}}(r) = \int_\infty^r dr''\mathsf{E}(r'') = Ke^2\int_\infty^r dr''\frac{1}{(r'')^2}\int_0^{r''} dr'\sigma(r'). \tag{2.116}$$

This can be rewritten as a parts integration $u\,dv = d(uv) - v\,du$ using $u \equiv \int_0^{r''} dr'\sigma(r')$ and $dv \equiv dr''/(r'')^2$ to obtain

$$V_{\text{eff}}(r) = Ke^2\left[\frac{1}{r}\int_0^r dr'\sigma(r') + \int_r^\infty dr''\frac{\sigma(r'')}{r''}\right]. \tag{2.117}$$

From this we can define two useful quantities: an internal screening parameter

$$S(r) = \int_0^r dr'\sigma(r'); \tag{2.118}$$

and an external screened energy

$$W(r) = Ke^2 \int_r^\infty dr'' \frac{\sigma(r'')}{r''}. \tag{2.119}$$

This same process can be repeated for any number of passive electrons. We simply label each with an index, hold one fixed, and sum over the others to obtain

$$S_i(r_i) = \sum_{j \neq i} \int_0^{r_i} dr \sigma_j(r) \tag{2.120}$$

$$W_i(r_i) = Ke^2 \sum_{j \neq i} \int_{r_i}^\infty dr \frac{\sigma_j(r)}{r}. \tag{2.121}$$

With this specification, each electron in the atom can then be described by a momentum

$$p_i(r_i) = \sqrt{2m \left(E_i - \frac{Ke^2[Z - S_i(r_i)]}{r_i} + W_i(r_i) \right) - \frac{\left(\ell_i + \frac{1}{2}\right)^2 \hbar^2}{r_i^2}}. \tag{2.122}$$

This quantity occurs not only in the EBK quantization equation

$$\left(n_i - \ell_i - \frac{1}{2}\right) \hbar = \frac{1}{\pi} \int_{A_-}^{A_+} dr_i \ p_i(r_i), \tag{2.123}$$

but also in the expression for the position probability density

$$\sigma_i(r_i) dr_i = \frac{dr \ [p_i(r_i)]^{-1}}{\int dr_i' \ [p(r_i')]^{-1}}. \tag{2.124}$$

Equations 2.122, 2.123, and 2.124 permit a self-consistent iteration to be set up for any atom in terms of its nuclear charge Z and the configurations $\{n_i, \ell_i\}$ of its orbital electrons. Since the eccentricity of each orbital is determined by its value for n_i and ℓ_i, its orbit and hence its charge distribution is completely specified by its semimajor axis (or equivalently, its energy E_i, since both depend on n^2). This set of values for the $\{\sigma_i(r)\}$ can be obtained either from an initial estimate of the $\{E_i\}$, or as a value deduced from a previous iteration of this process. Then a new set of eigenenergies can be determined by separately considering each electron in the effective field of all of the others, and solving the EBK quantum condition of Eq. 2.123 for E_i by numerical inversion (using a numerical technique such as the Newton–Raphson method). After repeating this process for each E_i, the effective charge density can be recomputed, and the entire process can be repeated until convergence to a self-consistent set of values is obtained.

From the self-consistent values for the charge distributions it is possible to compute, for each orbital, expectation values for the various contributions to the energy. For the ith electron, the kinetic energy is given by

$$T_i = \int_0^\infty dr_i \sigma_i(r_i) \frac{[p_i(r_i)]^2}{2m}, \tag{2.125}$$

the potential energy due to the attraction to the nucleus is

$$V_i^{(Z)} = -\int_0^\infty \mathrm{d}r_i \sigma_i(r_i) \frac{Ze^2}{r_i}, \tag{2.126}$$

and the potential energy due to the interelectron repulsion is

$$V_i^{(e)} = \sum_{j \neq i} \int_0^\infty \mathrm{d}r_i \sigma_i(r_i) \int_0^\infty \mathrm{d}r_j\, \sigma_j(r_j) \frac{e^2}{r_>}, \tag{2.127}$$

where $r_>$ is the greater of r_i and r_j. The eigenenergy obtained from the inversion of the quantization conditions is related to these energies by

$$E_i = T_i + V_i^{(Z)} + V_i^{(e)}, \tag{2.128}$$

but because of the double counting inherent in paired electron–electron interactions, the total energy of all of the electrons is given by

$$E = \sum_i \left(T_i + V_i^{(Z)} + \frac{1}{2} V_i^{(e)} \right) = \sum_i \left(E_i - \frac{1}{2} V_i^{(e)} \right). \tag{2.129}$$

This self-consistent procedure is predicated on the questionable assumption that the removal or excitation of one electron does not affect the charge distribution of the other electrons. In the corresponding quantum mechanical case this approximation is known as Koopmans's theorem [135]. [While developing this theorem in 1934 and obtaining his doctoral degree in Physics (under Hendrik Anton Kramers) in 1936, Tjalling Koopmans developed an interest in Economics. In 1975 he shared the Nobel Memorial Prize in Economics for his "Optimum allocation theory."]

This formalism has been applied [62] to the sodiumlike (eleven electron) ions of phosphorus, iron, and molybdenum. (This is an example of an "isoelectronic sequence" in which ions with a fixed number of electrons are traced as a function of nuclear charge.) The results are compared with the results of a Hartree–Fock program in Table 2.4. Despite the apparent simplicity of this semiclassical self-consistent field calculation, there is close agreement between its results and those of the *ab initio* quantum mechanical model that improves with increasing stage of ionization. (With either method the central field approximation improves with higher effective central charge.) For this simple alkali-metallike system, the neglect of exchange and higher-order direct electron–electron interactions in the semiclassical approach seems to introduce only small discrepancies. Notice that the total energies quoted here are of the order of thousands of Rydberg units, whereas spectroscopic measurement accuracies are often to within 10^{-7} Rydberg units or better. This illustrates why the specification of energy differences in the quantum mechanical self-consistent field method is sometimes likened in an old adage to "determining the weight of a ship's captain by weighing the ship with and without the captain aboard."

Figure 2.4 displays a plot of the SCSCF and Hartree–Fock calculations for radial distribution of the neonlike core as seen by the 3s electron in an Fe^{15+} ion. The classical distribution is much more sharply peaked than the quantum mechanical case because of the poles of the periapses and apoapses of the individual orbits that are summed on this plot.

Table 2.4. Comparison of a quantum mechanical (QM) Hartree–Fock calculation with the SCSCF method for contributions to the ground configuration energies (in Rydbergs) for ions in the Na isoelectronic sequence. The electron–electron interaction is denoted by (dir) for direct monopole interaction, by (dir') for direct higher-order interactions, and by (exc) for the exchange interaction. The QM subtotal excludes exchange and higher-order direct moments. (From Ref. [62].)

		P^{4+}		Fe^{l5+}		Mo^{31+}	
		QM	Class.	QM	Class.	QM	Class.
1s	T	212.48	213.22	651.98	645.57	1723.10	1720.32
	$V^{(Z)}$	−437.26	−437.47	−1327.70	−1317.38	−3486.80	−3482.38
	$V^{(e)}$ (dir)	61.70	61.44	119.24	116.62	202.60	205.46
	$V^{(e)}$ (exc)	−1.17	–	−2.79	–	−5.21	–
2s	T	31.93	27.57	123.81	94.22	365.67	349.70
	$V^{(Z)}$	−85.20	−79.51	−289.75	−255.15	−803.58	−786.07
	$V^{(e)}$ (dir)	36.75	35.93	72.73	68.22	124.87	124.50
	$V^{(e)}$ (exc)	−2.63	–	−5.18	–	−8.84	–
2p	T	29.80	27.52	120.71	109.57	361.00	356.48
	$V^{(Z)}$	−81.55	−78.78	−285.38	−272.78	−797.75	−792.79
	$V^{(e)}$ (dir)	38.51	38.51	78.62	77.37	133.92	137.52
	$V^{(e)}$ (dir')	−0.71	–	−1.49	–	−2.61	–
	$V^{(e)}$ (exc)	−0.99	–	−2.05	–	−3.59	–
3s	T	5.83	5.72	39.15	34.25	135.74	134.52
	$V^{(Z)}$	−25.12	−24.44	−109.24	−102.27	−326.94	−325.07
	$V^{(e)}$ (dir)	14.80	14.32	35.20	33.25	63.90	63.62
	$V^{(e)}$ (exc)	−0.25	–	−0.77	–	−1.51	–
	Subtotal	−664.46	−658.61	−2296.00	−2279.20	−6445.90	−6430.97
	Total	−673.47	–	−2315.00	–	−6479.30	–

It is clear that the quantum numbers of these states are too small to expect the quantum mechanical distribution to approach the correspondence limit. Nevertheless, the high level of agreement exhibited in Table 2.4 indicates that the expectation values obtained from the two distributions are nearly the same. Thus it appears that the correspondence limit is reached much more easily for expectation values than for wave functions.

2.4.13 Relativistic formulation of angle-action integrals

Using the relationships for energy and momentum from Special Relativity, the total energy for a velocity-independent potential $V(r)$ is given by

$$E + mc^2 = \sqrt{(pc)^2 + (mc^2)^2} + V(r). \qquad (2.130)$$

Here $V(r)$ is assumed to be part of the fourth component of the momentum four-vector.

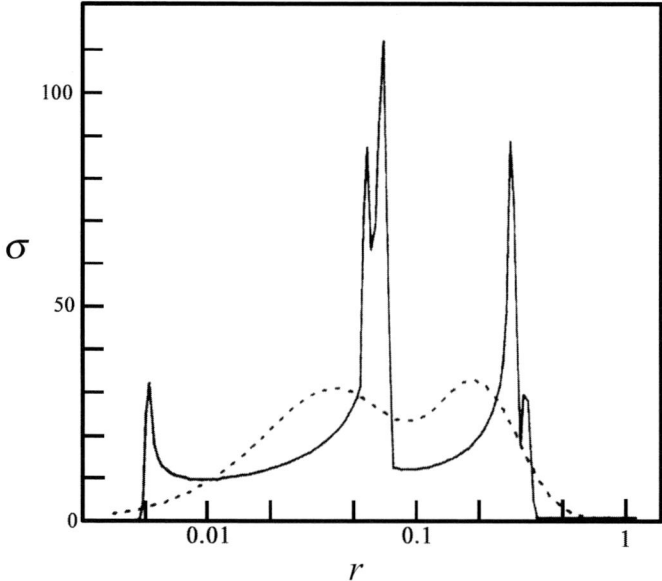

Fig. 2.4. Position probability densities for $1s^2 2s^2 2p^6$ core of the Fe^{15+} ion. The solid curve represents the SCSCF calculation and the dashed curve represents the Hartree–Fock calculation. (From Ref. [62].)

Regrouping, isolating the square root, and squaring both sides

$$(pc)^2 + (mc^2)^2 = \{[E - V(r)] + mc^2\}^2. \tag{2.131}$$

Expanding the squared binomial eliminates the $(mc^2)^2$ terms. Solving the resulting expression for p^2 yields

$$p^2 = p_r^2 + L^2/r^2 = 2m[E - V(r)] + [E - V(r)]^2/c^2. \tag{2.132}$$

For a potential $V(r) = -\kappa/r$ this becomes, after expansion and factorization in powers of $1/r$,

$$p_r = \sqrt{2mE\left(1 + \frac{E}{2mc^2}\right) + 2m\kappa\left(1 + \frac{E}{mc^2}\right)\frac{1}{r} - \left(1 - \frac{\kappa^2}{c^2 L^2}\right)\frac{L^2}{r^2}} \tag{2.133}$$

which is of the form

$$p_r = \sqrt{A + B/r - C/r^2}. \tag{2.134}$$

with

$$A \equiv 2mE\left(1 + \frac{E}{2mc^2}\right)$$

$$B \equiv 2m\kappa\left(1 + \frac{E}{mc^2}\right) \tag{2.135}$$

$$C \equiv L^2\left(1 - \frac{\kappa^2}{c^2 L^2}\right).$$

(Note that, since the system is bound, $E = -|E|$.) The orbit will have classical turning points

$$A_{\pm} = -\frac{B}{2A} \pm \sqrt{\left(\frac{B}{2A}\right)^2 + \frac{C}{A}}. \tag{2.136}$$

Using Eq. 2.39, the semimajor axis is

$$a = -\frac{B}{2A} = -\frac{\kappa}{2E} \frac{(1 + E/mc^2)}{(1 + E/2mc^2)}, \tag{2.137}$$

and the semiminor axis is

$$b = \sqrt{-\frac{C}{A}} = \sqrt{-\frac{L^2}{2mE} \frac{(1 - \kappa^2/c^2 L^2)}{(1 + E/2mc^2)}}. \tag{2.138}$$

Notice that the semiminor axis becomes imaginary if

$$\kappa/cL > 1 \tag{2.139}$$

which produces a breakdown in the definition of periapsis. For an electrostatic potential $\kappa = \zeta k e^2$ and the EBK quantization $L = (\ell + \frac{1}{2})\hbar$, these quantities can be written in terms of the fine structure constant $\alpha \equiv ke^2/\hbar c$, and this condition becomes

$$\alpha\zeta > \left(\ell + \frac{1}{2}\right). \tag{2.140}$$

Thus the small but finite periapsis for $\ell = 0$ s-states caused by the Maslov index vanishes (becomes pure imaginary) in this relativistic formulation when

$$\zeta > 1/2\alpha \cong 68.5. \tag{2.141}$$

Analogous breakdowns occur at high Z in relativistic quantum theory whereby the Coulomb potential becomes too strong to permit stationary states to occur. In the case of the Dirac equation (describing the spin $\frac{1}{2}$ electron) this occurs for $\alpha Z > 1$. For the Klein–Gordon equation (describing a spinless boson, or force carrier) it occurs at $\alpha Z > \frac{1}{2}$. The cause of this breakdown in the presence of very strong Coulomb fields can be attributed [102] to the delocalization of the wave function produced by Zitterbewegung. The intense absorption and emission of virtual photons causes the size of the atom to be dominated by Zitterbewegung, and thus cannot lead to a stationary state. In the spinless case (which is the approximation used in the relativistic SCSCF approach) the breakdown occurs at a lower value of Z that in the spin $\frac{1}{2}$ case.

2.4.14 The EBK quantization

With this relativistic expression for the momentum, the overall form of the quantization integral for a particle in a $1/r$ potential remains the same

$$n_r + \frac{1}{2} = \frac{1}{2\pi} \oint dr \sqrt{A + B/r - C/r^2} = -\sqrt{C} - B/\sqrt{-A}, \tag{2.142}$$

except that the values for the constants A, B, and C assume the values given in Eqs. 2.135, yielding

$$n_r + \frac{1}{2} = -L\sqrt{1 - \frac{\kappa^2}{c^2 L^2}} - \kappa\sqrt{-\frac{2m}{E}\frac{(1 + E/mc^2)}{\sqrt{1 + E/2mc^2}}}.$$ (2.143)

This leads to a quadratic equation that can be solved to obtain E. The angular phase integrals are unchanged, yielding $L = (\ell + \frac{1}{2})\hbar$ so that $n_r = n - \ell - 1$. If we use the electrostatic potential $\kappa = \zeta K e^2$ and note that $\alpha \equiv K e^2/\hbar c$, this yields

$$E + mc^2 = mc^2 \left(1 + \frac{(\alpha\zeta)^2}{\left[n - (\ell - \frac{1}{2}) + \sqrt{(\ell + \frac{1}{2})^2 - (\alpha\zeta)^2}\,\right]^2}\right)^{-1/2}.$$ (2.144)

This is the Sommerfeld relativistic correction to the hydrogen atom. Since it assumes an electrostatic potential (no magnetic effects) it includes only the effects of the relativistic momentum, and not the spin–orbit interaction. However, once again here, the correct quantum mechanical solution to the Dirac equation (which of course neglects the effects of quantum electrodynamics) yields this same equation with the substitution $\ell \to j$.

2.4.15 Relativistic extension of SCSCF

The extension of the semiclassical self-consistent field method described in Section 2.4.12 only requires using the relativistic expression for the momentum, as prescribed in Eq. 2.132, to obtain

$$[p_i(r_i)]^2 = 2m \left[E_i - \frac{K e^2 [Z - S_i(r_i)]}{r_i} + W_i(r_i)\right] - \frac{(\ell_i + \frac{1}{2})^2 \hbar^2}{r_i^2}$$

$$+ \frac{1}{c^2}\left[E_i - \frac{K e^2 [Z - S_i(r_i)]}{r_i} + W_i(r_i)\right]^2$$ (2.145)

in the phase integral and charge density calculation. The relativistic expression for radial speed

$$\frac{1}{c}\frac{d\mathbf{r}}{dt} = \frac{\mathbf{p}c}{\sqrt{(pc)^2 + (mc^2)^2}}$$ (2.146)

can be rewritten in terms of the radial and angular portions as

$$\frac{dr}{dt} = \frac{p_r c^2}{\sqrt{(p_r c)^2 + (Lc/r)^2 + (mc^2)^2}}$$ (2.147)

and used to compute the quantity

$$\sigma(r)dr = \frac{2dt}{T} = \frac{dr}{T\,dr/dt} = \frac{\sqrt{(p_r c)^2 + (Lc/r)^2 + (mc^2)^2}\,dr}{T\,p_r\,c^2}$$ (2.148)

which specifies the charge distribution. The self-consistent calculation can be carried out in the same manner as was done for the nonrelativistic case.

Table 2.5. Isoelectronic comparison of the SCSCF orbital turning points using both the nonrelativistic and the relativistic formulations. The periapses (A_-) and apoapses (A_+) are in units a_0/Z. (After Ref. [62].)

| | Nonrelativistic | | | | Relativistic | | | |
| | 1s | | 2p | | 1s | | 2p | |
Z	A_-	$A+$	A_-	A_+	A_-	A_+	A_-	A_+
42	0.134	1.89	1.45	7.52	0.088	1.64	1.41	7.39
54	0.135	1.89	1.42	7.24	0.049	1.40	1.37	7.13
74	0.133	1.85	1.41	7.10	0	0.96	1.27	6.81

The relativistic formulation contains some interesting features not present in the nonrelativistic formulation. This is illustrated in Table 2.5, which displays the turning points of the 1s and 2p orbitals obtained from the nonrelativistic and relativistic SCSCF calculations for the of the ground configuration energy levels for the copperlike ions Mo^{+13}, Xe^{+25}, and W^{+45}. The periapses and apoapsis are listed in the charge-scaled units a_0/Z to elucidate the relative shifts along the sequence.

As described earlier, in the relativistic case the apoapsis of the s-states collapses for $Z > 68$, but there are also other interesting occurrences. Notice that the apoapsis of 1s is outside the periapsis of the 2p for $Z = 42$, but shrinks inside the 2p by $Z = 74$, with the crossing occurring in the vicinity of $Z = 60$. This observation is valid for the entire np manifold: the 1s apoapsis shrinks inside the locus of the np periapses near $Z = 60$. This crossing is a purely relativistic effect and the nonrelativistic SCSCF calculation shows no such phenomenon. The exact position of the crossing is difficult to determine because the self-consistent convergence is disturbed by it, and results become very sensitive to the initial trial charge distribution. When the 1s apoapsis axis is inside the np periapses it screens them from nuclear attraction and they move further out. When the 1s is outside the np periapses the screening is reduced and they move further in. Thus the classical crossing cannot occur in a smooth isoelectronic manner, but rather as a sudden and discontinuous jump from one distribution of charge to another. If the standard isoelectronic approach is taken whereby the final self-consistent values for one ion are used as initial starting values for a neighboring ion, there is a hysteresis effect, in which the crossing occurs at a different value of Z when approached from below than it does when approached from above.

It is interesting to note that departures from regularity in semiempirical parametrizations of measured data have been observed [62] in the region above $Z = 60$, which suggest that there may be some quantum mechanical counterpart to this semiclassical occurrence.

2.5 Semiclassical formulation of the decay meanlife

The study of atomic structure through spectroscopy is carried out not only through high-resolution wavelength measurements to determine energy-level structures, but also through relative intensity and time-resolved lifetime measurements to determine transition probabilities. The time-dependent process is not a subject usually emphasized in quantum mechanics courses. This de-emphasis probably occurs because of difficulties inherent in presenting the subject in a rigorous and transparent manner. Unlike other applications in atomic physics that can be presented in an elegant exposition, the time dependence (and its manifestation in the exponential decay law) is not a rigorous consequence of quantum mechanics, but rather the result of a series of delicate approximations made in order to conform to experimental observations [155]. There is a great deal of support for the validity of the exponential law in tests over a wide range of time-scales at a precision level of a few percent. However, on a time-scale commensurate with that of the meanlife itself, there exists neither theoretical nor experimental evidence that precludes distortions superimposed on the exponential behavior at the level of parts per thousand [72].

2.5.1 The Wien model

In the consideration of time-dependent processes, it is interesting to note that in 1919 Wilhelm Wien proposed [32] a theoretical model to predict the atomic meanlives that he had measured using time-of-flight methods with "canal rays." Using only classical electrodynamics and the Bohr model of the atom, Wien obtained a specific equation by which he predicted the meanlife of the 3d level in hydrogen to within 10 percent. Upon modern re-examination [32] it has been shown that Wien's lifetime formula is indeed correct to within the limits he prescribed, and yields very nearly correct lifetimes for an even larger class of atomic transitions.

Wien began with Larmor's formula [125] for the energy loss by charge with a time-dependent energy and an acceleration a

$$-\frac{\mathrm{d}E}{\mathrm{d}t} = \frac{2Ke^2a^2}{3c^3}.$$ (2.149)

He assumed that the radiation arises from the centripetal acceleration of an orbiting charge, but he found it necessary to depart from a purely classical picture. To avoid the variation in frequency that would occur from an orbiting charge spiraling in, Wien assumed that the electron first makes the quantum jump, and then radiates the quantum of energy liberated at the constant centripetal acceleration that is characteristic of the lower orbit. For simplicity, Wien restricted his consideration to yrast [112] orbits ($\ell = n - 1$, corresponding to the minimum eccentricity for a given n). In the limit of an ideally circular orbit, the centripetal acceleration is given by

$$a = v^2/r = \omega_e^2 r$$ (2.150)

where $\omega_e = v/r$ is the orbital frequency. He assumed that the quantum of energy was given by the Planck quantum condition

$$E = \hbar\omega_R$$ (2.151)

where ω_R is the frequency of the emitted radiation. Combining these expressions

$$\frac{1}{\tau_W} = -\frac{1}{E}\frac{\mathrm{d}E}{\mathrm{d}t} = \frac{2Ke^2\omega_e^4 r^2}{3\hbar c^3 \omega_R}.$$ (2.152)

To take the correspondence limit, Wien equated the frequencies of the electron and the radiation so they could be specified by the emitted wavelength λ, hence

$$2\pi c / \lambda = \omega_R \approx \omega_e. \tag{2.153}$$

Using the substitution $Ke^2/\hbar = \alpha$, this becomes

$$\frac{1}{\tau_W} = (2\pi)^3 \frac{2\alpha c}{3} \frac{r^2}{\lambda^3}. \tag{2.154}$$

Consistent with the circular orbit assumption, Wien limited the consideration to unbranched orbit jumps that diminish n and ℓ each by one unit. He associated the quantity r^2 with the square of the average of the radius of the lower (final) orbit (Table 2.1)

$$\langle r \rangle_f^2 = \left(\frac{a_0 n_f^2}{\zeta} \right)^2, \tag{2.155}$$

but the average of the square in the case of a transition among yrast states (Table 2.1, $\ell = n - 1$)

$$\langle r^2 \rangle_f = \left(\frac{a_0}{\zeta} \right)^2 \frac{n_f^4}{2} \left[5 - 3\frac{\left(n_f - \frac{1}{2}\right)^2}{n_f^2} \right] \tag{2.156}$$

yields the same result in the correspondence limit of large n_f. Grouping the constants with forethought to standard quantum mechanical definitions

$$(2\pi)^3 \left(4\alpha c a_0^2 / 3 \right) = (1265.38)^3 \; \text{Å}^3/\text{ns}, \tag{2.157}$$

the Wien formula is given by

$$\frac{1}{\tau_W(ns)} = \left[\frac{1265.38}{\lambda(\text{Å})} \right]^3 \frac{n_f^4}{2\zeta^2}. \tag{2.158}$$

For a hydrogenlike system, the wavelength is proportional to $1/\zeta^2$, so the reciprocal lifetime is proportional to ζ^4. For complex atoms, the wavelength factors contain properties of the atom that can improve the reliability of the predictions of the formula.

Using physical wavelengths in alkali-metal and alkali-metallike atoms, this formula predicts quite well the meanlives of the yrast transitions. A sampling of cases calculated by the Wien formula and by quantum mechanical methods [32] is presented in Table 2.6.

If the Wien formula is rewritten using the Balmer formula for hydrogen (Eq. 2.62, noting that the initial principal quantum number is $n_i = n_f + 1$), it becomes

$$\frac{1}{\tau_W} = \frac{2\alpha^4 c \; \zeta^4}{3a_0} \left[\frac{\left(n_f + \frac{1}{2}\right)^3}{n_f^2 \, (n_f + 1)^6} \right] \tag{2.159}$$

whereas the quantum mechanical expression is

$$\frac{1}{\tau_{QM}} = \frac{2\alpha^4 c \; \zeta^4}{3a_0} \left[\frac{(n_f + 1)^{2n_f - 2}(n_f)^{2n_f}}{\left(n_f + \frac{1}{2}\right)^{4n_f + 3}} \right]. \tag{2.160}$$

Table 2.6. Comparison of predictions of the formula of Wien with quantum mechanical values. (From Ref. [32].)

Atom or Ion	Transition	$\lambda(\text{Å})$	τ_W(ns)	τ(ns)
H I	1s–2p	1 215	1.77	1.60
	2p–3d	6 563	17.4	15.5
	3d–4f	18 751	80	73
	4f–5g	40 512	256	235
	5g–6h	74 578	655	608
Li I	2p–3d	6 104	14	14
Be II	2p–3d	1 512	0.85	0.88
B III	2p–3d	677	0.172	0.176
C IV	2p–3d	384	0.056	0.056
Na I	3d–4f	18 465	76.7	71.4
Mg II	3d–4f	4 481	4.39	4.44
Al III	3d–4f	1 936	0.80	0.82

Although these two expressions appear superficially to be quite different, an expansion of these two results yields

$$\frac{\tau_W}{\tau_{QM}} = \frac{(1 + 1/n_f)^{2n_f+4}}{(1 + 1/2n_f)^{4n_f+6}} \approx 1 + 1/2n_f. \tag{2.161}$$

It is interesting to note that this ratio is maximum for $n_f = 2$. (In an initial violation of the correspondence principle, the ratio is 1.110 for $n_f = 1$, increases to 1.127 for $n_f = 2$, decreases to 1.107 for $n_f = 3$, and continues to decrease thereafter.) The only data available to Wien were the 2p–3d elements of the Balmer H-alpha visible manifold (ultraviolet and infrared transitions were not accessible to his apparatus), and this is the least accurate case of the formula.

2.5.2 Connection with the quantum mechanical formulation

This semiclassical development can be connected to the quantum mechanical case by the substitution

$$r^2 \rightarrow 2 \, |\langle f|r|i\rangle|^2 \tag{2.162}$$

where $\langle f|r|i\rangle$ is the quantum mechanical dipole matrix element connecting the initial and final states. In the quantum mechanical case the energy levels may be degenerate, consisting of a manifold of states with the same energy, but a different projection of its angular momentum along a prescribed z axis. The number of degenerate states for the levels i and f are denoted g_i and g_f. The concept of the matrix element is then extended to the definition of the line strength factor S_{if} (defined in Chapter 6), which is summed over the degenerate states

$$S_{if} = \sum_{m_i} \sum_{m_f} |\langle f|r|i\rangle|^2 .$$ (2.163)

With these substitutions made in Eq. 2.154, the result yields the correct expression for the unbranched transition probability

$$\frac{g_i}{\tau_i(ns)} = \left[\frac{1265.38}{\lambda(\text{Å})}\right]^3 S_{if} .$$ (2.164)

These relationships will be developed in more detail in Chapter 6.

3

Semiempirical parametrization of energy-level data

What seems like black and white drabness, can reveal hidden color sublime; you need
only to look at it deeply, and see it one part at a time.

3.1 Historical development

The study of optical radiation, dispersed to reveal its frequency content, has a long and
venerable history. However, the fact that this radiation consists of a continuous distribution
of colors when emitted by free ions in a dense plasma or solid, and of a discrete distribution
of lines of color when emitted by an atomic gas, was long unnoticed. The first recorded
observation of the dispersed solar (ark) spectrum is usually attributed to Noah, who beheld
the rainbow after the flood. In Genesis 9:13, God is reported to have said "I have placed my
rainbow in the clouds." Regrettably, no revelation of the Fraunhofer lines was reported.

The first published observation of a dispersed solar spectrum using a slit and a prism
was by Isaac Newton in his 1666 treatise on optics. Again, Newton made no mention of
observing dark lines superimposed on the continuous "Phænomena of Colours." The first
recorded observation of a line spectrum was by Thomas Melvill [154] in 1752. Melvill
inserted a piece of sea-salt into a flame and allowed the emitted radiation to pass through
a slit onto a prism. He noted a "constancy of refrangeability" of the bright yellow sodium
light.

The observation of seven dark lines superimposed on the solar spectrum was noted [195]
by William Wollaston in 1802, and that number was increased to several hundred by Joseph
Fraunhofer [99] in 1814. In 1859 Robert Bunsen and Gustav Kirchhoff [20] combined
the experiments of Melvill and Fraunhofer to launch the field of laboratory astrophysics.
They allowed both sunlight and light from a sodium vapor lamp to pass simultaneously,
in controllable amounts, through the slit of their spectrometer thus superimposing these
spectra. By balancing intensities they were able to fill in Fraunhofer's "D" dark gaps in the
solar spectrum with Melvill's bright yellow sodium light. In this manner they discovered that
the element sodium is present in the cooler outer envelope of the Sun, and absorbs light at
these characteristic frequencies from the continuum radiation emitted from the Sun's plasma
interior. It is interesting that in 1835 the French positivist philosopher August Comte had
cited the chemical constitution of the stars as an example of a type of information that
would be eternally inaccessible to man. Comte modeled his theory of science on Laplacian

determinism, and rejected the inclusion of properties that were imperfectly or incompletely measured. Comte died in 1857, and was thus denied knowledge of the power of atomic spectroscopy.

In Uppsala, Anders Jonas Ångström [1] had anticipated the work of Kirchhoff in concluding that a cool gas absorbs just those wavelengths of light that it emits (in transitions to the ground state) when it is hot. Ångström not only extended the use of spectroscopy in laboratory astrophysics, but also placed the wavelength measurements on a firm and precise scale in terms of the unit that bears his name (the measurements of Kirchhoff and Bunsen were on an arbitrary scale based only on markings on the spectrometer in their Heidelberg laboratory).

In 1885, the precise measurements of the visible spectrum of the hydrogen atom by Ångström attracted the attention of Johann Jakob Balmer. Balmer was a sixty-year-old teacher of projective geometry in a girl's school in Basel, and applied an interesting geometric construction [7] to his study of the spacing of these spectral lines. In this way he discovered [6] that the spacings of these lines bear relationships that involve the squares of integers. This resulted in the Balmer formula which, although based on only the visible transitions to the $n = 2$ level, also predicted the wavelengths of many other lines in the then-unobserved ultraviolet and infrared regions.

Although the Balmer formula was applicable only to single-electron atoms, in 1889 Janne Rydberg showed [173] that (as initially presented in an 1887 Research Proposal) the Balmer formulation can be extended to describe wide classes of complex atoms. Rydberg's method required that the integer values be shifted slightly by a constant, which we now call the quantum defect δ. (The origin of the success of this method was discussed in Section 2.4.6.)

With increased accuracies of measurement, the specification of spectra in terms of extracted quantum defects soon revealed that these quantities themselves have deeper regularities. It was shown [192] in 1903 by Ritz that these quantities could accurately be expressed by a low-order power series in the quantity $1/(n - \delta)^2$. Semiempirical approaches such as these were polished and extended [87] to a high level of sophistication by Bengt Edlén. These methods now permit interpolative and extrapolative predictions that match the extremely high precision of the measured database.

An essential aspect of the analysis of spectra is the Ritz combination principle, which was recognized by Rydberg and subsequently codified [170] by Ritz in 1908. This states that it is possible to find pairs of spectral lines such that the sums of their reciprocal wavelengths correspond to the reciprocal wavelength of another observed line. This principle is now obvious since we know that spectra are due to transitions between energy levels whose difference is proportional to the reciprocal of the emitted wavelength. Thus atomic states with energy-ordered labels 1, 2, and 3 can decay in a single step 3–1, or in two steps 3–2 and 2–1, with the sum of the reciprocal wavelengths of the latter equaling that of the former. For this reason, spectroscopic energy-level compilations are usually given in the units of cm^{-1} (Kaysers). Conversions to SI energy units would be possible, but would introduce constants ($\hbar$ and c) that may be known to lower precision than the wavelength measurements. (For optical wavelengths measured in air the values must be converted to vacuum values using the index of refraction of air.) Thus, energy levels in the examples to follow will be given in cm^{-1}.

The nomenclature of spectroscopy labels states with $\ell = 0, 1, 2, 3$ by the letters s, p, d, f, corresponding to the words sharp, principal, diffuse, fundamental. The origin of this notation lies in the appearance of the lines emitted by alkali-metal atoms. As has been discussed [151] by McGucken, by 1890 three types of alkali spectral series had been characterized according to their distinctive appearance. The reasons for these characterizations can now be understood in terms of whether: the states were simple or composed of multiple sublevels; the states had a transition to the ground state; the spacings of the sublevels could be resolved by then-existing methods.

Alkali transitions involving an upper state with $\ell = 1$ are dominated by the doublet resonance transitions to the $\ell = 0$ ground state, which are strong, distinct, and easily absorbed (and, with the exception of lithium, the doublet was resolved in 1890). Thus they were characterized as "principal" by Kayser and Runge and by Rydberg. Alkali transitions involving upper levels with $\ell = 2$ exhibit fine structures in both the upper and lower states due to electron spin, and this was unresolved in 1890. Thus these were characterized as "diffuse" by Liveing and Dewar and by Rydberg. Alkali transitions involving an upper state with $\ell = 0$ owe their fine structure entirely to the splitting of the lower $\ell = 1$ levels, which was sufficiently large to be resolved (except for lithium) in early work. Thus they were characterized as "sharp." Additional lines were observed which did not belong to these three readily distinguishable groups, many of which involved transitions from an upper state with $\ell = 3$. Much later, Bergmann characterized these lines as "fundamental." This notation was extended and standardized at an informal meeting of spectroscopists in 1928, as reported in Ref. [172]. Details of this and other aspects of spectroscopic notation are given in Chapter 5.

3.2 Description of complex atoms by two-body collective modes

Certainly a complex atom must be considered a many-body problem, and must, therefore, be treated by approximation methods. However, the definition of a "complex atom" here must be considered in the context of the following oft-repeated [150] quotation:

> ...the many-body problem may be defined as *the study of the effects of interaction between bodies on the behaviour of a many-body system.* (It might be noted here, for the benefit of those interested in exact solutions, that there is an alternative formulation of the many-body problem, i.e., how many bodies are required before we have a problem? G. E. Brown points out that this can be answered by a look at history. In eighteenth-century Newtonian mechanics, the three-body problem was insoluble. With the birth of general relativity around 1910 and quantum electrodynamics around 1930, the two- and one-body problems became insoluble. And within modern quantum field theory, the problem of zero bodies (vacuum) is insoluble. So, if we are after exact solutions, no bodies at all is already too many!)

3.2.1 The nonrelativistic two-body problem

The description of a one-electron atom with a nucleus of finite mass is achieved by the well-known reduced mass transformation. In this procedure the two interacting masses m and M

are replaced by two collective modes. One is a fictional particle of mass $(m + M)$ moving at the center-of-mass, and the other is a fictional particle of mass $mM/(m + M)$ moving about the center-of-mass. When consideration is restricted to two particles their center-of-mass is a property of the entire system, hence the solution is obtained by consideration the $mM/(m + M)$ collective mode alone. Thus, in the case of hydrogen and hydrogenlike single-electron ions, the solution can be obtained simply by expressing energies and lengths in reduced-mass values for the Rydberg energy

$$R_Z \equiv Ry/(1 + m/M) \tag{3.1}$$

and the Bohr radius

$$a_Z \equiv a_0(1 + m/M). \tag{3.2}$$

(Recall that $Ke^2 = 2Ry\, a_0$ and $\hbar^2/m = 2Ry\, a_0^2$.) It is conceptually instructive to notice that the distance from the electron to the center-of-mass is independent of the nuclear mass. In the case of an infinite nuclear mass, the nucleus resides at the center-of-mass, whereas for finite mass the nucleus retreats to an orbital position behind the center-of-mass. This increases the electron–nucleus separation, and thus reduces the attractive Coulomb energy between them. These facts have been incorrectly presented in some textbooks [152].

3.2.2 Relativistic nonseparability of the two-body problem

Even for this simple binary case, the energy independence of these collective modes occurs only in the nonrelativistic approximation. In the relativistic expressions, values for the energy and momenta in the center-of-momentum (CM) system can be written as

$$E = E_m + E_M$$
$$(Pc)^2 = E_m^2 - (mc^2)^2 = E_M^2 - (Mc^2)^2, \tag{3.3}$$

where E is the total energy, E_m and E_M are the energies of the masses m and M, and P is the magnitude of the momentum of either particle (since in the CM system the momenta are equal and opposite). These can be solved for the CM energies of the individual particles by considering

$$E_M^2 = (E - E_m)^2 = E^2 - 2EE_m + E_m^2. \tag{3.4}$$

Subtracting $(Pc)^2$ from both sides of this equation and noting the triangle relationship relating energy, momentum, and mass, this becomes

$$(Mc^2)^2 = E^2 - 2EE_m + (mc^2)^2. \tag{3.5}$$

This can be solved for E_m to obtain

$$E_m = \frac{E^2 + (mc^2)^2 - (Mc^2)^2}{2E}. \tag{3.6}$$

The momentum can be obtained from this using

$$Pc = \sqrt{E_m^2 - (mc^2)^2}$$
$$= \frac{1}{2E}\sqrt{[E^2 + (mc^2)^2 - (Mc^2)^2]^2 - (2Emc^2)^2}. \tag{3.7}$$

This expression can be expanded and refactored to obtain the form

$$Pc = \frac{1}{2E}\sqrt{(E - mc^2 - Mc^2)(E - mc^2 + Mc^2)(E + mc^2 - Mc^2)(E + mc^2 + Mc^2)}. \tag{3.8}$$

The kinetic energies can be computed using

$$T_m = E_m - mc^2 = \frac{1}{2E}[E^2 + (mc^2)^2 - (Mc^2)^2 - 2Emc^2]. \tag{3.9}$$

This can also be expanded and refactored to obtain the form

$$T_m = \frac{1}{2E}[(E - mc^2 - Mc^2)(E - mc^2 + Mc^2)]. \tag{3.10}$$

The total energy E can be written in terms of the available kinetic energy Q, and its partition, T_m and T_M, between the particles

$$Q = T_m + T_M = E - mc^2 - Mc^2. \tag{3.11}$$

In terms of Q the momentum is given by

$$P = \sqrt{\frac{2Q(m + Q/2c^2)(M + Q/2c^2)(m + M + Q/2c^2)}{(m + M + Q/c^2)^2}}, \tag{3.12}$$

and the kinetic energies are given by

$$T_m = \frac{Q(M + Q/2c^2)}{(m + M + Q/c^2)} \; ; \quad T_M = \frac{Q(m + Q/2c^2)}{(m + M + Q/c^2)}. \tag{3.13}$$

In the approximation of small Q the momentum reduces to

$$P \approx \sqrt{2Q\frac{mM}{m + M}}, \tag{3.14}$$

and the kinetic energies reduce to

$$T_m \approx Q\frac{M}{m + M} \; ; \quad T_M \approx Q\frac{m}{m + M}. \tag{3.15}$$

This approximation yields the familiar nonrelativistic energy partition of Q that depends only on the masses. It should be noted, however, that in the relativistic case the partition of Q depends not only on the masses but also on the value of Q itself. Thus, at least in principle, every different energy level in an atom possesses a slightly different CM energy partition. Fortunately, the corrections for this nonseparability are negligible in most practical cases.

3.2.3 Extension to complex atoms

For multielectron atoms, the many-body problem is invariably modeled as a sum of two-body pairings, and the motion of each individual electron is transformed to a separate center-of-mass representation with the nucleus. However, the motion of the nucleus occurs in response to all of the electrons in the atom, so the fictional $(m + M)$ center-of-masses of all of the individual pairings move relative to each other and relative to the center-of-mass of the entire system. In a gravitational analogue, the center-of-mass of the Earth–Sun binary moves relative to the center-of-mass of the Jupiter–Sun binary, in a way not fully described by the Earth–Sun binary reduced mass.

In the study of isotope shifts in hyperfine structure, the characterization of the relative motion of these many pairwise center-of-mass systems is referred to [136] as the "coupling effect". These effects differ for various energy levels in the same atom, and are very difficult to calculate for complex atoms. The effect may operate in either the same or the opposite direction as the simple nuclear motion effect. This depends on whether the electrons move in the same or mutually opposite directions, thus causing an increase or a decrease in the nuclear motion to keep the center-of-mass of the entire system at rest. The effects of nuclear motion are usually much smaller when applied to the gross structure (dependent on n and ℓ only) and the fine structure (dependent on electron angular momentum coupling) of the atom than they are for specification of the hyperfine structure (dependent on nuclear properties). Nevertheless, some care should be exercised in carrying over the reduced-mass concept into the specification of multielectron systems.

3.3 The Rydberg formula and the Ritz parametrization

A Rydberg series consists of the set of values for the binding energies of the electron for various values of n with a fixed value of ℓ. In the Rydberg formula, these binding energies are described by a hydrogenlike Balmer formula with an effective value for the principal quantum number that is deduced from the measured data. The deviation of this effective quantum number from the corresponding hydrogenlike integer quantum number (the quantum defect) is then expanded in powers of the measured binding energy (a Ritz parametrization). This simple algorithm has been found to have immense predictive power with a small number of free parameters, producing accuracies that are virtually unrivaled in the physical sciences.

Despite its long history and phenomenological origins, the importance of the Rydberg–Ritz formulation is undiminished, and it remains a primary tool of modern ultrahigh-precision energy-level spectroscopy [101]. The Rydberg constant is now known to better than 11 significant figures, placing it among the most accurately known quantities in nature. The Rydberg–Ritz formulation permits precise optical measurements of energy levels in a Rydberg series to be extrapolated with great accuracy to the ionization limit, permitting the determination of ionization potentials in atoms and ions with accuracies in excess of nine significant figures.

Although the accuracy of the Rydberg–Ritz formulation is well known, the reasons for its success can be concealed by its empirical origins. A semiclassical exposition is presented below that provides some insight into the reasons for the form and success of this algorithm.

3.3.1 Semiclassical derivation of the Rydberg–Ritz formula

As discussed in Section 2.4.6, all perturbations to a pure Coulomb potential that vary as r^k with $k \leq -2$ result in a correction with a leading term that is proportional to $1/n^3$ (see Eq. 2.70). Such corrections can be taken into account in the energy by introducing an effective quantum number $n^* = n - \delta$ into the Balmer equation, thus transforming it into the Rydberg formula

$$E = -\frac{Ry\, \zeta^2}{(n - \delta)^2}.$$

(3.16)

In the quantum mechanical formulation, δ corresponds to a phase shift $r' \to r + \delta\pi$ in the asymptotic form of the radial wave function at large r.

Since δ is the only free parameter in the Rydberg formula, it provides a convenient one-to-one mapping parameter by which the raw energies can be re-expressed as an equivalent quantity that is a slowly varying function of the other quantum numbers. We shall examine here the expected behavior of δ as a function of the principal quantum number n, if all other quantities are held fixed.

In the EBK quantization for a hydrogenic system, the radial quantization integral is given by

$$\left(n - \ell - \frac{1}{2}\right)\hbar = \frac{1}{2\pi} \oint dr\, p(r).$$

(3.17)

In the case of a pure Coulomb potential, the orbit closes and the contour integral can be written in terms of a real integral between the periapsis and the apoapsis

$$\left(n - \ell - \frac{1}{2}\right)\hbar = \frac{1}{\pi} \int_{A_-}^{A_+} dr \sqrt{2m\left[E + \frac{K\zeta e^2}{r} - \frac{\left(\ell + \frac{1}{2}\right)^2 \hbar^2}{r^2}\right]}.$$

(3.18)

For an atom possessing a core of "running" electrons in addition to the "jumping" electron, the potential will be modified from the pure Coulombic form for small values of r. These deviations are often characterized in terms of the polarization and penetration of the core by the jumping electron. In a system for which the potential $V(r)$ deviates slightly from the Coulombic form the orbit will precess, and the conversion of the contour integral to the form above cannot rigorously be made. However, a useful approach is to retain the form of the real integration, and include the effects of the correction for orbit precession by moving its factor to the left side of the equation, as the quantum defect parameter $\delta\hbar$,

$$\left(n - \delta - \ell - \frac{1}{2}\right)\hbar = \frac{1}{\pi} \int_{A_-}^{A_+} dr \sqrt{2m\left[E - V(r) - \frac{\left(\ell + \frac{1}{2}\right)^2 \hbar^2}{r^2}\right]}.$$

(3.19)

In this form, the quantum defect can be obtained by subtracting Eq. 3.19 from Eq. 3.18 to obtain

$$\delta = \frac{\sqrt{2m}}{\pi\hbar} \int_{A_-}^{A_+} dr \left[\sqrt{E + \frac{K\zeta e^2}{r} - \frac{(\ell + \frac{1}{2})^2 \hbar^2}{r^2}} - \sqrt{E - V(r) - \frac{(\ell + \frac{1}{2})^2 \hbar^2}{r^2}} \right].$$

(3.20)

If we assume that the deviations of $V(r)$ from pure Coulombic form occur only in the inner core near the periapsis, then

$$V(r) \to -\frac{K\zeta e^2}{r} \qquad (r \gg A_-)$$

(3.21)

and the two square roots in the integral for δ effectively cancel beyond an upper limit r_{core}. In this inner region of small r the kinetic and potential energies are both large, and dominate over the binding energy, which is the small difference between these two large numbers. We can denote and characterize this by defining the quantities W_0 and W

$$W_0 \equiv \frac{K\zeta e^2}{r} - \frac{(\ell + \frac{1}{2})^2 \hbar^2}{r^2} \geq |E|$$

$$W \equiv -V(r) - \frac{(\ell + \frac{1}{2})\hbar^2}{r^2} \geq |E|.$$

(3.22)

Since these quantities dominate over the binding energy in the square roots, the integrals can be Taylor-expanded

$$\delta = \frac{\sqrt{2m}}{\pi\hbar} \int_{A_-}^{r_{core}} dr \left[\sqrt{W_0} \left\{ 1 + \frac{1}{2}\left(\frac{E}{W_0}\right) - \frac{1}{8}\left(\frac{E}{W_0}\right)^2 + \cdots \right\} \right.$$
$$\left. - \sqrt{W} \left\{ 1 + \frac{1}{2}\left(\frac{E}{W}\right) - \frac{1}{8}\left(\frac{E}{W}\right)^2 + \cdots \right\} \right].$$

(3.23)

If we express E in terms of the Rydberg formula $E = -Ry\,\zeta^2/(n-\delta)^2$, this expression can written

$$\delta = a + \frac{b}{(n-\delta)^2} + \frac{c}{(n-\delta)^4} + \cdots.$$

(3.24)

and thus yields the Ritz formula. The constants can be formally expressed as the integrals

$$a \equiv \frac{\sqrt{2m}}{\pi\hbar} \int_{A_-}^{r_{core}} dr[\sqrt{W_0(r)} - \sqrt{W(r)}]$$

(3.25)

$$b \equiv \frac{\sqrt{2m}}{\pi\hbar} Ry\,\zeta^2 \int_{A_-}^{r_{core}} dr \left[\frac{1}{2\sqrt{W_0(r)}} - \frac{1}{2\sqrt{W(r)}} \right]$$

(3.26)

$$c \equiv -\frac{\sqrt{2m}}{\pi\hbar} (Ry\,\zeta^2)^2 \int_{A_-}^{r_{core}} dr \left[\frac{1}{8[W_0(r)]^{3/2}} - \frac{1}{8[W(r)]^{3/2}} \right]$$

(3.27)

and numerical values for these quantities can be experimentally determined directly from measured data.

The Ritz expansion is very useful for the semiempirical systematization of certain types of measured energy-level data. The measured data can be converted to a corresponding set of empirical values δ, and the fitting parameters a, b, c, etc., are deduced by their least-squares adjustment to the Ritz expansion. The Ritz expansion can then be used to interpolate and extrapolate the data base. The quantity δ appears on both sides of the Ritz equation, but the predictive use of the equation can be done very effectively through successive iteration.

3.3.2 Use of the Rydberg–Ritz quantum defect parametrization of spectroscopic data

The Rydberg formula describes the binding energy for the outermost electron in an atomic system. However, the binding energy is not the quantity that is usually measured. The energy-level scheme is constructed from the energy differences between the levels by observation of the radiation emitted in the various transitions. This system is usually formulated in terms of excitation energies defined relative to the ground state, and the position of the ionization limit is not directly measurable to the high spectroscopic accuracy that characterizes the excitation energy measurements.

The ionization limit (or ground-state ionization energy) of an atom or ion is defined as the amount of energy required to remove the most weakly bound electron from the system in the ground state, while leaving the ion subsequently formed also in the ground state. The ionization potential is the corresponding energy/charge expressed in volts. Ionization limits can be measured in a number of ways, both directly and indirectly. Direct methods include electron-impact studies of the Franck–Hertz type, as well as photoionization measurements. However, owing to the very high accuracy that can be achieved through optical measurements, indirect spectroscopic methods offer significant advantages. Through the measurement of transition wavelengths for a finite number of levels along a Rydberg series, the ionization limit can be specified with great precision through the extrapolation of the Rydberg–Ritz formula to infinite n.

Thus, the ionization limit I_0 is often left as an additional fitting parameter to be evaluated from the measured data. The applicability of this method requires that the ionization limit be a well-defined physical quantity, which may not be the case in situations where the levels are affected by quantum mechanical mixing between two or more electron configurations that converge to different ionization limits.

The quantum defect formulation employs a binary representation, in which a single electron is modeled as interacting with another single entity that consists of the nucleus dressed with the core electrons. As long as the application is restricted to an excited electron that is well separated from the core electrons, the binary approximation should be valid. The orbital period of the outer electron will be much longer than that of the core electrons, thus averaging out the coupling effect. The reduced-mass transformation can be used, and the quantum defects extracted will reflect the shift in phase of the radial wave function relative to a true hydrogenic atom of the same nuclear mass. The energies are expressed in reduced Rydbergs R_Z and the distances are expressed in center-of-mass modified Bohr radii a_Z.

In terms of these quantities, the Rydberg formula for a level with quantum numbers n, ℓ, j (principal, orbital angular momentum, and total angular momentum) is given by

$$I_0 - E_{n\ell j} = \frac{R_Z \zeta^2}{(n - \delta_{n\ell j})^2}. \tag{3.28}$$

Here both the ionization limit and the level energy $E_{n\ell j}$ are excitation energies above the ground state, so the quantity $I_0 - E_{n\ell j}$ is the binding energy of the level. The quantity $\zeta = Z - N_e + 1$ is the net charge of the nucleus and core electrons. (In later sections various charge-screening parametrizations will be developed, but in the quantum defect formulation it is essential to parametrize only the effective quantum number so as to obtain a unique one-to-one mapping of the data.)

The quantity ζ also characterizes the number of the spectrum, with $\zeta = 1$ for the neutral atom, $\zeta = 2$ for the singly charged ion, $\zeta = 3$ for the doubly charged ion, etc. Two alternative notations are used to specify the degree of ionization. One denotes ζ by a roman numeral, the other uses a superscript to indicate the number of electrons removed. Thus, a neutral magnesium atom is denoted as Mg I, whereas the Mg^{+1} ion is denoted as Mg II, the Mg^{+2} ion by Mg III, etc.

The measured data are mapped into quantum defects using Eq. 3.28 and then fitted to a Ritz expansion

$$\delta_{n\ell j} = a_{\ell j} + \frac{b_{\ell j}}{(n - \delta_{n\ell j})^2} + \frac{c_{\ell j}}{(n - \delta_{n\ell j})^4} + \cdots. \tag{3.29}$$

The remapping of the data is done by considering the dimensionless quantities

$$t_{n\ell j} \equiv \frac{1}{(n - \delta_{n\ell j})^2} = \frac{I_0 - E_{n\ell j}}{R_Z \zeta^2} \tag{3.30}$$

$$\delta_{n\ell j} = n - 1/\sqrt{t_{n\ell j}}. \tag{3.31}$$

For a given Rydberg series (various values for n and fixed values for ℓ and j), the fitting parameters I_0 and $a_{\ell j}, b_{\ell j}, c_{\ell j}, \ldots$ can be obtained by a least-squares adjustment (a nonlinear search for the ionization limit, a linear regression for the Ritz parameters).

When the best-fitted value of I_0 has been obtained (since this quantity is characteristic of the atom or ion, its value can combine the results of analysis of many Rydberg series of various ℓ and j in that system), the quality of the fit can be displayed in an exposition of $\delta_{n\ell j}$ vs $t_{n\ell j}$.

An example is given in Fig. 3.1. Experimental characterization of a spectrum involves identifying the quantum numbers corresponding to the measured levels. The energy levels are then reduced to values for the quantum defect using the Rydberg formula. The points (denoted by circles, beginning with $n = \ell + 1$ in the upper right portion of the plots) represent the measured values. Notice that in this type of plot both the ordinate and the abscissa involve measured quantities, and differ from the "dependent" and "independent" variables that occur when a mathematical function is plotted versus a continuous variable. Error bars on this type of plot are not vertical, since a shift in δ alters both coordinates.

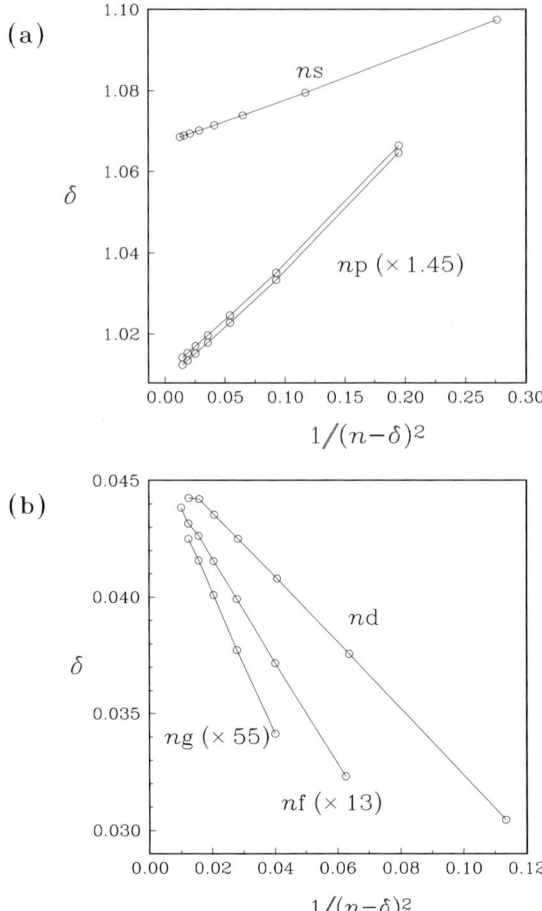

Fig. 3.1. Rydberg–Ritz plot for Mg II.

The upper panel of Fig. 3.1 shows the ns, np($j = \frac{1}{2}$), and np($j = \frac{3}{2}$) levels, and the lower panel shows the nd($j = \frac{3}{2}$), nf($j = \frac{5}{2}$), and ng($j = \frac{7}{2}$) levels. Since the abscissa is in reciprocal units, the plots increase in n from right to left, and reveal a characteristic trend that warrants consideration.

Although there are many possible phenomenological origins for the deviations from the pure Coulombic potential that are embodied in the quantum defect, these are usually characterized in terms of effective values for the polarization and the penetration of the core by the orbital electron. Notice that the quantum defects decrease with increasing n for the s- and p-orbitals, and increase with increasing n for the d-, f-, and g-orbitals. This provides an empirical operational definition of whether the nonCoulombic corrections are dominated by core polarization or by core penetration, since it is characterized by the sign of the second Ritz parameter $b_{\ell j}$. A negative value for $b_{\ell j}$ indicates that penetration effects are large and thus dominate over the ever-present effects of polarization. A positive value for

$b_{\ell j}$ indicates that penetration effects are negligible, so that the weaker polarization effects can be observed. (Notice that the various Rydberg series in Fig. 3.1 have been multiplied by overall scaling factors so they can appear on a single plot. Thus the plot indicates the slopes more effectively than the overall strengths of the corrections.) The reasons for this signature criterion can be made plausible in terms of classical arguments.

The penetration effects are dominated by monopole corrections due to the increase in the effective central charge ζ_{eff} seen by the running electron along the segment of its orbit where it penetrates the core ($\zeta_{eff} > Z - N_e + 1$, the fully screened limit). As was shown in Section 2.4.4, the periapsis and apoapsis of the orbiting electron are approximately

$$A_- \cong a_0(\ell + 1/2)^2/2\zeta_{eff}; \qquad A_+ \cong 2a_0 n^2/\zeta_{eff}. \tag{3.32}$$

For fixed ℓ the periapsis is nearly independent of n, whereas the apoapsis increases strongly with n. If the radius vector of the orbital motion is to sweep out equal areas in equal times (with an increased total area but no increase in the distance of closest approach), then the time spent inside the core must decrease with increasing n. Thus the quantum defect due to penetration effects must decrease with increasing n.

The polarization effects are primarily due to the distortion of the core in the presence of the electric field of the orbital electron. For the dipole polarizability it was shown in Section 2.4.10 that this is given by

$$\langle \Delta E \rangle = -\frac{\alpha_d Ry\, a_0}{2n^3 \ell_{op}^5} \left(\frac{\zeta}{a_0}\right)^4 \left[3 - \frac{\ell_{op}^2}{n^2}\right]. \tag{3.33}$$

By expansion of the Rydberg formula it was also shown in Section 2.4.10 that the quantum defect can be written in terms of a perturbation as

$$\delta = -\frac{n^3 \Delta E}{2Ry\, \zeta^2} \left[3 - \frac{\ell(\ell + 1)}{n^2}\right] \tag{3.34}$$

(here the quantum mechanical expression $\ell_{op} = \ell(\ell + 1)$ has been inserted), so the second Ritz parameter for this polarization correction is given by

$$b_{\ell j} = -\ell(\ell + 1)a_{\ell j}/3. \tag{3.35}$$

Thus the polarization contribution to the second Ritz coefficient is negative, and is small for low ℓ (where penetration is large because of the high eccentricity of the orbit) and becomes larger for higher ℓ (where the more nearly circular orbit does not penetrate the core). As can be seen from Table 2.1 in Chapter 2, the second terms of the expectation values of all powers of r are negative, since this is a property of the Legendre polynomials. Thus the quantum defect due to polarization alone must increase with increasing n.

This pure polarization result differs from the case where penetration effects are significant. As discussed earlier, for fixed ℓ the periapses of the orbits are nearly independent of n, but speed at periapsis increases with increasing n. Thus as n increases, the penetrating electron spends less time inside the core, hence ζ_{eff} decreases. The lower effective charge leads to less binding and a smaller quantum defect. Thus the quantum defect due to penetration decreases with increasing n.

Table 3.1. Rydberg–Ritz systematization of the energies
(in cm^{-1}) of the np series in Mg II.

n	np ($j = 1/2$)		np ($j = 3/2$)	
	Obs.	Rydberg[a]	Obs.	Rydberg[b]
3	35 669.31	35 669.23	35 760.88	35 760.80
4	80 619.50	80 619.46	80 650.02	80 649.99
5	97 455.12	97 455.10	97 468.92	97 468.90
6	105 622.34	105 622.32	105 629.72	105 629.71
7	110 203.58	110 203.55	110 207.99	110 207.96
8	113 030.25	113 030.23	113 033.09	113 033.06
9	114 896.79	114 896.75	114 898.72	114 898.69
10		116 193.66		116 195.04
11		117 131.28		117 132.29
12		117 831.06		117 831.83
13		118 367.17		118 367.76
14		118 786.93		118 787.40
15		119 121.74		119 122.12
16		119 393.07		119 393.38
17		119 616.01		119 616.26
18		119 801.41		119 801.63
19		119 957.27		119 957.45
20		120 089.52		120 089.68
∞		121 267.61		121 267.61

[a] $\delta(j = 1/2) = 0.697\,025\,9 + 0.169\,484\,6t + 0.130\,371t^2 + 0.064\,51t^3$
[b] $\delta(j = 3/2) = 0.695\,758\,8 + 0.170\,099\,5t + 0.128\,363t^2 + 0.072\,47t^3$

A Rydberg–Ritz systematization of the np $j = \frac{1}{2}$ and $\frac{3}{2}$ levels in the Mg II example of
Fig. 3.1 is shown in Table 3.1. Data are taken from Ref. [87].

3.4 The core polarization model

The Rydberg–Ritz quantum defect approach permits the prediction of levels of arbitrary n in
a Rydberg series of fixed quantum numbers ℓ and j. This parametrization requires a separate
set of Ritz parameters for each Rydberg series in an atom or ion, and the formulation can
be applied only to those Rydberg series for which a suitable database exists.

For nonpenetrating high n and ℓ states, another formalism exists [15, 87] that can extract
fundamental atomic structure parameters from the available database. This formalism can
be used to predict *all* of the energy levels in the system that satisfy the nonpenetrating
criterion.

For sufficiently high principal and orbital angular momentum quantum numbers, the orbit
of the highly excited outermost electron is large and nearly circular, and the core penetration
becomes negligible. In such cases, the active electron and passive core are coupled only by

central electrostatic interactions, and can be described by the so-called "core polarization model." In this model [44], the core is represented by a point nucleus surrounded by a deformable cloud of negative charge. The magnitude of the core charge is denoted by ζ, its dipole and quadrupole polarizabilities are denoted by α_d and α_Q, and the nonadiabatic correlation factor (a correction for the inability of the core to instantaneously respond to the motion of the orbital electron) is denoted by 6β (the factor six corresponds to a commonly used normalization definition for this quantity). The range of ℓ that can be included can sometimes be extended by defining [46] a penetration correction f_ℓ that is a constant for a given value of ℓ, that decreases with ℓ, and that vanishes for sufficiently large ℓ.

As discussed in Section 2.4.10, the classical formulation of the polarization corrections is based on the definitions of the induced vector dipole moment $p_i = \alpha_d' E_i$ and induced tensor quadrupole moment $Q_{ij} = \alpha_Q' (dE_i/dx_j)/4$, where E_i is the external field. Here the primes indicate the use of SI units (m^3/K for α_d', m^5/K for α_Q', where K is the Coulombic electrostatic constant). The unprimed symbols α_d and α_Q will later denote the use of atomic units. The interactions between these moments and the field yield

$$\Delta E = -\frac{1}{2}\alpha_d' E^2 - \frac{1}{8}\alpha_Q' \left(\frac{dE}{dr}\right)^2. \tag{3.36}$$

In the atomic case the external field affecting the core arises from the Coulomb field produced by the orbital electron is $E = -Ke/r^2$, and this interaction becomes

$$\Delta E = -\frac{1}{2}\alpha_d' \left(-\frac{Ke}{r^2}\right)^2 - \frac{1}{8}\alpha_Q' \left(\frac{2Ke}{r^3}\right)^2 = -\frac{K^2 e^2}{2}\left[\frac{\alpha_d'}{r^4} + \frac{\alpha_d'}{r^6}\right]. \tag{3.37}$$

By similar arguments it can be shown that the nonadiabatic correlation factor [22] and the corrections to the Coulomb energy for penetration effects [41] are also proportional to r^{-6}. In atomic energy units the energy shift in Eq. 3.37 can be rewritten using $Ke^2 = 2Ry\, a_0 = 2R_Z a_Z$ and defining the quantities

$$\alpha_d \equiv \alpha_d' K/a_Z^3 ; \qquad \alpha_Q \equiv \alpha_Q' K/a_Z^5, \tag{3.38}$$

and

$$\rho \equiv \zeta r/a_Z \tag{3.39}$$

(here ρ is a dimensionless distance). Notice that these commonly used atomic units for the polarizabities are not actually constants, but contain element-specific corrections for the reduced mass.

The contribution to the binding energy is obtained by taking the expectation values of these expressions. The term values (freed of magnetic fine structure and exchange effects by an appropriate configuration average) are then given by

$$I_0 - E_{n\ell} = T_{n\ell}^{(H)} + R_Z a_Z [\alpha_d \zeta^4 \langle \rho^{-4}\rangle_{n\ell} + (\alpha_Q - 6\beta + f_\ell)\zeta^6 \langle \rho^{-6}\rangle_{n\ell}] \tag{3.40}$$

where $T_{n\ell}^{(H)}$ is the energy of a hydrogenlike atom or ion of the same central charge (see Eq. 2.86)

$$T_{n\ell}^{(H)} \equiv \frac{R_Z \zeta^2}{n^2}\left[1 + \left(\frac{\alpha\zeta}{n}\right)^2 \left(\frac{n}{\ell + \frac{1}{2}} - \frac{3}{4}\right)\right], \tag{3.41}$$

Table 3.2. Theoretical values for the polarizabilities of the Ne and Na sequence ground states.

Ion	Charge	α_d	α_Q	Charge	α_d	α_Q	β
		Ne sequence			Na sequence		
Na	1+	0.946	1.521				
Mg	2+	0.470	0.518	1+	33.9	137.	104.
Al	3+	0.265	0.216	2+	13.8	30.8	28.1
Si	4+	0.162	0.102	3+	7.22	10.4	11.0
P	5+	0.106	0.0531	4+	4.30	4.36	5.24
S	6+	0.0721	0.0296	5+	2.79	2.12	2.84
Cl	7+	0.0509	0.0175	6+	1.92	1.14	1.67
Ar	8+	0.0371	0.0108	7+	1.38	0.648	1.05
K	9+	0.0277	0.0069	8+	1.02	0.393	0.689
Ca	10+	0.0211	0.0046	9+	0.782	0.252	0.472
Sc	11+	0.0164	0.0031	10+	0.605	0.164	0.332
Ti	12+	0.0129	0.0022	11+	0.478	0.110	0.239
V	13+	0.0103	0.0015	12+	0.384	0.078	0.176
Cr	14+	0.0083	0.0011	13+	0.312	0.055	0.132
Mn	15+	0.0068	0.0008	14+	0.257	0.039	0.101
Fe	16+	0.0056	0.0006	15+	0.215	0.030	0.078

and, using Tables 2.1 and 2.2, the expectation values are given by

$$\langle \rho^{-4} \rangle_{n\ell} = \frac{3 - \ell(\ell+1)/n^2}{2n^3\left(\ell - \frac{1}{2}\right)\ell\left(\ell + \frac{1}{2}\right)(\ell+1)\left(\ell + \frac{3}{2}\right)} \tag{3.42}$$

$$\langle \rho^{-6} \rangle_{n\ell} = \frac{35 - 5[6\ell(\ell+1) - 5]/n^2 + 3(\ell-1)\ell(\ell+1)(\ell+2)/n^4}{8n^3\left(\ell - \frac{3}{2}\right)(\ell-1)\left(\ell - \frac{1}{2}\right)\ell\left(\ell + \frac{1}{2}\right)(\ell+1)\left(\ell + \frac{3}{2}\right)(\ell+2)\left(\ell + \frac{5}{2}\right)}. \tag{3.43}$$

Here the quantum numbers have been made explicit in the notation for the expectation values through subscripting. The nonpenetrating spectrum can be specified from this equation if empirical values for the quantities α_d, $(\alpha_Q - 6\beta)$, and f_ℓ can be evaluated from the existing data.

These quantities can also be computed theoretically, but not to the spectroscopic accuracy that can be obtained by empirical fitting. A sampling of theoretical calculations [38, 40, 128] for the neonlike core of the sodium spectroscopic sequence and the sodiumlike core of the magnesium spectroscopic sequence is given in Table 3.2. Notice that the polarizabilities are much smaller for the closed neonlike core than for the sodiumlike core, which contains an out-of-shell electron.

The nonadiabatic correlation factor 6β detracts from the binding energy of the dipole polarizability, but depends on $\langle \rho^6 \rangle$ rather than $\langle \rho^4 \rangle$, hence it appears as a negative correction to the effective value of α_Q. The penetration factor f_ℓ is a correction that increases the attraction in the primary Coulomb attraction (since the effective central charge is increased if penetration occurs). Since it has been found to be approximately proportional to $\langle\rho^6\rangle$ it is a positive correction to the effective value of α_Q. While an overall best estimate of the quantity α_d for a given ion can be obtained by combining the results obtained from several of its Rydberg series, the results obtained for the fitted value of the composite quantity $(\alpha_Q - 6\beta + f_\ell)$ usually vary significantly among the various Rydberg series. However, if data for several series of successively high ℓ show a constancy in this coefficient, it can be assumed that $f_\ell \approx 0$ for these series. The value for $(\alpha_Q - 6\beta)$ extracted therefrom can then be used to determine the values of f_ℓ for the individual series of lower ℓ.

The core polarization formula can be used to fit the measured data in the form it is written above, but a convenient alternative exposition of the results was suggested [87] by Edlén. This involves a rewriting of the equation in terms of a linear expression

$$y = A + Bx \tag{3.44}$$

where

$$y \equiv \left(I_0 - E_{n\ell} - T_{n\ell}^{(H)}\right)\big/R_Z\langle\rho^{-4}\rangle_{n\ell} \tag{3.45}$$

$$x \equiv \langle\rho^{-6}\rangle_{n\ell}/\langle\rho^{-4}\rangle_{n\ell} \tag{3.46}$$

$$A \equiv \alpha_d\zeta^4 \tag{3.47}$$

$$B \equiv (\alpha_Q - 6\beta + f_l)\zeta^6. \tag{3.48}$$

A plot of these quantities for measured data for P IV (the fourth spectrum of phosphorus, P^{3+}) in the Mg isoelectronic sequence in shown in Fig. 3.2. The quantities contained in the

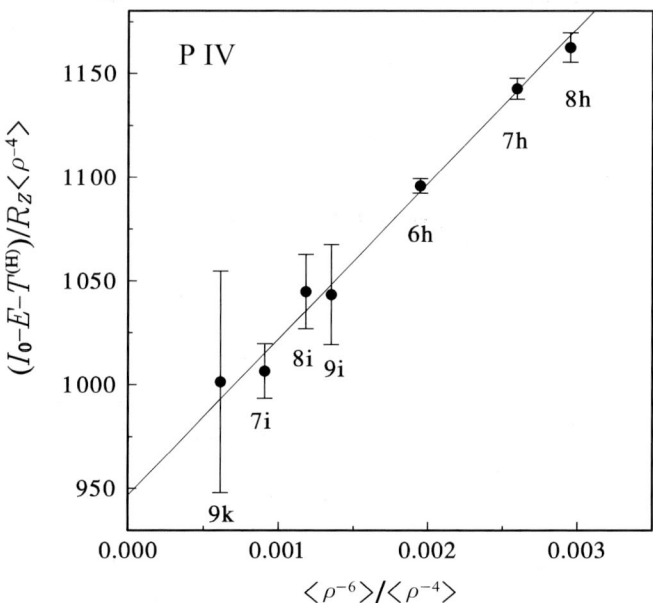

Fig. 3.2. Edlén plot for P IV, yielding $\alpha_d = 3.70$. The labels h, i, and k indicate the $l = 5$, 6, and 7 Rydberg series.

fitting parameters A and B can be extracted as the intercept and slope of the straight-line fit shown on this plot. The labels h, i, and k indicate the $\ell = 5$, 6, and 7 Rydberg series. The fitted value for the dipole polarizability is $\alpha_d = 3.70$. Since h-series energies are probably affected somewhat by penetration, and there are not sufficient numbers of points in the i and k series to determine slopes independently, the quantity $\alpha_Q - 6\beta + f = 18.3$ must be regarded as only an effective value.

3.5 Screening parametrizations

As has been demonstrated in the sections above, the predictive systematization of precision measurements for properties of complex atoms is often accomplished through the semiempirical parametrization of simple models. While the procedure is usually described in terms of a semiclassical picture, these methods also have a basis in a corresponding parametrization of the quantum mechanical formulation.

3.5.1 Characterizing the regions of small and large r

Two of the most successful types of systematizations are the quantum defect and the screening parameter methods, and each has both a semiclassical empirical and a quantum mechanical interpretation. The quantum defect method involves the replacing of a quantum number in a hydrogenic expression with an effective quantum number to be determined from measured data. The screening parametrization involves replacing the central charge in a hydrogenic expression with an effective central charge determined from measured data.

The quantum mechanical correspondence can be understood by considering the natural units in which the radial wave equation [for $\psi = r R(r)$] is expressed for a single-electron atom with a nucleus of charge Z

$$E\psi = -\frac{\hbar^2}{2m} \left[\frac{d^2}{dr^2} - \frac{\ell(\ell+1)}{r^2} \right] \psi - \frac{ZKe^2}{r} \psi. \tag{3.49}$$

If the constants are rewritten using

$$\hbar^2/2m = Ry \, a_0^2; \qquad Ke^2/2 = Ry \, a_0, \tag{3.50}$$

and we make the substitutions

$$\epsilon \equiv \frac{E}{Ry \, Z^2}; \qquad \rho \equiv \frac{Zr}{a_0}, \tag{3.51}$$

then the Schrödinger equation becomes

$$\epsilon\psi = -\frac{d^2\psi}{d\rho^2} + \frac{\ell(\ell+1)}{\rho^2}\psi - \frac{2}{\rho}\psi, \tag{3.52}$$

and the solutions for any specific value of the nuclear charge can be obtained by a scaling of the radial coordinate.

There are two solutions to this equation: one is regular at $\rho = 0$ and the other is irregular and diverges as $\rho \to 0$. For hydrogenlike atoms, the irregular solution is rejected. For an atom with a core the hydrogenlike solution is valid only outside the core, since the potential inside the core is modified from the Coulombic form. Thus boundary conditions for $\rho \to 0$ are not relevant, and the external wave function can be modified by a phase shift, since a cutoff radius must be imposed to exclude the region where the mathematical divergences would appear. Thus the solution is

$$\psi \left(\frac{Zr}{a_0} \right) \to \psi \left(\frac{Zr}{a_0} + \delta \pi \right), \tag{3.53}$$

where the quantum defect is the phase shift in units of π. Clearly this approach is most likely to be successful in describing quantities that are sensitive to the portion of wave function at large values of r, such as the gross binding energy (which varies as $1/r$) and electric dipole transition probabilities (which vary as transition integrals over r). Quantities that depend on higher reciprocal powers of r (such as the spin–orbit energy and the relativistic kinetic energy) will be sensitive to the inner part of the wave function, where the phase shift would introduce divergences. For quantities such as these, a screening parametrization is more suitable.

The screening parametrization maintains the phase of the wave function, but alters its radial scaling

$$\psi \left(\frac{Zr}{a_0} \right) \to \psi \left(\frac{(Z - S)r}{a_0} \right). \tag{3.54}$$

This approach has been shown to be very successful in parametrizing magnetic fine structure, electron–electron direct and exchange energies, quantum electrodynamic corrections, etc., which are sensitive to the inner part of the wave function.

The screening parameter approach began with Moseley's adaptation of the Balmer formula to describe x-ray line series and Sommerfeld's relativistic formulation of the regular doublet law of x-ray fine structure splittings.

3.5.2 The regular and irregular doublet laws

Even in the earliest attempts at the use of screening parametrizations, it was clear that the nature of the charge screening of the Balmer energy is quite different from that of the Sommerfeld energy. In x-ray spectra, studies were made of intershell and intrashell splittings which indicated screening mechanisms that differed both in magnitude and in Z dependence. In order to use parametrized hydrogenlike formulae to describe complex atoms, initial attempts used an equation

$$I_0 - E_{n\ell j} = \frac{R_Z (Z - S_0)^2}{n^2} + \frac{R_Z \alpha^2 (Z - S_1)^4}{n^4} \left[\frac{n}{j + \frac{1}{2}} - \frac{3}{4} \right], \tag{3.55}$$

where S_0 and S_1 are the effective Balmer and Sommerfeld screening parameters that depend on n, ℓ and Z, but (for purposes of this phenomenological modeling) are assumed not to depend on j.

The so-called "irregular doublet law" treats the electrostatic fine structure levels, which have the same n and j but different ℓ. These levels are called "screening doublets" since their separations result primarily from variations in the core penetration, corresponding to the differences in the eccentricities of the classical orbits. Under the assumptions of Eq. 3.55, S_0 and S'_0 differ for these levels, and the Balmer energy dominates the energy separation. The dependence on Z for this expression reduces from a quadratic to a linear form by an expansion and refactoring

$$E_{n\ell'j} - E_{n\ell j} = \frac{R_Z}{n^2}\left[(Z^2 - 2S_0 Z + S_0^2) - (Z^2 - 2S'_0 Z + S_0'^2)\right]$$
$$= \frac{2R_Z(S'_0 - S_0)}{n^2}\left[Z - \left(\frac{S'_0 + S_0}{2}\right)\right]. \tag{3.56}$$

Thus a plot of this energy difference vs the nuclear charge yields an approximately straight line. An example of the application of this approach to the sodium isoelectronic sequence is given in Fig. 3.3. Here the screening-doublet energy separations corresponding to the transitions $3s\,^2S_{1/2}$–$3p\,^2P_{1/2}$ and $3p\,^2P_{3/2}$–$3d\,^2D_{3/2}$ are plotted vs Z. With the exception of the region very near the neutral end of the sequence (where the centrality of the interaction is weaker), the overall linearity of the plot is apparent. A linear regression yields values $18.05(Z - 10.64)$ for the $j = \frac{1}{2}$ splittings, and $22.91(Z - 9.55)$ for the $j = \frac{3}{2}$ splittings. However, small deviations from exact linearity do occur in the use of this formalism, and additional fitting terms are usually included in high-precision work [168].

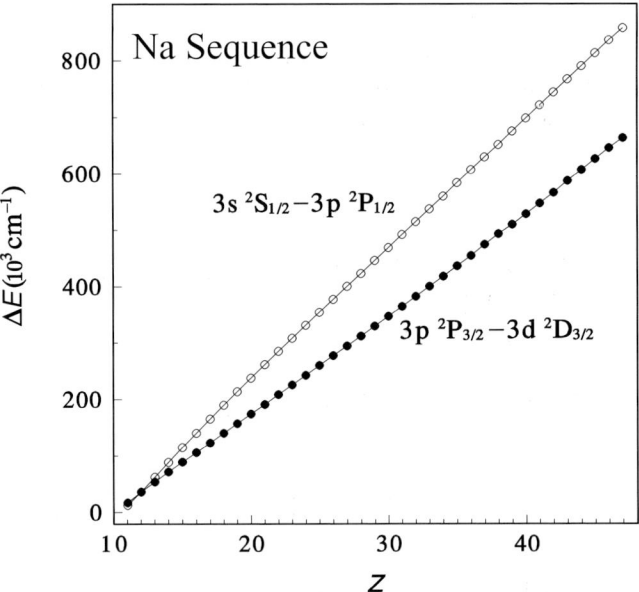

Fig. 3.3. Application of the irregular doublet law to the 3s–3p and 3p–3d intervals in the Na sequence. (Energies are in kilokaysers.)

If the ionization limit is known and the Sommerfeld terms are negligible, an alternative systematization used by Moseley and Hertz can be accomplished simply by subtracting the square roots of the two energies

$$\sqrt{\frac{I_0 - E_{n\ell'j}}{R_Z}} - \sqrt{\frac{I_0 - E_{n\ell j}}{R_Z}} \approx \frac{S_0 - S_0'}{n} \tag{3.57}$$

which, for a given x-ray doublet, was found to be a constant nearly independent of Z.

While the irregular doublet law is sometimes useful in interpolating, extrapolating, and critically evaluating ℓ-dependent screening doublet data, in general the Balmer energy is more precisely specified by a quantum defect formulation than by a screening parametrization. In contrast, the j-dependent fine structure can be very precisely characterized by a screening parametrization.

The so-called "regular doublet law" treats the magnetic fine structure levels which have the same n and ℓ but different values for j. Neither S_0 nor the Balmer energy depends on j, so the gross energies cancel leaving only the Sommerfeld terms. Since S_1 is independent of j, the expression becomes

$$E_{n\ell j'} - E_{n\ell j} = \frac{R_Z \alpha^2 (Z - S_1)^4}{n^4} \left[\frac{n}{j' + \frac{1}{2}} - \frac{n}{j + \frac{1}{2}} \right]$$

$$= \frac{R_Z \alpha^2 (Z - S_1)^4}{n^3 \ell(\ell + 1)} \tag{3.58}$$

where the final form uses the fact that $j' = \ell + \frac{1}{2}$ and $j = \ell - \frac{1}{2}$. In order to make optimum use of this expression it is necessary to use higher-order terms in the expansion. A method for quantitatively specifying these higher-order terms will be developed in the next section.

3.6 Screening parametrization of the fine structure

It is desired to make an isoelectronic reduction of the fine structure intervals along an isoelectronic sequence (a fixed number of electrons in the ion, with a increasing value of the atomic number and hence the charge of the nucleus) which maps the measured data into a parameter that has a regular and slowly varying behavior over the sequence. There are two commonly used expressions for the fine structure, one fully relativistic, the other a nonrelativistic series expansion. If the entire isoelectronic sequence is to be considered, nonrelativistic and relativistic regimes must be smoothly connected. If the relativistic expression is used, the fine structure splitting is a small difference between two large numbers that include the electron mass. If the nonrelativistic expansion is used, then a large number of terms must be included in the expansion so that it will remain valid in the relativistic limit. Thus it is desirable to develop a procedure for obtaining the coefficients of the terms in this expansion to very high order.

3.6.1 Expansion of the Dirac–Sommerfeld formula

As discussed in Sections 2.4.9 and 2.4.14, the expression for the energy predicted by the relativistic Dirac equation [82] can be obtained from the Sommerfeld formulation of Eq. 2.144 with the substitution $\ell \to j$. Thus, the Dirac energy of a one-electron atom is given by

$$\frac{E(n, j, Z)}{mc^2} = \left(1 + \frac{(\alpha Z)^2}{\left[n - (j + \frac{1}{2}) + \sqrt{(j + \frac{1}{2})^2 - (\alpha Z)^2}\right]^2}\right)^{-1/2} \tag{3.59}$$

where E is the relativistic energy, which includes rest energy mc^2.

Since this exact expression is dominated by the electron mass, its series expansion is of more practical use in accurately specifying the binding energy. A systematic method has been developed [35] for obtaining such an expansion quickly and reliably. The number of parameters in this equation can be reduced from three to two through the substitutions

$$x \equiv (\alpha Z/n)^2; \qquad b \equiv n/(j + 1/2) \tag{3.60}$$

in terms of which the equation becomes

$$\frac{E}{mc^2} = \left(1 + \frac{x}{[1 - (1 - \sqrt{1 - b^2 x})/b]^2}\right)^{-1/2}. \tag{3.61}$$

This consists of a nesting of four binomial quantities

$$E/mc^2 \equiv (1 + xv)^{-1/2} \tag{3.62}$$

$$v \equiv (1 - u/b)^{-2} \tag{3.63}$$

$$u \equiv (1 - t) \tag{3.64}$$

$$t \equiv (1 - b^2 x)^{1/2} \tag{3.65}$$

which can each be expanded by the binomial theorem, to become

$$\frac{E}{mc^2} = \sum_{p=0}^{\infty} \binom{-1/2}{p} (xv)^p \tag{3.66}$$

$$v^p = \sum_{q=0}^{\infty} \binom{-2p}{q} (-u/b)^q \tag{3.67}$$

$$u^q = \sum_{r=0}^{\infty} \binom{q}{r} (-t)^r \tag{3.68}$$

$$t^r = \sum_{s=0}^{\infty} \binom{r/2}{s} (-b^2 x)^s \tag{3.69}$$

where the binomial coefficients have their standard definition

$$\binom{A}{B} = \frac{A!}{B!(A - B)!}. \tag{3.70}$$

If these binomial expansions are nested, the coefficients of the powers of x and b are collected, and the orders of summation over q and r are interchanged with s, they form

$$\frac{E}{mc^2} = \sum_{p=0}^{\infty} \sum_{s=0}^{\infty} x^{p+s} \sum_{q=0}^{\infty} b^{2s-q} \binom{-1/2}{p} \binom{-2p}{q} \sum_{r=0}^{\infty} \binom{q}{r} \binom{r/2}{s} (-1)^{q+r+s}. \tag{3.71}$$

To obtain the numerical coefficients for specific powers of x and b we shift the p sum to $P = p + s$, interchange the orders of the s and q summations, and shift the q sum to $Q = 2s - q$. After paying back the x and b substitutions, this yields

$$\frac{E(n, j, Z)}{mc^2} = \sum_{P=0}^{\infty} \left(\frac{\alpha Z}{n}\right)^{2P} \sum_{Q=0}^{2P} \left(\frac{n}{j + \frac{1}{2}}\right)^{Q} C_{PQ}, \tag{3.72}$$

where

$$C_{PQ} = \sum_{s=s_{\min}}^{Q} \binom{-1/2}{P-s} \binom{-2P+2s}{2s-q} \sum_{r=0}^{2s-Q} \binom{2s-Q}{r} \binom{r/2}{s} (-1)^{r-s-Q}, \tag{3.73}$$

with $s_{\min} = Q/2$ for Q even, and $s_{\min} = (Q + 1)/2$ for Q odd. Notice that the C_{PQ} coefficients involve only finite sums and correspond to a set of rational fractions that are independent of n, j and Z. These coefficients are given for $P = 0$–9 and $Q = 0$–15 in Table 3.3.

Using these coefficients the expression can be expanded to yield

$$\frac{E(n, j, Z)}{mc^2} = 1 - \frac{1}{2}\left(\frac{\alpha Z}{n}\right)^2 - \frac{1}{2}\left(\frac{\alpha Z}{n}\right)^4 \left[\left(\frac{n}{j + \frac{1}{2}}\right) - \frac{3}{4}\right]$$

$$- \frac{1}{8}\left(\frac{\alpha Z}{n}\right)^6 \left[\left(\frac{n}{j + \frac{1}{2}}\right)^3 + 3\left(\frac{n}{j + \frac{1}{2}}\right)^2 - 6\left(\frac{n}{j + \frac{1}{2}}\right) + \frac{5}{2}\right] + \cdots. \tag{3.74}$$

In the power series in $(\alpha Z/n)$ on the right-hand side, the zeroth-order term is the rest energy, the second-order term is the Balmer energy, the fourth-order term is the Dirac–Sommerfeld energy, and the sixth-order term is the first of a series of higher-order corrections. (Since $b \geq 1$, the series is written in its descending powers.) All terms past the rest energy are contributions to the binding energy, and therefore have negative signs. It is clear from this expression that the definition of the Rydberg constant ($Ry \equiv \alpha^2 mc^2/2$) conceals the true role of α. It is not *just* a "fine structure constant," but is rather the coupling constant of the electromagnetic interaction and occurs in all orders above zeroth.

The quantum number j enters first in the $(\alpha Z/n)^4$ term, hence the rest energy and the Balmer energy cancel from the expression for the fine structure splitting $\sigma(n, l, Z)$, which

Table 3.3. Coefficients C_{PQ}. The rational fraction is obtained by dividing the numerator, given in the table, by the denominator, given beneath it in the row labeled Den   (From Ref. [35].)

$P =$	0	1	2	3	4	5	6	7	8	9
$Q = 0$	1	−1	3	−5	35	−63	231	−429	6435	−12155
1			−4	12	−120	280	−1260	2772	−48048	102960
2				−6	120	−420	2520	6930	144144	−360360
3				−2	−8	180	−1960	7770	−210672	648648
4					−24	80	0	−2520	129360	−576576
5					−8	−24	504	−1736	15120	121968
6						−30	140	518	−39312	152460
7						−10	−84	690	−11664	−33660
8							−84	140	9072	−58608
9							−28	−140	7920	−8624
10								−126	1200	16632
11								−42	−1872	11880
12									−1584	1320
13									−528	−3168
14										−2574
15										−858
Den.	1	2	8	16	128	256	1024	2048	32768	655326

is given by

$$\sigma(n, l, Z) \equiv E(n, \ell + 1/2, Z) - E(n, \ell - 1/2, Z). \tag{3.75}$$

This subtraction yields

$$\frac{\sigma(n,l,Z)}{mc^2} = \sum_{\beta=0}^{\infty}\left(\frac{\alpha Z}{n}\right)^{2\beta+4}\sum_{\gamma=0}^{2\beta}\left[\left(\frac{n}{\ell+1}\right)^{\gamma+1}-\left(\frac{n}{\ell}\right)^{\gamma+1}\right]C_{\beta+2,\gamma+1}. \tag{3.76}$$

This can be expanded

$$\frac{\sigma(n,l,Z)}{mc^2} = \frac{(\alpha Z)^4}{2n^3\ell(\ell+1)}\left\{1+\frac{(\alpha Z)^2}{4}\left[\frac{3\ell^2+3\ell+1}{\ell^2(\ell+1)^2}+\frac{3(2\ell+1)}{n\ell(\ell+1)}-\frac{6}{n^2}\right]+\cdots\right\}. \tag{3.77}$$

These equations can be used to generate an expansion to arbitrary order for the splitting of a term of any given n and ℓ.

3.6.2 Application to complex atoms

The screening parametrization of a set of measured fine structure intervals $\sigma(n, \ell, Z)$ consists of mapping those numbers onto the effective charge $Z_{\text{eff}} \equiv (Z - S)$ of a corresponding quasi-hydrogenlike atom of the same quantum numbers that would have that splitting. The

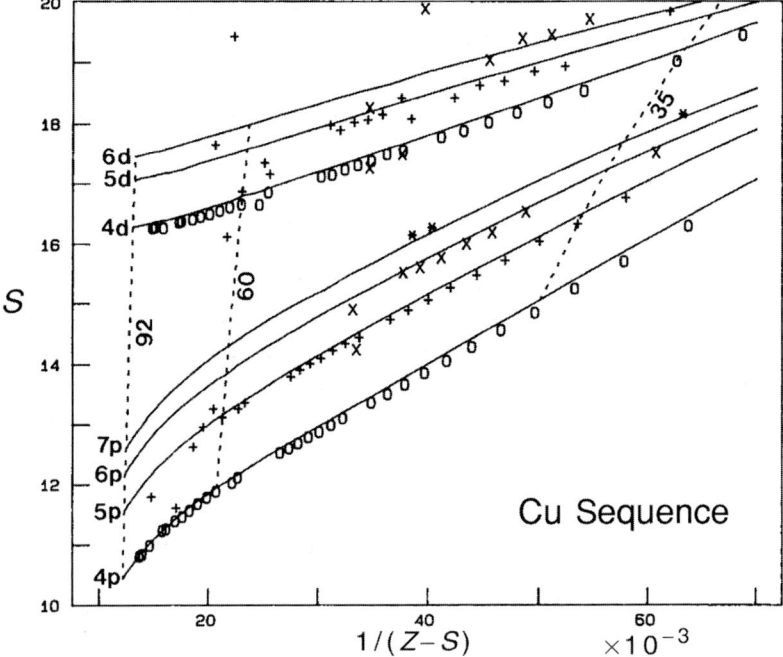

Fig. 3.4. Screening parameter plot for np and nd fine structures of the Cu sequence. (From Ref. [47].)

extracted effective screening parameters are then presented in a graphical exposition that displays either a linearity or a slowly varying regularity that expedites interpolation, extrapolation, and the identification of flawed data. An empirical description [87] that has been found to exhibit a very regular behavior for many isoelectronic systems is given by

$$S = A + B/(Z - S). \qquad (3.78)$$

An example of the application of this method to the fine structure splittings of the 4p–7p and 4d–6d terms of the 29-electron ions in the copper isoelectronic sequence is shown in Fig. 3.4

The quantities S are experimentally measured, obtained by a one-to-one mapping of each of the $\sigma(n, \ell, Z)$ values into screening parameter space. Notice that here, as in the case of the Ritz expansion, the quantity S appears on both sides of the fitting equation, and in both the ordinate and the abscissa of the plot. The fact that both axes depend on the values of S means that a given value of the abscissa does not correspond to the same value of Z for data of differing $n\ell$. As a guide in Fig. 3.4, dashed lines trace the corresponding ions of the specific chemical elements having $Z = 35$ (Br^{6+}), $Z = 60$ (Nd^{31+}), and $Z = 92$ (U^{63+}) among the various terms. The quantity S that is predicted by the fitting parameters A and B in Eq. 3.78 can be easily obtained through successive iteration (e.g., by a feedback loop whereby $S_{\mathrm{try1}} = A + B/(Z - A)$, $S_{\mathrm{try2}} = A + B/(Z - S_{\mathrm{try1}})$, etc.).

The expression obtained in the previous section is the expansion of the exact solution to the Dirac equation for an electron with the Dirac magnetic g-factor $g_{\mathrm{e}} = 2$, and a nucleus of infinite mass. Neither of these assumptions is rigorously valid for a physical atom or ion, and neither effect can be exactly included.

A first approximation to a correction for the finite mass can be made by the use of the reduced-mass correction $m \rightarrow mM/(m + M)$, but the reduced mass is a nonrelativistic concept, and since the fine structure is already a relativistic correction to the Balmer energy, higher-order corrections may be needed in high-precision work. It is clear from the semiclassical presentation of the spin–orbit interaction in Chapter 2 that a proper description of the fine structure requires quantum electrodynamic corrections to the g-factor of the electron, but the properties of the free electron are not sufficient, since higher-order QED processes are also present in the bound atomic system. For purposes here we shall include the finite mass of the nucleus through the use of the reduced-mass Rydberg constant (because of the high precision obtainable in optical spectroscopy, Ry is known to higher accuracy than either the mass of the electron or the fine structure constant), and an approximation to the electron g-factor [43].

3.6.3 Relative importance of higher-order terms

Although the details contained in the corrections of the previous section may seem excessive in view of the semiempirical nature of the formulation, studies indicate that the inclusion of these higher-order corrections produces a set of screening parameters of greater isoelectronic regularity, and thus of greater predictive power. Spectroscopic accuracies of the order of ± 0.001 cm^{-1} are typical in many types of spectroscopic measurements, and can provide stringent tests of a data-based predictive formulation.

To further emphasize the significance of including these seemingly small effects, a breakdown of the various contributions to the 4p fine structure of the copperlike Nd^{31+} ion is given in Table 3.4, and scaled to gravitational potentials at the surface of the Earth. The values for the Nd^{31+} energy contributions were obtained by reducing the measured fine structure interval to a corresponding screening parameter in a hydrogenlike equation that includes all of these interactions. The contribution of each separate term was then computed using this effective screening parameter.

The analogous gravitational problem can be written in terms of the radius of the Earth R and a height h above its surface.

$$\frac{GM}{R+h} \cong \frac{GMm}{R} \left[1 - \frac{h}{R} + \cdots \right]. \tag{3.79}$$

Thus the quantity h/R corresponds to the ratio of a change in the gravitational potential energy of an object near the Earth to the binding energy of that object to the Earth. Table 3.4 considers the analogous ratio of a given perturbation to the fine structure of the 4p level in the Nd^{31+} ion to the total binding (ionization) energy of the Balmer centroid of that term. For a conceptual comparison, Table 3.4 expresses the perturbation in the atomic case as a proportional shift in height of an object in the analogous gravitational case.

It is clear from this example that the Hamiltonian formulation of quantum mechanics (which optimizes the accuracy of the knowledge of the ionization energy due to the $1/r$

Table 3.4. Contributions to the 4p fine structure of the Cu-like ion Nd^{31+}, expressed both as cm^{-1} and as relative gravitational heights above the surface of the Earth. (From Ref. [49].)

Phenomenological origin	Energy contribution (cm^{-1})	Earth's gravity analogue
Binding energy		
4p (Balmer centroid)	9 282 570	6 378 km
Sommerfeld fine structure		
4th Order	244 282	168 km
6th Order	18 800	13 km
8th Order	1 463	1 km
10th Order	123	85 m
12th Order	11	8 m
14th Order	1	69 cm
16th Order	0.1	7 cm
18th Order	0.01	7 mm
Quantum electrodynamics		
$g - 2$ (free electron)	567	390 m
higher-order QED	−52	−36 m
Nuclear effects		
reduced mass	0.9	62 cm
higher order recoil	−0.004	−3 mm
penetration	14	10 m
Total	265 210	182 km

potential and treats all other interactions perturbatively) has significant limitations in high-precision specifications of separations between levels in atoms. If we navigated on Earth by this process, the neglect of the nuclear penetration correction would be equivalent to ignoring the effect of the sidewalk below in a fall off a two-story building at sea level. Since the risk of injury is restricted to the last few millimeters of the fall, only a few parts in 10 thousand of the orbit are affected by sidewalk penetration. Thus the experience of an electron penetrating the nucleus has similarities to that of a gravitational object dropped from a great height penetrating the ground below.

This example indicates some of the reasons why, despite the fact that quantum mechanical methods can provide reliable predictions on the level of 1%, semiempirical methods are needed when accuracies become better than parts in 10^8.

3.7 Screening parametrization of transition rates

Progress in atomic spectroscopy has been greatly enhanced by the development of light sources that permit measurements of lifetimes in highly ionized atoms. This provides a database that permits studies of transition rates for systems with the same number of electrons but variable charge on the nucleus. An extensive base of these isoelectronic data is now available in systems corresponding to neutral atoms of alkali metals, alkaline earths, and inert gases. Unfortunately, at present these data are primarily limited to $\Delta n = 0$ transitions to the ground state. This is because these transitions possess only a single decay channel, so the transition rate is equal to the reciprocal lifetime. For higher lying levels multiple decay channels usually exist, and the transition rates cannot be deduced from lifetime measurements without accompanying measurements for the individual branching fractions. At present, virtually no branching fraction data exist for multiply charged ions.

3.7.1 Parametrizing line strengths

In order to characterize these isoelectronic lifetime measurements in a manner that permits accurate interpolation, extrapolation, and identification of flawed data, it is necessary to map the raw data into a quantity that exhibits a slow and regular isoelectronic variation. It has been found that the line-strength factor fulfills these criteria very well. As was described semiclassically in Section 2.5, and will be developed quantum mechanically in Chapter 6, the line-strength factor for an unbranched decay can be deduced from the lifetime τ_u, the transition wavelength $\lambda_{u\ell}$, and the degeneracy g_u of the upper level u by using Eq. 2.164, which can be written in the form

$$S_{u\ell} = \left[\frac{\lambda_{u\ell}(\text{Å})}{1265.38} \right]^3 \frac{g_u}{\tau_u(\text{ns})}. \tag{3.80}$$

It has been observed [88] that this quantity, scaled (like the corresponding hydrogenlike value) by the square of the nuclear charge, has an almost linear dependence on a suitably chosen reciprocal change of constant screening C

$$Z^2 S_{u\ell} = A + B/(Z - C). \tag{3.81}$$

An example of such a parametrization for the lowest resonance transitions of the sodium isoelectronic sequence is shown in Fig. 3.5. It is clear from this plot that, not only is the variation very nearly linear, but the extrapolation of the trend to infinite Z yields the value for the corresponding transition in hydrogen. The plot contains experimental data for ions from neutral Na to Au^{68+}. For the sodium sequence shown in this figure, the hydrogenic limits are indicated by diamonds at $S_\text{H} = 108$ and 216 for the upper levels $j = \frac{1}{2}$ and $\frac{3}{2}$.

The overall linearity of this plot is quite striking, but since the model by which the data reduction was made is based on the nonrelativistic Schrödinger equation, further semiempirical investigation of the very high Z region is warranted. A suitable case for this study is provided by the Li isoelectronic sequence, for which a measurement of the lifetime of the 2p $^2\text{P}_{1/2}$ lifetime exists for lithiumlike uranium. A reduction of the data for the Li isoelectronic sequence is shown in Fig. 3.6(a). As can be seen on this plot, the U^{89+} point exhibits a significant dip from the linear extrapolation to the value 18 at the high-Z hydrogenic limit. A clear question that can be investigated is whether this dip occurs because of relativistic corrections to the single-electron model, or indicates a deviation due to a high-Z

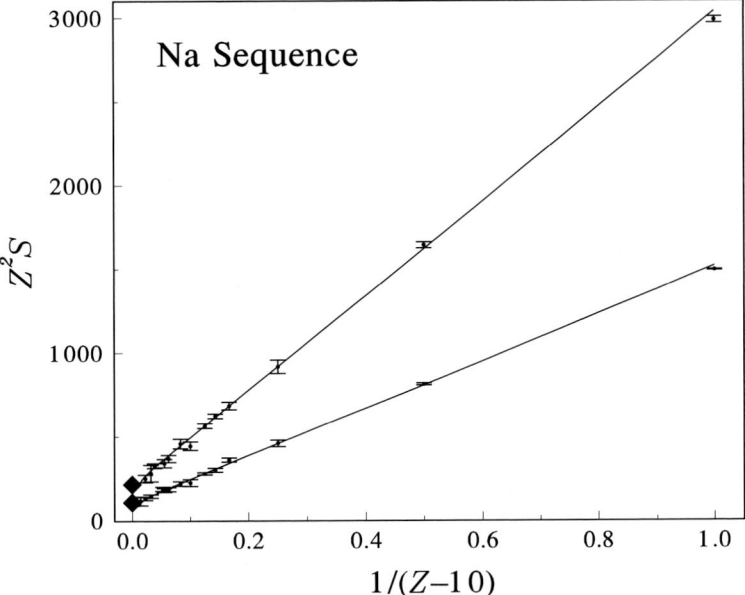

Fig. 3.5. Plot of the Z-scaled line strength of the Na-like resonance lines vs the reciprocal screened charge. The upper line represents the $\frac{1}{2}-\frac{3}{2}$ transition and the lower line represents the $\frac{1}{2}-\frac{1}{2}$ transition. Diamonds indicate the hydrogenlike value approached at infinite Z. (After Ref. [69].)

redistribution within the inner core charge. To investigate this question, the relativistic form of the hydrogenic line strength will be developed below.

3.7.2 Relativistic form of the hydrogenic line strength

Nonrelativistically, the hydrogenic line-strength factor is a constant that is given, for an ns–np transition, by [14]

$$S_{\text{H}}(1) = \frac{3}{4}n^2(n^2 - 1)(2J_u + 1). \tag{3.82}$$

In the Schrödinger approximation for a hydrogenlike atom the ns and np levels are degenerate, so this line-strength factor is not relevant to normal E1 transition probability calculations. However, the quantity can be useful for semiempirical applications in complex atoms. Here J_u is the total angular momentum of the upper level, and involves an extension of the Schrödinger model to include electron spin. For hydrogen $J_u = \frac{1}{2}$ or $\frac{3}{2}$, but for more complex systems other values for J_u can occur.

The nonrelativistic Z dependence for the hydrogenlike isoelectronic sequence is given simply by $Z^2 S_{\text{H}}(Z) = S_{\text{H}}(1)$, but the corresponding relativistic value $S_{\text{H}}^{\text{R}}(Z)$ is of the form

$$Z^2 S_{\text{H}}^{\text{R}}(Z) = S_{\text{H}}(1)\left[1 - \sum_i a_i(\alpha Z)^{2i}\right]. \tag{3.83}$$

The dipole line strengths have been computed [69] as a function of Z employing

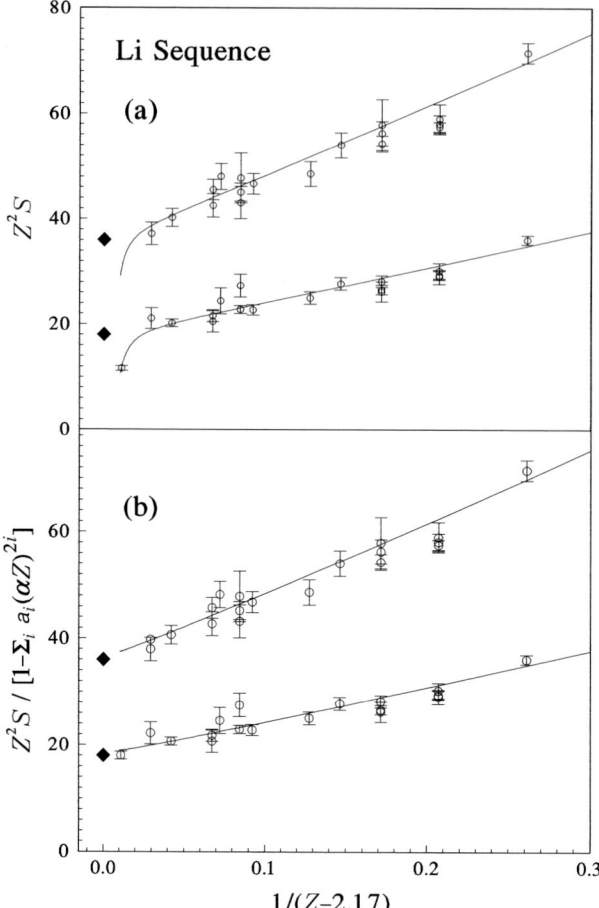

Fig. 3.6. Comparison of the high-Z region reduced with the (a) nonrelativistic and (b) relativistic hydrogenlike formulations. (After Ref. [69].)

hydrogenlike Dirac wave functions [14], employing a symbolic algebra package. This calculation neglects quantum electrodynamic and retardation effects, and assumes no variation in the electromagnetic field over the dimensions of the atom. For the $n\mathrm{s}_{1/2}-n\mathrm{p}_{1/2}$ transitions, the result can be written in a particularly simple form which is given in Ref. [69]. Numerical values for $S_\mathrm{H}(1)$ and a_i are presented in Table 3.5. For the $n\mathrm{s}_{1/2}-n\mathrm{p}_{3/2}$ transition, the a_i coefficients for $i > 1$ are irrational, involving the Riemann zeta function and factors of π, and are presented in decimal form.

Therefore the Li isoelectronic sequence data were reparametrized by defining the relativistic quantity

$$\frac{Z^2 S}{1 - \sum_i a_i (\alpha Z)^{2i}} = S_\mathrm{H}(1) \left[A' + \frac{B'}{Z - C} \right]. \tag{3.84}$$

A plot of this new parametrization is shown in Fig. 3.6(b). Clearly these relativistic

Table 3.5. Constants for relativistic hydrogenlike line strengths. (From Ref. [69].)

Transition	$S_H(1)$	a_1	a_2	a_3	a_4	a_5
$2s_{1/2}-2p_{1/2}$	18	5/6	$-1/48$	1/96	7/768	11/1536
$2s_{1/2}-2p_{3/2}$	36	1/3	0.110 187	0.059 476	0.037 032	0.024 925
$3s_{1/2}-3p_{1/2}$	108	7/12	5/144	7/288	37/2304	53/4608
$3s_{1/2}-3p_{3/2}$	216	19/72	0.139 267	0.082 219	0.050 840	0.033 291
$4s_{1/2}-4p_{1/2}$	360	9/20	3/64	17/640	87/5120	123/10240
$4s_{1/2}-4p_{3/2}$	720	103/480	0.147 237	0.093 466	0.058 853	0.038 459
$5s_{1/2}-5p_{1/2}$	900	11/30	29/600	31/1200	157/9600	221/19200
$5s_{1/2}-5p_{3/2}$	1800	9/50	0.150 012	0.100 133	0.064 068	0.041 970

corrections to the hydrogenlike line strength have removed the dip, and the linearity displayed on this exposition can be used to interpolate and extrapolate the measured database for the corresponding ions of all chemical elements in the isoelectronic sequence.

These semiempirical methods will be applied to more complex systems in Chapter 7.

4

The vector model of angular momentum

Yr, yrare, yrast.

4.1 The Schrödinger approximation

Dirac has recounted [82] an interesting story told to him by Schrödinger regarding his development of the wave equation. In trying to generalize the ideas of DeBroglie regarding waves associated with particles, Schrödinger considered a mathematical operator constructed from the relativistic energy relationships governing the Coulomb potential. He began with what Dirac calls "Schrödinger's first wave equation"

$$\left(E + mc^2 + \frac{Ke^2}{r} \right)^2 \psi = [(mc^2)^2 + (pc)^2]\psi. \tag{4.1}$$

(Here the relativistic mass energy is written as $W = E + mc^2$ to make the connection with the nonrelativistic energy E, and the fourth component of the four-momentum is assumed to be $E - V$, where $V = -Ke^2/r$). Schrödinger immediately applied this equation to the behavior of the electron in the hydrogen atom, but he obtained results that did not agree with experiment. This disagreement was a great disappointment and caused him to abandon the work for several months.

Schrödinger later returned to this study and rewrote the equation in an approximate way, neglecting the refinements required by relativity. By taking the square root of the operators on both sides of the equation above, and expanding the right-hand side in powers of $(p/mc)^2$, he obtained what Dirac calls "Schrödinger's second wave equation"

$$\left(E + \frac{Ke^2}{r} \right) \psi = \left[\frac{p^2}{2m} \right] \psi. \tag{4.2}$$

To his surprise, Schrödinger found that the results obtained using this rough approximation were in agreement with observation (if the effects of electron spin are not resolved).

Both equations are flawed in that they do not contain the intrinsic spin of the electron, but they omit this crucial element in different ways. "Schrödinger's first equation" is now known as the Klein–Gordon equation, and it describes a spinless particle. "Schrödinger's second equation" is the normal Schrödinger equation, and it represents a factorization of the spatial and spin portions of the wave function. In both cases the loss of the spin information occurs because the mathematical operator is written as a squared quantity (a dot product either of two four-vectors or two three-vectors), and is thus only a magnitude from which the vector information has been removed.

Instead of a square root, Dirac's formulation rewrites Schrödinger's first equation as a complex factorization

$$\left(E + mc^2 + \frac{Ke^2}{r} \right) \psi = [\mathbf{p}c + imc^2]\psi. \tag{4.3}$$

The product of this with its complex conjugate recovers the Klein–Gordon equation. The solution to this equation is not a scalar wave function but rather a four-component vector, spanning the space of an electron and a positron, each with two possible spin orientations. Schrödinger's first and second equations thus represent two different limits of the Dirac equation. The first retains Lorentz covariance but loses electron spin. The second forfeits Lorentz covariance, but retains the possibility of including electron spin as a multiplicative patch.

The Dirac wave equation includes the intrinsic spin of the electron as a dynamical consequence of the four-dimensional formulation, and its solution yields results that describe most aspects of the hydrogen atom or a single-electron ion with high accuracy. However, even for hydrogen, it is an approximation since it does not include quantum electrodynamic effects, and implies that the gyromagnetic g-factor of the electron is exactly two. For atoms with more than one electron, there are additional problems, since the spin–own-orbit interaction is automatically included as a dynamical contribution, whereas spin–spin, spin–other-orbit, and other types of couplings must be included [14] perturbatively through the Breit interaction.

For the purposes of an introductory course in quantum mechanics, a formulation that involves the Schrödinger model with its factorization of a space and a spin wave function has certain advantages, and can describe many atomic systems with great accuracy. For few-electron systems the spatial wave function can be obtained as a solution to the scalar Schrödinger equation, and the electron spin can be formulated by the Pauli representation [163], variously coupled through appropriate angular momentum (Racah [167]) algebra.

For the characterization of the structure of complex many-electron atoms it is less clear that this approximation is optimal. In these systems the orbitals plunge deeply into the core, introducing large magnetic (relativistic) effects, their proximity to the highly charged nucleus causes intense virtual photon fields to exist, and standard formulations of elementary quantum mechanics can be confining. For example, the language by which atomic levels are described is based on hydrogenic quantum numbers, and the characterization of spin and space properties is made as if they were the independent quantities that are obtained for single-electron systems of low nuclear charge. These concepts and notational assumptions must be continually re-examined in complex atomic systems.

4.2 The intrinsic angular momentum and magnetic moment of the electron

The ratio of the magnetic moment to the angular momentum of an electron is approximately twice the so-called "classical" value $e/2m$ that occurs when mass and charge are assumed to have identical distributions. There is really nothing "nonclassical" about such a value, since the same gyromagnetic ratio of two occurs for any uniform solid cylinder of mass that spins about its axis, and has a uniform charge confined to its outer cylindrical surface. However, attempts to apply this macroscopic model to the electron invariably lead to self-contradictory results. The minimum radius for a mechanically spinning electron model that will yield a value for the electromagnetic inertia that does not exceed its observed mass (the so-called "classical electron radius") leads to a tangential velocity much greater than the speed of light. Thus, any attempt to gain conceptual insights by considering the electron as anything other than a point particle are ill-conceived and counter-pedagogic. However, a very attractive model does exist that provides a clear conceptual picture of a mechanism by which a point particle can possess both an angular momentum and a magnetic moment. This lies in the formulation of the Foldy–Wouthuysen transformation.

In 1950, L. L. Foldy and S. A. Wouthuysen reported [98] on a phenomenon similar to that encountered in Schrödinger's first and second equations, in that it involved differing choices in evaluating a nonrelativistic limit. They noted that different operators occur in the Dirac Hamiltonian (which includes both the positive and negative energy states) and the nonrelativistic limit of the Pauli Hamiltonian (with its position and spin operators). When the proper transformation is made from the Dirac Hamiltonian to the Pauli Hamiltonian in the presence of an electromagnetic field, the behavior of a point electron exhibits some properties characteristic of a particle of finite extension. In this transformation, a point particle that moved smoothly in the Dirac space–time coordinates acquires a jittery motion (or "Zitterbewegung") in the Pauli representation, in which it dances about under the influence of its continual absorption and re-emission of virtual photons from the electromagnetic field. Since photons possess angular momentum, this dancing motion can produce a non-vanishing precession of the electron motion, which leads to a circulation of both mass and charge when the motion of the point particle is averaged over time.

In the old representation, the Dirac particle interacted with the electromagnetic field only at its position. In the new representation the interaction between the particle and the electromagnetic field is expressed in terms of electromagnetic field quantities at its mean-position variable, which is spread out in space over a region of dimensions of the order of its Compton wavelength. In addition to presenting a mathematical development of the canonical transformation, this work [98] provides a clear conceptional picture for the mechanism by which a point particle shows a behavior characteristic of a particle of finite extension, including an angular momentum, a magnetic moment, and the Darwin term [78] (as discussed in Section 2.4.9, this is the nonvanishing s-state spin–orbit interaction in hydrogen).

4.3 The Pauli spin matrices

The spin of the electron is conveniently described in terms of spinor matrix algebra. In this formulation the basis states $m_s = \pm\frac{1}{2}$ are designated by the "up" and "down" spin vectors

$$|\uparrow\rangle \equiv \begin{pmatrix} 1 \\ 0 \end{pmatrix}, \quad |\downarrow\rangle \equiv \begin{pmatrix} 0 \\ 1 \end{pmatrix}. \tag{4.4}$$

The spin wavefunction for an arbitrary state $\chi_{1/2}^{m_s}$ can be represented as an admixture of the up and down basis states

$$\chi_{1/2}^{m_s} = \cos\alpha \mid \uparrow\rangle + \sin\alpha \mid \downarrow\rangle \tag{4.5}$$

where the unitarity of the transformation permits the amplitudes to be written as a mixing angle α. However, since "up" and "down" are apart by $180°$ in configuration space, whereas the transformation angle for a two-dimensional orthogonal coordinate system is $90°$, the wave function can also be rewritten in terms of the physical half-angle $\alpha = \theta/2$

$$\chi_{1/2}^{m_s} = \cos\frac{\vartheta}{2} \mid \uparrow\rangle + \sin\frac{\vartheta}{2} \mid \downarrow\rangle = \begin{pmatrix} \cos(\vartheta/2) \\ \sin(\vartheta/2) \end{pmatrix}. \tag{4.6}$$

Thus, in order to make a transformation for a spin-$\frac{1}{2}$ particle that returns to the starting point, it is necessary to make two complete revolutions ($720°$). This is in contrast to a spin-1 particle, which requires a $360°$ rotation, and a spin-2 particle, which requires a $180°$ rotation. For the spin-1 and spin-2 cases, Hawking has made [118] an analogy to a deck of cards, in which the Ace of Spades must be rotated $360°$ to look the same, but the Queen of Hearts looks the same after only a $180°$ rotation. A $720°$ conceptual analogue can also be made by considering the "Philippine Wine Dance" [10]. In this folk dance, a symbolic waiter lifts a tray containing a glass of wine with the palm of his hand up, then rotates the tray about a vertical axis first past his waist, and then outward and around for one $360°$ rotation, and then (with a twist of his body) over his head for a second $360°$ rotation, which returns both the tray and his arm to their initial positions.

The spin operator is defined by $\mathbf{S} \equiv \sigma\hbar/2$ where σ denotes the Pauli spin matrices

$$\sigma_x \equiv \begin{pmatrix} 0 & 1 \\ 1 & 0 \end{pmatrix}, \quad \sigma_y \equiv \begin{pmatrix} 0 & -i \\ i & 0 \end{pmatrix}, \quad \sigma_z \equiv \begin{pmatrix} 1 & 0 \\ 0 & -1 \end{pmatrix}. \tag{4.7}$$

Products of these matrices have the properties

$$\sigma_x^2 = \sigma_y^2 = \sigma_z^2 = \begin{pmatrix} 1 & 0 \\ 0 & 1 \end{pmatrix} \tag{4.8}$$

and

$$\begin{aligned} \sigma_x\sigma_y &= -\sigma_y\sigma_x = i\sigma_z \\ \sigma_y\sigma_z &= -\sigma_z\sigma_y = i\sigma_x \\ \sigma_z\sigma_x &= -\sigma_x\sigma_z = i\sigma_y. \end{aligned} \tag{4.9}$$

These relationships can be compactly summarized (denoting the coordinates x, y, and z by the indices 1, 2, and 3) as

$$\sigma_j\sigma_k = \delta_{jk}\begin{pmatrix} 1 & 0 \\ 0 & 1 \end{pmatrix} + i\varepsilon_{jkl}\sigma_l. \tag{4.10}$$

Here δ_{jk} is the Kronecker tensor ($\delta_{jk} = 1$ for $j = k$; $\delta_{jk} = 0$ otherwise) and ε_{jkl} is the Levi–Civita tensor, which is antisymmetric under exchange of any two of its indices (i.e., $\varepsilon_{123} = \varepsilon_{231} = \varepsilon_{312} = 1$; $\varepsilon_{321} = \varepsilon_{213} = \varepsilon_{132} = -1$; $\varepsilon_{jkl} = 0$ otherwise).

By the rules of addition of angular momenta, the sum of two spin-$\frac{1}{2}$ angular momenta can yield $S = 0$ or $S = 1$. These two cases can be written as

$$\chi_0^0 = [|\uparrow_1\rangle|\downarrow_2\rangle - |\downarrow_1\rangle|\uparrow_2\rangle]/\sqrt{2}$$
$$\chi_1^1 = |\uparrow_1\rangle|\uparrow_2\rangle$$
$$\chi_1^0 = [|\uparrow_1\rangle|\downarrow_2\rangle + |\downarrow_1\rangle|\uparrow_2\rangle]/\sqrt{2}$$
$$\chi_1^{-1} = |\downarrow_1\rangle|\downarrow_2\rangle \tag{4.11}$$

4.4 Internal magnetic fields

In the approximation of separability of the space and spin portions of the wave function inherent in the use of the Schrödinger equation, the radial wave function is mathematically (although it is incorrect physically) independent of the spin and total angular momentum of the system. Thus the angular momentum portion can be treated separately, and various coupling schemes can be considered.

The so-called spin–orbit coupling of the electron is clearly not a coupling of the spin and orbital angular momentum of the electron, since these are properties of the inertial mass. It is the coupling of the intrinsic magnetic moment of the electron to the magnetic field generated by the apparent motion of the nucleus as seen by the electron as a result of its orbital motion. Compared to macroscopic fields that are normally generated in the laboratory this is a very large field, as can be seen from some calculations made with Bohr orbit parameters. To simplify the model, consider an ideally circular orbit, for which the radius r is given by

$$r = a \approx b = a_0 n^2/\zeta, \tag{4.12}$$

where ζ is the effective central charge ($Z - N_e + 1$ for the outermost electron). The tangential speed is given by

$$v = \sqrt{2E/m} = \sqrt{2(Ry\, Z^2/n^2)/m} = \alpha c \zeta/n \tag{4.13}$$

(since $Ry = \frac{1}{2}mc^2\alpha^2$). The current I corresponding to a nucleus of charge Ze orbiting an electron (as seen from the noninertial frame of the electron) in such a Bohr orbit is of the order

$$I = \frac{Ze}{T} = \frac{Zev}{2\pi r} = \frac{Ze}{2\pi}\frac{\alpha c\zeta/n}{a_0 n^2/\zeta} \approx (10^{-3}\text{ amps})\frac{Z\zeta^2}{n^3} \tag{4.14}$$

which corresponds to a magnetic field

$$B = \frac{\mu_0 I}{2r} = \frac{\mu_0 Ze}{4\pi}\frac{\alpha c\zeta/n}{(a_0 n^2/\zeta)^2} \approx (10\text{ teslas})\frac{Z\zeta^3}{n^5}. \tag{4.15}$$

These are very strong fields (a 10-tesla field is more than an ordinary saturated iron magnet can bear). Notice also how rapidly the field increases with increasing central charge, and how rapidly it decreases with increasing n.

For an electron in a light atom such as lithium ($Z = 3$), currents and fields felt by an electron due to the apparent motion of the nucleus are less significant than those produced by the relative motions of the other electrons. Since the electron possesses a g-factor approximately equal to 2, their spins tend to precess in the presence of a magnetic field with a frequency $\omega = eB/m$, twice the Larmor frequency with which their orbits precess.

Thus, in light atoms, the electron moments tend to couple to each rather than to the field produced by the relative motion of the nucleus. Moreover, just as the spin magnetic moments of the electrons tend to couple among themselves, so do their orbital magnetic moments. This leads to the phenomenon known as LS, or Russell–Saunders coupling [171]. (Henry Norris Russell has been called the "Dean of American Astronomers." Frederick A. Saunders subsequently became a leading researcher on the acoustics of the violin.)

For an electron in a heavy atom such as mercury ($Z = 80$), the currents and fields felt by an electron due to the apparent motion of the nucleus are much greater than those produced by the relative motions of the other electrons. Here the relative motions of the individual electrons and the nucleus correspond to currents of the order of amperes, leading to fields of the order of kiloteslas (megagauss).

Thus in heavy atoms, the spin magnetic moment of each electron tends to couple to the magnetic moment generated its own orbital motion about the nucleus. This leads to the phenomenon known as jj coupling.

4.5 Coupling approximations

Quantum mechanical angular momenta are restricted to values that are either integer or half-integer multiples of $\hbar$.

The rules of addition of two angular momenta $\mathbf{J}_1$ and $\mathbf{J}_2$

$$\mathbf{J}_{12} = \mathbf{J}_1 + \mathbf{J}_2 \tag{4.16}$$

are the same as those for non quantum mechanical vectors, namely

$$|J_1 - J_2| \le J_{12} \le J_1 + J_2 \tag{4.17}$$

with the additional restriction that the values of the angular momenta occur in integer steps of $\hbar$ from the lower limit to the upper limit.

When three angular momenta are added, the mathematical procedure is to first add two of them to form a pair-resultant, then add that pair-resultant to the third to form the final resultant. Clearly there are three choices for doing this: combining 1 and 2 to obtain (12) and then combining that to obtain (12)3; combining 1 and 3 to obtain (13) and then combining that to obtain (13)2; and combining 2 and 3 to obtain (23) and then combining that to obtain (23)1. For higher numbers of angular momenta the possibilities increase accordingly. Thus the mathematical coupling scheme that one selects is dependent upon the actual physical couplings that exist in the system, as described in the considerations above, and any scheme that is adopted is likely to be only an approximation.

Two systems that have asymptotic applicability are the LS and the jj coupling models.

4.5.1 *LS* coupling

For electrons in an atom of low-Z, where spins tend to couple to other spins and orbits tend to couple to other orbits, the rotational dynamics can be described by the LS coupling approximation. In this limit, the atom as a whole possesses both a total spin angular momentum and a total orbital angular momentum

$$\mathbf{S} = \sum_{\text{electrons}} \mathbf{S}_i, \qquad \mathbf{L} = \sum_{\text{electrons}} \mathbf{L}_i. \tag{4.18}$$

In this approximation (approached in light multielectron atoms, but truly attained only in the degenerate case of hydrogenlike atoms) the spin angular momenta of the individual electrons have eigenvalues for the lengths

$$\mathbf{S}_i^2 = \frac{1}{2}\left(\frac{1}{2}+1\right)\hbar^2 = \frac{3}{4}\hbar^2, \qquad \mathbf{L}_i^2 = \ell_i(\ell_i+1)\hbar^2. \tag{4.19}$$

Their projections precess about their **L** and **S** resultants (vector sums) and not about an external magnetic field. Similarly, their resultants have eigenvalues for their lengths

$$\mathbf{S}^2 = S(S+1)\hbar^2; \qquad \mathbf{L}^2 = L(L+1)\hbar^2, \tag{4.20}$$

but in the presence of a moderate external magnetic field the spin and orbital moments will not separately precess about the field, but will first couple to each other to form a total angular momentum **J**

$$\mathbf{J} = \mathbf{L} + \mathbf{S}. \tag{4.21}$$

and **J** precesses as a whole about an external magnetic field **B**. Only the presence of a huge external magnetic field that is commensurate with the large internal fields characterized by Eq. 4.15 can cause L and S to decouple. In such a case they precess (independently and with frequencies that differ by a factor 2) with eigenvalues M_s and M_ℓ along the direction of the field (this is known as the Paschen–Back effect, and will be discussed later). In moderate magnetic fields, **J** will have the eigenvalues

$$\mathbf{J}^2 = J(J+1)\hbar^2; \qquad \mathbf{J}_z = M_J\hbar, \tag{4.22}$$

where

$$|L-S| \le J \le L+S. \tag{4.23}$$

and the state of the system is described by the quantum numbers $\{\ell_1, \ell_2, \dots\}$ L, S, J, and M_J. A diagram illustrating this coupling is presented in Fig. 4.1(a).

4.5.2 *jj* coupling

For electrons in an atom of high-Z, their magnetic interactions with the nucleus dominate, and each electron tends to couple its spin magnetic moment to the magnetic moment generated by the relative motion of its orbit with the nucleus, leading to a fixed value for its

(a) *LS* coupling

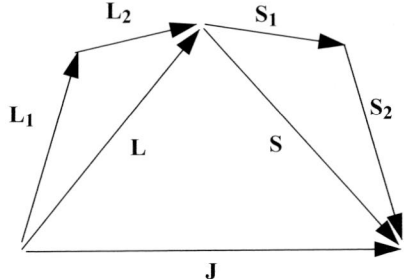

(b) *jj* coupling

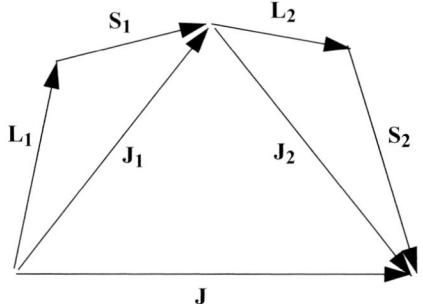

Fig. 4.1. *LS* and *jj* coupling schemes.

total angular momentum

$$\mathbf{J}_i = \mathbf{L}_i + \mathbf{S}_i. \tag{4.24}$$

Its eigenvalues are

$$|\ell_i - 1/2| \leq j_i \leq \ell_i + 1/2, \tag{4.25}$$

again in integer steps. The orbital and spin angular momenta of the individual electrons precess about their resultant, but not about any external field.

The total angular momentum of the system is given by

$$\mathbf{J} = \sum_{\text{electrons}} \mathbf{J}_i, \tag{4.26}$$

and the total state of the system is described by $\{j_1, j_2, \dots\}$ J, M_J. A diagram illustrating this coupling is presented in Fig. 4.1(b).

4.6 Quantum mechanical vector coupling of angular momenta

Quantum mechanical wave functions for the coupling schemes that have been described semiclassically herein by the vector model can be obtained from basis states and vector coupling coefficients. For example, if two angular momenta J_1 and J_2 couple to form a resultant J_{12}, the corresponding wave function can be written

$$|J_1 J_2, J_{12} M_{12}\rangle = \sum_{M_1} \sum_{M_2} (J_1 M_1, J_2 M_2 | J_{12} M_{12}) | J_1 M_1\rangle | J_2 M_2\rangle, \tag{4.27}$$

where J_{12} has integer or half-integer values and satisfies the triangle relationship

$$|J_1 - J_2| \leq J_{12} \leq J_1 + J_2. \tag{4.28}$$

The symbol $(J_1 M_1 J_2 M_2 | J_{12} M_{12})$ is a number called the Clebsch–Gordan coefficient, and is generally a square root of a rational fraction. Tabulations of the numerical values and expositions of the symmetry properties of these coefficients are available in many sources [29, 30, 167], as well as in equivalent alternative formulations such as the Wigner 3-j symbol. Because of the properties of the Clebsch–Gordan coefficients the double sum is only apparent. At the sacrifice of symmetry, the expression can be rewritten as

$$|J_1 J_2, J_{12} M_{12}\rangle = \sum_{M_1} (J_1 M_1, J_2 M_{12} - M_1 | J_{12} M_{12}) | J_1 M_1\rangle | J_2 M_2\rangle. \tag{4.29}$$

If this procedure is continued to couple the resultant J_{12} to a third angular momentum J_3, this can be written

$$|(J_1 J_2) J_{12} J_3, J_{123} M_{123}\rangle = \sum_{M_{12}} \sum_{M_3} (J_{12} M_{12}, J_3 M_3 | J_{123} M_{123}) | J_1 J_2, J_{12} M_{12}\rangle | J_3 M_3\rangle. \tag{4.30}$$

However this wave function is not uniquely formed from the vectors J_1, J_2, and J_{12}, since a resultant could also have been formed in the alternative coupling schemes

$$|(J_2 J_3) J_{23} J_1, J_{123} M_{123}\rangle = \sum_{M_{23}} \sum_{M_1} (J_{23} M_{23}, J_1 M_1 | J_{123} M_{123}) | J_2 J_3, J_{23} M_{23}\rangle | J_1 M_1\rangle, \tag{4.31}$$

and

$$|(J_3 J_1) J_{31} J_2, J_{123} M_{123}\rangle = \sum_{M_{31}} \sum_{M_2} (J_{31} M_{31}, J_2 M_2 | J_{123} M_{123}) | J_3 J_1, J_{31} M_{31}\rangle | J_2 M_2\rangle. \tag{4.32}$$

The connections among these three coupling schemes is determined by another set of quantities, the Wigner 6-j or Racah W coefficients. These quantities play a very important role in the theory of complex spectra (e.g., in LS, jj, and intermediate coupling). The manipulation of these quantities provides many of the relationships that will be used later in this text to characterize empirical data.

4.7 The connection between spin and statistics

Much of the verbiage concerning "wave–particle duality" that pervades elementary physics textbooks could be reduced to a single reference to this misleading historical artifact, if a qualitative discussion of the connection between spin and statistics were presented at the outset.

Entities that possess intrinsic spins that are integer multiples of $\hbar$ have symmetric wave functions. This permits ensembles of such entities to exist in a common state where all do the same thing at the same time. Thus their macroscopic behavior mimics their microscopic behavior, masking their individualities and revealing their periodic coherences. Early workers used words such as "fields" and "waves" to describe what we now call "bosons."

Entities that possess intrinsic spins that are half-odd-integer multiples of $\hbar$ have antisymmetric wave functions. This precludes ensembles of such particles from doing the same thing at the same time. Thus their macroscopic behavior differentiates their individualities and averages out their periodicities. Early workers used words such as "particles" to describe what we now call "fermions."

The examples cited in elementary textbooks as illustrative of one or the other of these "duality" aspects (ocean waves, sound waves, billiard balls, human beings, etc.) are themselves constructed granularly from atoms, but simultaneously possess (both individually and collectively) basic periodic frequency modes. Thus the dichotomy of duality may be more a conceptual impediment than a pedagogical tool, since it introduces a counter-experiential distinction for the express purpose of its subsequent refutation.

The reason that the connections between bosons and fermions and their Bose–Einstein and Fermi–Dirac distributions is deferred until advanced study probably lies in the lack of an accepted elementary explanation of the connection between spin and statistics. A simple intuitive model for this connection was recently presented [80] by Deck and Walker.

Rather than framing the presentation in terms of an operator that exchanges the identities of two arbitrarily labeled particles, this formulation utilizes coordinate rotation operators that accomplish the same task. They demonstrate that the exchange of two identical particles with particular spin orientations can be brought about by a rotation through an angle π of a coordinate system defined by the two particles. Since each such rotation introduces a minus sign, an even number of rotations produces symmetric wave functions, while an odd number of rotations produces antisymmetric wave functions. As a consequence of the connection between the generator of the space rotation and the spin angular momentum operator, the effects of the particle exchange are shown to depend on the integer or half-odd-integer characteristic of the particle spins. From this they conclude that the connection between the spins of the particles and the statistics that derive from the symmetry of the individual identical particle system under particle exchange can be explained in a most fundamental way, by the equivalence between the exchange of two identical particles and a rotation in configuration space.

4.8 The Landé interval rule

In Section 2.4.9, it was shown that the spin–orbit interaction is proportional to the quantity $\mathbf{L} \cdot \mathbf{S}$. In the approximation of LS coupling, projection of $\mathbf{S}$ along $\mathbf{L}$ (or vice versa, depending on which is larger) can be obtained from

$$\mathbf{J}^2 = \mathbf{L}^2 + \mathbf{S}^2 + 2\mathbf{L} \cdot \mathbf{S}, \tag{4.33}$$

so

$$\mathbf{L} \cdot \mathbf{S} = [J(J+1) - L(L+1) - S(S+1)]\hbar^2/2. \qquad (4.34)$$

For a group of levels that differ only in the quantum number J, the fine structure splitting between a level labeled by J and the adjacent level $J - 1$ is

$$\langle E_J \rangle - \langle E_{J-1} \rangle \propto J(J+1) - (J-1)J = 2J. \qquad (4.35)$$

This permits one to immediately draw a scaled picture of the fine structure splitting as a function of J for any system, since

$$\frac{\langle E_J \rangle - \langle E_{J-1} \rangle}{\langle E_{J-1} \rangle - \langle E_{J-2} \rangle} = \frac{J}{J-1} \qquad (4.36)$$

and the ratio of adjacent fine structure splittings is proportional to the Js of their upper levels. This is known as the Landé interval rule [139] and is valid only in the limit of pure LS coupling. Deviations from the rule provide a test of the presence of intermediate coupling or configuration interaction, and examples of these effects will be given in Chapter 5.

4.9 External magnetic fields

The behavior of the atom in an external magnetic field depends upon the strength of that field compared to the internal atomic fields. These are described in terms of the Zeeman and Paschen–Back limits.

4.9.1 Zeeman effect

Extending the development in Section 2.4.9, it can be shown that the simple vector addition $\mathbf{J} = \mathbf{L} + \mathbf{S}$ describes the circulation of mass, but the description of the circulation of charge is more complicated. Because of the g-factor of the electron, the magnetic moment differs from the angular momentum not only in relative magnitude, but also in direction. Assuming $g_e = 2$ exactly, the energy associated with the coupling of the magnetic moment to an external magnetic field $\mathbf{B}$ is given by

$$\boldsymbol{\mu}_J \cdot \mathbf{B} = \frac{e}{2m}(\mathbf{L} + 2\mathbf{S}) \cdot \mathbf{B} = \frac{e}{2m}(\mathbf{J} + \mathbf{S}) \cdot \mathbf{B}. \qquad (4.37)$$

A diagram illustrating these vector relationships is given in Fig. 4.2.

We can use this vector model to find the projection of $\mathbf{J} + \mathbf{S}$ along $\mathbf{J}$, and the projection of $\mathbf{J}$ along $\mathbf{B}$

$$(\mathbf{J} + \mathbf{S}) \cdot \mathbf{B} = \left[\frac{(\mathbf{J} + \mathbf{S}) \cdot \mathbf{J}}{|\mathbf{J}|} \right] \left[\frac{\mathbf{J} \cdot \mathbf{B}}{|\mathbf{J}|} \right], \qquad (4.38)$$

where the quantity $\mathbf{S} \cdot \mathbf{J}$ can be obtained from

$$\mathbf{L}^2 = (\mathbf{J} - \mathbf{S})^2 = \mathbf{J}^2 + \mathbf{S}^2 - 2\mathbf{J} \cdot \mathbf{S}. \qquad (4.39)$$

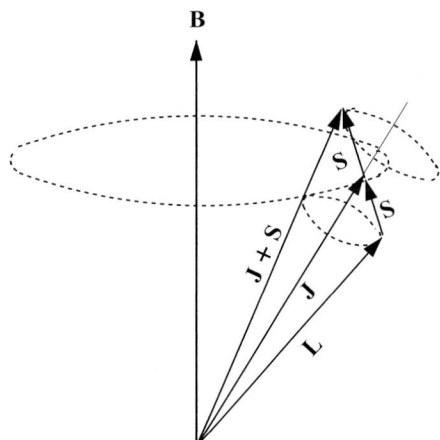

Fig. 4.2. Projections of angular momenta and magnetic moments in LS coupling.

Recognizing $\mathbf{J} \cdot \mathbf{B}$ in Eq. 4.38 as the magnetic projection quantum number M_J, this (together with the definitions of Eq. 4.20) yields

$$\langle \boldsymbol{\mu}_J \cdot \mathbf{B} \rangle = g_J \frac{e\hbar B}{2m} M_J, \qquad (4.40)$$

where g_J is the Landé g-factor [138]

$$g_J = \frac{3J(J+1) + S(S+1) - L(L+1)}{2J(J+1)}. \qquad (4.41)$$

The constant $e\hbar/2m$ is known as the Bohr magneton, μ_{B}. The expectation value in Eq. 4.40 corresponds to the Wigner–Eckart theorem in the quantum mechanical formulation.

Thus a level of energy E_0 placed in a magnetic field B will be shifted to the new energy

$$E_B = E_0 + g_J \mu_{\mathrm{B}} B \, M_J, \qquad (4.42)$$

and the energy of the photon emitted in a transition between this and another level is given by

$$\Delta E = E_0 - E_0' + \mu_{\mathrm{B}} B (g_J M_J - g_J' M_J'). \qquad (4.43)$$

The observation of a frequency change in light emitted by a spectroscopic light source in the presence of a magnetic field was made in 1896 in Leiden by Pieter Zeeman, and the results were published [200] in 1897. Shortly after their observation, the Leiden theoretician H. A. Lorentz informed his former student Zeeman of a possible explanation of this phenomenon on the basis of his corpuscular theory of electricity [145]. As discussed in the next section, the theory of Lorentz predicted a splitting into three components.

It can be seen from Eqs. 4.41 and 4.43 that the prediction of Lorentz is satisfied for the special case of singlet levels. For such levels $S = 0$, so $J = L$ and Eq. 4.41 yields $g_J = 1$. In this case Eq. 4.43 becomes

$$\Delta E = E_0 - E_0' + \mu_{\mathrm{B}} B (M_J - M_J'). \qquad (4.44)$$

The selection rules $\Delta M_J = 0, \pm 1$ yield three energies: one unshifted and two shifted by a constant amount $\pm \mu_B B$.

This splitting is called the "Lorentz triplet" and the interaction is called the "normal Zeeman effect." Unfortunately it is hardly normal, since singlet transitions comprise only a small fraction of the levels studied in atomic spectroscopy. However, at the time it was a significant discovery. It permitted an early determination of the charge-to-mass ratio, and it has been argued [151] that this influenced J. J. Thomson, who announced his discovery of the corpuscle (subsequently called the electron) one month after the publication of Zeeman's paper. For this work Zeeman and Lorentz shared the Nobel Prize in Physics in 1902.

4.9.2 Classical model for the Zeeman effect

In the absence of external fields, Newton's law for a circular orbit equates the Coulomb attraction to the centripetal acceleration

$$\frac{KZe^2}{r^2} = m\frac{v^2}{r} \equiv m\omega^2 r. \tag{4.45}$$

If a magnetic field B is applied perpendicular to the plane of the orbit, the electron will experience an additional radial force $\mathbf{F} = q\mathbf{v} \times \mathbf{B}$. This force will be inward if the electron's orbital moment detracts from the external field, and outward if the electron's orbital moment adds to the external field. The Newton's law equation for this situation becomes

$$\frac{KZe^2}{r^2} \pm e\omega_\pm r B = m\omega_\pm^2 r. \tag{4.46}$$

Here we have assumed that the effect of the magnetic field is small compared to the electrostatic attraction, so that the radius is essentially unchanged by the field ($r_0 \approx r_+ \approx r_-$). If we subtract the equations for the two orbital directions from each other, the Coulomb forces cancel. Dividing this difference by r, we are left with

$$eB(\omega_+ + \omega_-) = m(\omega_+^2 - \omega_-^2) = m(\omega_+ + \omega_-)(\omega_+ - \omega_-). \tag{4.47}$$

Dividing by the factor involving the frequency sums, this yields

$$(\omega_+ - \omega_-) = eB/m. \tag{4.48}$$

If ω_0 is defined to be the average of these frequencies

$$\omega_0 \equiv (\omega_+ + \omega_-)/2, \tag{4.49}$$

then the two frequencies can be written

$$\omega_\pm = \omega_0 \pm \frac{eB}{2m}. \tag{4.50}$$

If this result is substituted into Eq. 4.46, expanded, refactored, and compared with Eq. 4.45, the average frequency in the presence of the field can be written in terms of the field-free case as

$$\omega_0 = \sqrt{\omega^2 + \left(\frac{eB}{2m}\right)^2}. \tag{4.51}$$

This classical formulation predicts that the frequency of a spectral line viewed at right angles to the magnetic field will split into three components, one of higher, one of lower, and one of nearly unchanged frequency. (If the radiation is viewed along the axis of the field, only the higher and lower frequencies will be observed.) This is the "Lorentz triplet" of the "normal Zeeman effect," which does not include the effects of electron spin, but is valid for singlet states.

This phenomenon has many applications. Applied to known spectra it can provide a measure of the strength of the magnetic field. In this manner, Zeeman maps are made of the surface of the Sun (revealing, e.g., the strengths of magnetic fields in sunspots) and of stars. In an atom possessing electrons with paired clockwise and counterclockwise orbitals, this process is responsible for the diamagnetic polarizability. Unlike an atom with an unpaired electron, where the paramagnetic polarizability enhances the applied magnetic field, this detracts from the field. For paired clockwise and counterclockwise electron orbitals, Lenz's law increases the frequency of the orbit that generates a secondary field opposing the external field, and decreases the frequency of the orbit that generates a secondary field enhancing the field. Thus the net effect is an induced field that detracts from the external field.

4.9.3 Paschen–Back effect

The validity of LS coupling requires that the external field be small compared to the internal fields (shown by classical arguments earlier in this chapter to be of the order of $(10\,\mathrm{T})Z\zeta^3/n^5$ for a quasi-hydrogenlike atom). If the external field strength dominates over that of the LS coupling interaction, then the $\mathbf{L}$ and $\mathbf{S}$ vectors will break apart, and each precesses independently (at its own characteristic frequency) about the direction of the external field. In such a case $\mathbf{L}$ will precess at the Larmor frequency, whereas $\mathbf{S}$ will precess (assuming $g_e = 2$ exactly) at twice that frequency. The vector sum of $\mathbf{L}$ and $\mathbf{S}$ will thus oscillate in magnitude over the cycle, so J is no longer a constant of the motion.

This high-field limit is known as the Paschen–Back effect, and corresponds to the separate independent precessions of the L and S vectors. Although J is not a constant of the motion in this case, the projections of L and S along the direction of the field are constants of the motion, and their sum can be interpreted as a projection of the total angular momentum. Thus

$$M_J = M_L + M_S \tag{4.52}$$

is a constant for all regions, encompassing the low-, intermediate-, and high-field cases.

In this case the energy of interaction with the field is given by

$$\boldsymbol{\mu}_J \cdot \mathbf{B} = \frac{e}{2m}(\mathbf{L} + 2\mathbf{S}) \cdot \mathbf{B}, \tag{4.53}$$

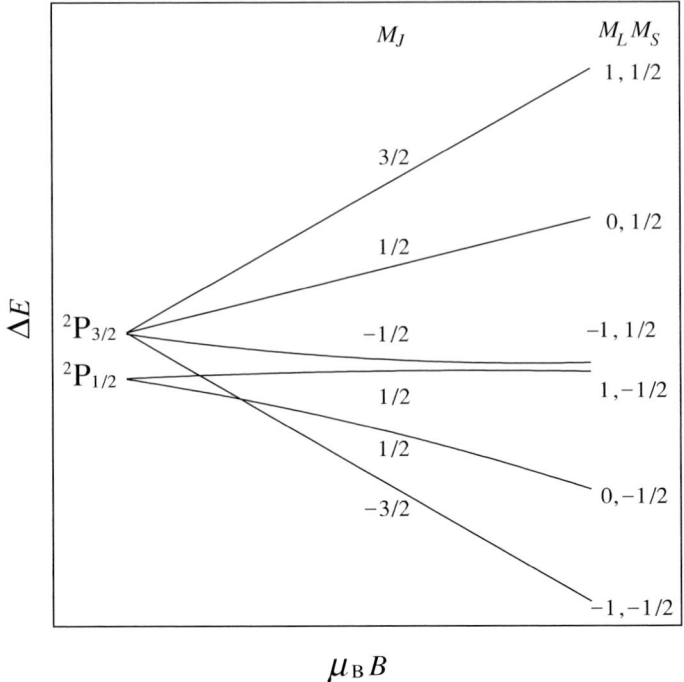

Fig. 4.3. The dependence upon external magnetic field of energies of the magnetic substates in a ^{2}P term. (After Ref. [198])

so

$$\langle \boldsymbol{\mu}_J \cdot \mathbf{B} \rangle = \mu_B \, B(M_L + 2M_S). \tag{4.54}$$

A plot of the dependence upon external magnetic field of energies of the magnetic substates in a ^{2}P term is shown in Fig. 4.3. The low-field region illustrates the Zeeman effect, whereas the high-field region illustrates the Paschen–Back effect. In the intermediate region there is partial decoupling and it is necessary to solve a secular equation for the energy levels.

As an aid to interpreting Fig. 4.3, a comparison of the relevant quantum numbers for the low-field and high-field limits is given in Table 4.1 for an np configuration. For the Zeeman case the energy is proportional to $g_J M_J$ whereas for the Paschen–Back case it is proportional to $M_L + 2M_S$, so those two quantities are also tabulated.

In the region of intermediate field strength, the interaction mixes states of the same M_J. During the transition from the Zeeman limit to the Paschen–Back limit, sublevels of the same M_J never cross. (This can be understood in the context of the off-diagonal elements that occur between these states in the perturbation matrix that removes the zero-field degeneracy [30].)

In the np configuration of Table 4.1, the $M_J = \pm\frac{3}{2}$ states have no partners with which to mix, and thus increase linearly with B for all field strengths. In contrast, the two $M_J = \frac{1}{2}$

Table 4.1. Zeeman and Paschen–Back limits of the specification of the M_J substates for the $np\,^2P_J$ levels in an external magnetic field.

M_J	J	g_J	$g_J M_J$	M_L	M_S	$M_L + 2M_S$
		Low field			High field	
3/2	3/2	4/3	2	1	1/2	2
1/2		4/3	2/3	0	1/2	1
−1/2		4/3	−2/3	−1	1/2	0
−3/2		4/3	−2	−1	−1/2	−2
1/2	1/2	2/3	1/3	1	−1/2	0
−1/2		2/3	−1/3	0	−1/2	−1

states and the two $M_J = -\frac{1}{2}$ states each diverge from each other. The ($M_L = 1$, $M_S = -\frac{1}{2}$) states and the ($M_L = -1$, $M_S = \frac{1}{2}$) states both approach the same limit of zero magnetic splitting for large field.

4.9.4 Extremely strong fields

While the Paschen–Back region is representative of the high-field limit for most laboratory situations, there is justification for examining the behavior of systems in the presence of even more severe conditions [201]. Although a laboratory field of 10 T must be considered as very strong, fields as high as 10^8 T have been estimated to exist in neutron stars. Similarly, because the Coulomb interactions of exiton states in semiconductors are very weak, extreme-strong-field approximation conditions can be achieved with very modest absolute field strengths.

The Hamiltonian of an atom in an electromagnetic field is given [107] by

$$H = (\mathbf{p} - e\mathbf{A})^2/2m + V(\mathbf{r}), \tag{4.55}$$

where $V(\mathbf{r})$ is the electrostatic potential, perhaps modified from the Coulombic case, and the cross term $\mathbf{p} \times \mathbf{A}$ introduces a velocity dependence to the interaction (corresponding to the dependence on velocity of the Lorentz force on a moving charge in a magnetic field). For a uniform magnetic field $\mathbf{B}$, the vector potential $\mathbf{A}$ is given by

$$\mathbf{A} = \frac{1}{2}(\mathbf{r} \times \mathbf{B}). \tag{4.56}$$

In this case the cross term in Eq. 4.55 involves

$$2(\mathbf{p} \cdot \mathbf{A}) = \mathbf{p} \cdot (\mathbf{r} \times \mathbf{B}) = -\mathbf{B} \cdot (\mathbf{r} \times \mathbf{p}) = -\mathbf{B} \cdot \mathbf{L}. \tag{4.57}$$

Consider a uniform magnetic field in the z-direction (with which the radius vector $\mathbf{r}$ makes an angle θ) and a Coulomb electrostatic potential. In this case Eq. 4.57 becomes

$$2(\mathbf{p} \cdot \mathbf{A}) = -BL_z, \tag{4.58}$$

and the quadratic term involves

$$\mathbf{A} \cdot \mathbf{A} = (Br \sin\theta/2)^2. \tag{4.59}$$

For an atom or ion of effective central charge ζ, the Hamiltonian is given by

$$H = \frac{1}{2m} \left[p_r^2 + \frac{L^2}{r^2} + eBL_z + \frac{1}{4}e^2B^2r^2 \sin^2\vartheta \right] - \frac{K\zeta e^2}{r}. \tag{4.60}$$

If the term involving B^2 is neglected the linear approximation is obtained, which (if the effects of electron spin are incorporated into the model) leads to the normal and anomalous Zeeman effect, and the Paschen–Back effect.

When the magnetic field is further increased, a "quadratic regime" is entered in which the system can still be studied by perturbation theory, but the term in B^2 must be included. This corresponds to the term in B^2 in the classical case of Eq. 4.51. Beyond that, semiclassical calculations indicate [81] that the classical orbits do not retain their orderly periodic behavior, but become irregular or chaotic. It has been suggested that the quantum mechanical energy spectrum will also be irregular in this regime (cf. [81] and references therein). If the magnetic field is increased to such a degree that it dominates over the electrostatic term, then a whole new class of solutions emerges, involving spiralling helical orbits.

Application

As an exercise, estimates can be made of the external field strengths that would correspond to the transitions through these various regimes, namely the Zeeman, Paschen–Back, quadratic, chaotic, and helical regions. Consider an atom or ion in a Rydberg state of principal quantum number n. In Eq. 4.15 it was estimated that the internal spin–orbit coupling field of such a system is approximately $(10\ \mathrm{T})\ Z\zeta^3/n^5$, providing an estimate of the field strength corresponding to transition from the normal Zeeman to the Paschen–Back region. The contribution of the linear term in Eq. 4.58 to Eq. 4.55 corresponds to the energy

$$E_1 = eL_z B/2m = \mu_{\mathrm{B}} B M_L \tag{4.61}$$

(here M_L is replaced by $M_L + 2M_S$ when electron spin is included in the Paschen–Back effect). The contribution of the quadratic term in Eq. 4.59 to Eq. 4.55 is given by the energy

$$E_2 = (eBr \sin\vartheta)^2/8m. \tag{4.62}$$

If we consider an ideally circular orbit with $r = a_0 n^2/\zeta$, and note that the constants can be written

$$\hbar/ea_0^2 = Ry/\mu_{\mathrm{B}} = 2.35 \times 10^5 \text{ tesla}, \tag{4.63}$$

then this becomes (for the maximum projection $\sin\vartheta = 1$)

$$E_2 = \frac{(\mu_{\mathrm{B}}B)^2}{Ry} \frac{n^4}{4\zeta^2}. \tag{4.64}$$

The approach to the quadratic regime is thus characterized by the ratio of the quadratic and linear energies

$$\frac{E_2}{E_1} = \frac{\mu_B B}{Ry} \frac{n^4}{4M_L \zeta^2} = \left(\frac{B}{2.35 \times 10^5 \text{ T}}\right) \frac{n^4}{4M_L \zeta^2}. \tag{4.65}$$

Similarly, if the quadratic energy not only exceeds the linear energy, but also the Coulomb binding energy of the atom, this regime is characterized by the ratio of the quadratic energy to the Balmer energy $E_{\text{Balmer}} = Ry\, \zeta^2/n^2$, so

$$\frac{E_2}{E_{\text{Balmer}}} = \left(\frac{\mu_B B}{Ry}\right)^2 \frac{n^6}{4\zeta^4} = \left(\frac{B}{2.35 \times 10^5 \text{ T}}\right)^2 \frac{n^6}{4\zeta^4}. \tag{4.66}$$

To estimate the value of B for which the quadratic energy becomes significant compared to the linear energy, we can consider a value $E_2/E_1 \approx 1/10$, for which Eq. 4.65 yields

$$B \approx (9.4 \times 10^4 \text{ T}) M_L \zeta^2/n^4. \tag{4.67}$$

For the magnetic energy to dominate over the binding energy of the atom, we can consider a value $E_2/E_{\text{Balmer}} \approx 10$, for which Eq. 4.66 yields

$$B \approx (4.7 \times 10^6 \text{ T})\zeta^2/n^3. \tag{4.68}$$

For a hydrogen or neutral lithium atom in the 2p level ($\zeta = 1$, $n = 2$), these considerations lead to the following estimates: the internal spin–orbit coupling would be commensurate with the external field for $B \approx 0.04$ T; its quadratic correction would be 1% of its linear magnetic energy for $B \approx 6 \times 10^3$ T, and its quadratic correction would exceed its Balmer energy by an order of magnitude for $B \approx 3 \times 10^5$ T.

Notice that the conditions for which the quadratic energy equals or exceeds the Balmer energy can be achieved in high Rydberg states, since the ratio in Eq. 4.66 is proportional to $n^6 B^2$. A measurement [104] of the principal Rydberg series in barium through $n = 80$ in a field of 3 T can be compared with measurement for an $n = 2$ level in a field of 2×10^5 T.

5

The intermediate coupling model

Sic transit gloria uni electroni.

5.1 Spectroscopic notation

The nomenclature that is used to describe measured quantities in atomic spectroscopy is very much governed by the approximations inherent in the Schrödinger equation. Two theoretical approximations are particularly important. One is the central-field approximation, in which a many-electron atom is described by wave functions that are constructed from products of one-electron states. Another is the nonrelativistic approximation, which leads to a separation of the space and spin portions of the wave function.

A one-electron atomic state is defined by the hydrogenic basis state of quantum numbers $| n \, \ell \, m_\ell \, m_s \rangle$, where states with a common value of n are denoted as a "shell" and states with common values of n and ℓ are denoted as a subshell. Since the electron–electron interaction is treated in an averaged manner by the central-field approximation, and the spin and space portions are treated as independent by the nonrelativistic approximation, electrons with the same value of n and ℓ are treated as "equivalent." As in the case of hydrogen, each electron is assigned a set of n and ℓ quantum numbers, to yield a "configuration." Here the numerical values of ℓ are replaced by letters according to the code s, p, d for $\ell = 0, 1, 2$ and f, g, h, etc., for $\ell = 3, 4, 5$ etc., (alphabetically from f with the letter j omitted), with a superscript to describe the number of equivalent electrons in each subshell. This notation was originally formalized [172] in a 1929 meeting that was attended by most of the leading spectroscopists of that era.

Since these single-particle wave functions are rigorously valid only for hydrogen or hydrogenlike ions, all physical multielectron atomic systems possess some degree of "configuration interaction" (CI), in that they require an admixture of these single-electron basis states for their description. In the standard notation the individual electron configurations in a complex atom or ion are nominally described in terms of their dominant single-electron configuration

$$\text{configuration} \equiv \{n\ell^i\} = (1\text{s}^2 2\text{s}^2 2\text{p}^6 \ldots). \tag{5.1}$$

An important property of the configuration is its parity $\mathcal{P}$, defined as

$$\mathcal{P} \equiv (-1)^{\Sigma \ell_i}. \tag{5.2}$$

The parity is said to be even or odd depending on whether the sum over all ℓ_i values is an even or an odd integer.

In addition to this nonrelativistically defined configuration notation (which is based only on spatial coordinates), a notation must be added to describe both the intrinsic spin of the individual electrons and their couplings to each other. For the individual electrons, the equivalent electrons are represented by $|\, n\,\ell\,j\,m_j\rangle$ rather than $|\,n\,\ell\,m_\ell\,m_s\rangle$, with $j = \ell \pm 1/2$. The LS or Russell–Saunders coupling scheme [171] is usually adopted. In this scheme the vector sums $\mathbf{L}, \mathbf{S}, \mathbf{J}, \mathbf{M}_J$ characterize the system and the nondegenerate magnitudes of the angular momenta are denoted by the symbol

$$^{2S+1}\mathrm{L}_J, \tag{5.3}$$

where $L = 0, 1, 2, \ldots$ is represented by the capital letter S, P, D, etc., according to the same code as used for lower-case letters in the configuration. The quantity $2S + 1$ is called the "multiplicity" and not only prescribes the total vector sum of the electron spin angular momenta, but also the number of different values of J that will occur if $S \leq L$. (If $S \geq L$, that number is given by $2L + 1$.) To include the parity in this LS notation, odd terms are indicated by a superscript $^\circ$ appended to the letter symbol for L.

The quantity $^{2S+1}L_J$ is characterized according to its value for $2S + 1 = 1, 2, 3, 4, 5, \ldots$ as a singlet, doublet, triplet, quartet, quintet, ... It has been pointed out [106] that this nomenclature mixes two musical notations. A solo, duet, trio, quartet, quintet, ... is a composition for 1, 2, 3, 4, 5, ... voices or instruments; a doublet, triplet, quadruplet, quintuplet, ... is 2, 3, 4, 5, ... notes played in one beat. Although not musically consistent, this usage is standardized in spectroscopy.

As noted earlier, the angular momenta of a complex atom are never in a pure state of LS coupling (and in many cases are closer to jj coupling), and therefore possess some degree of "intermediate coupling" (IC), in that they require an admixture of LS-coupling basis states for their description.

For a complex atom there may be many other quantum numbers that are needed to characterize the system, which will be symbolically denoted here by γ. In this notation, a "state" of the complex atom is described by the notation

$$\text{state} \equiv \{n\ell^i\} \,|\, \gamma S L J M_J\rangle. \tag{5.4}$$

Since, in the absence of an external magnetic field, the energies of states of the same M_J are degenerate, a "level" can be characterized as

$$\text{level} \equiv {}^{2S+1}L_J = \{n\ell^i\} \,|\, \gamma S L J\rangle. \tag{5.5}$$

In the Schrödinger picture, the quantum number J does not enter the radial wave function, hence the fine structure dependences must all be computed as perturbative corrections.

Table 5.1. Hierarchies of notation. (After Ref. [148].)

Hierarchy	Quantum numbers	Transitions
Configuration	$\{n\ell^i\}$	Transition array
Polyad	$\{n\ell^i\} \mid \gamma S\rangle$	Supermultiplet
Term	$\{n\ell^i\} \mid \gamma SL\rangle$	Multiplet
Level	$\{n\ell^i\} \mid \gamma SLJ\rangle$	Line
State	$\{n\ell^i\} \mid \gamma SLJM_J\rangle$	Component

This has led to the definition of the spectroscopic "term"

$$\text{term} \equiv {}^{2S+1}\mathrm{L} = \{n\ell^i\} \mid \gamma SL\rangle. \tag{5.6}$$

Another classification of quantities involves the "polyad" which encompasses the various terms of a given multiplicity that exist within a configuration. The various manifolds of transitions among these entities are given special names. A transition between two states is a component; the set of transitions between two levels is a line; the set of transitions between two terms is a multiplet; the set of transitions between two polyads is a supermultiplet; the set of transitions between two configurations is a transition array. These definitions are presented in Table 5.1.

It is clear that the restrictions imposed on the notation by the central-field and nonrelativistic approximations mean that this notation is often only nominal, and does not necessarily describe the quantum numbers of the complex atom in question. Indeed, in many situations the only quantities that truly characterize a level in an experimental situation are its measured energy, its total angular momentum, and its parity (which is here prescribed by whether ℓ is even or odd, but can also be defined in a more general manner in the case of intermediate coupling, or in the Dirac relativistic case where ℓ is not a "good quantum number").

As an illustrative example, consider a six-electron carbonlike atom or ion with the electron configuration $1s^2 2s^2 2p3p$ (an example of an "inner-shell-excited" or "hollow-atom" configuration that is copiously populated in, e.g., fast ion-beam excitation). Since the four s-electrons are in closed shells, we need consider only the 2p and 3p electrons. The possibilities for the vector sums of the angular momenta are $\mathbf{S} = \mathbf{1/2} + \mathbf{1/2} = \mathbf{0}, \mathbf{1}$ and $\mathbf{L} = \mathbf{1} + \mathbf{1} = \mathbf{0}, \mathbf{1}, \mathbf{2}$. Thus the polyads in the configuration include a singlet triad $({}^1\mathrm{S}\ {}^1\mathrm{P}\ {}^1\mathrm{D})$ and a triplet triad $({}^3\mathrm{S}\ {}^3\mathrm{P}\ {}^3\mathrm{D})$. These combine to give six terms $({}^1\mathrm{S}\ {}^3\mathrm{S}\ {}^1\mathrm{P}\ {}^3\mathrm{P}\ {}^1\mathrm{D}\ {}^3\mathrm{D})$. This corresponds to ten levels $({}^1\mathrm{S}_0\ {}^3\mathrm{S}_1\ {}^1\mathrm{P}_1\ {}^3\mathrm{P}_{0,1,2}\ {}^1\mathrm{D}_2\ {}^3\mathrm{D}_{1,2,3})$. Taking into account $2J+1$ degeneracies of the two $J=0$, four $J=1$, three $J=2$, and one $J=3$ levels, the total number of states is 36.

5.2 Two-valence-electron systems

Much of atomic structure involves interactions between electrons through processes involving intermediate coupling, configuration interaction, correlation, polarization, exchange, etc. In this sense the hydrogen atom is not at all the simplest example, but rather in a class by itself. The simplest case that provides these characteristics is a two-valence-electron system, which includes helium and heliumlike atoms, as well as systems dominated by configurations consisting of two electrons outside closed shells or subshells.

5.2.1 Solution of the unperturbed problem

The Schrödinger equation for a two-particle system in the presence of electrostatic and spin–orbit interactions can be written

$$H\psi_{n_1,n_2}(\mathbf{r}_1, \mathbf{r}_2) = \left[\frac{p_1^2}{2m} - \frac{KZe^2}{r_1} + \frac{p_2^2}{2m} - \frac{KZe^2}{r_2} \right] \psi_{n_1,n_2}(\mathbf{r}_1, \mathbf{r}_2)$$

$$+ \left[\frac{Ke^2}{|\mathbf{r}_1 - \mathbf{r}_2|} + \xi(r_1)\ell_1 \cdot \mathbf{s}_1 + \xi(r_2)\ell_2 \cdot \mathbf{s}_2 \right] \psi_{n_1,n_2}(\mathbf{r}_1, \mathbf{r}_2), \quad (5.7)$$

where the quantum number n_i is intended to include n_i, ℓ_i, m_{ℓ_i}, m_{s_i} implicitly. If we temporarily neglect the electron–electron and spin–orbit interaction terms in the second bracket to be treated later as perturbations, then the assumption of a product solution

$$\psi_{n_1,n_2}(\mathbf{r}_1, \mathbf{r}_2) = u_{n_1}(\mathbf{r}_1)u_{n_2}(\mathbf{r}_2) \quad (5.8)$$

leads to a separation of the variables, and a solution involving single-electron wave functions

$$u_n(\mathbf{r}) = R_{n\ell}(\mathbf{r})Y_\ell^{m_\ell}(\theta, \phi) \quad (5.9)$$

and single-electron eigenvalues. For equivalent electrons ($n_1\ell_1 = n_2\ell_2$) the choice of their labeling is inconsequential. For nonequivalent electrons the product wave functions must be constructed so as to be independent of the labeling.

Since the labeling of both the coordinates and the quantum numbers is arbitrary, the squared position probability must be independent of an interchange of the labels on the coordinates which retains the labeling of the quantum numbers. Thus

$$|\psi_{n_1,n_2}(\mathbf{r}_1, \mathbf{r}_2)|^2 = |\psi_{n_1,n_2}(\mathbf{r}_2, \mathbf{r}_1)|^2. \quad (5.10)$$

This is satisfied by two possibilities

$$\psi_{n_1,,n_2}(\mathbf{r}_2, \mathbf{r}_1) = \pm\psi_{n_1,n_2}(\mathbf{r}_1, \mathbf{r}_2), \quad (5.11)$$

so the wave functions must be an admixture of these two possibilities

$$\psi_{n_1,,n_2}^{\pm}(\mathbf{r}_1, \mathbf{r}_2) = \frac{1}{\sqrt{2}}[u_{n_1}(\mathbf{r}_1)u_{n_2}(\mathbf{r}_2) \pm u_{n_1}(\mathbf{r}_2)u_{n_2}(\mathbf{r}_1)]. \quad (5.12)$$

The plus sign corresponds to a spatial wave function that is symmetric under interchange of particle identities, whereas the minus sign corresponds to a spatial wave function that is antisymmetric under interchange of particle identities. Since we are dealing here with electrons, which obey Fermi–Dirac statistics, the total wave function must be antisymmetric under interchange of particles. However, the basic nonrelativistic solution to the Schrödinger equation omits the spin of the electron, and thus provides only the spatial portion of the total wave function.

As was shown in Section 4.3, the Pauli formulation of the two-electron problem adds a set of spin wavefunctions $\chi_S^{M_S}$ to the consideration. These consist of an antisymmetric singlet $(S = 0)$

$$\chi_0^0 = [|\uparrow_1\rangle|\downarrow_2\rangle - |\downarrow_1\rangle|\uparrow_2\rangle]/\sqrt{2}, \tag{5.13}$$

and a symmetric triplet $(S = 1)$ array

$$\chi_1^1 = |\uparrow_1\rangle|\uparrow_2\rangle$$
$$\chi_1^0 = [|\uparrow_1\rangle|\downarrow_2\rangle + |\downarrow_1\rangle|\uparrow_2\rangle]/\sqrt{2}$$
$$\chi_1^{-1} = |\downarrow_1\rangle|\downarrow_2\rangle. \tag{5.14}$$

The total antisymmetric wavefunction is thus given by the product of the space and spin portions

$$\Psi_{n_1,n_2,0,0}(\mathbf{r}_1, \mathbf{r}_2) = \psi_{n_1,n_2}^+(\mathbf{r}_1, \mathbf{r}_2)\chi_0^0$$
$$\Psi_{n_1,n_2,1,M_S}(\mathbf{r}_1, \mathbf{r}_2) = \psi_{n_1,n_2}^-(\mathbf{r}_1, \mathbf{r}_2)\chi_1^{M_S}. \tag{5.15}$$

By pairing symmetric space with antisymmetric spin and antisymmetric space with symmetric spin, a wave function of overall antisymmetric nature is obtained. The appropriate values for L and J can be constructed as admixtures of these wave functions. Corrections for other interactions can be computed using first-order perturbation theory.

5.2.2 Consequences of exchange symmetry for equivalent electrons

The factorization of spin and space in the Pauli formulation separates the interpretation of quantum numbers n_i, ℓ_i, m_{ℓ_i} from the quantum number m_{s_i}. The specification of a singlet or a triplet state sets the values of the m_{s_i}. As stated earlier, if the spatial quantum numbers n_i and ℓ_i of two electrons are the same, they are called "equivalent electrons" since they differ only in the quantum number m_{ℓ_i} that is degenerate in the single-particle model. When a configuration contains equivalent electrons, the designation of a singlet or a triplet spin state places severe restrictions on the possible values for the m_{ℓ_i} quantum numbers, and thus the Pauli exclusion principle rules out certain states.

For example, if a configuration consists of two $\ell = 0$ electrons (an ns^2 configuration), the only possibility is that both electrons have the value $m_\ell = 0$. Since all spatial quantum numbers are then the same, the antisymmetric spatial wave function will vanish, and no triplet can exist. Thus the $ns^2\,^3S_1$ state is ruled out by the Pauli exclusion principle.

It can be shown generally that for any configuration consisting of two equivalent electrons $n\ell^2$ the quantity $L + S$ must be even. Thus for an np^2 configuration, where $L = 0, 1, 2$ and $S = 0, 1$, there exist ^1S, ^3P, ^1D terms but no ^3S, ^1P, ^3D terms.

Whereas the antisymmetric spatial wave function vanishes entirely if the spatial quantum numbers of the two electrons are the same, it also becomes very small if the quantum numbers are different but the radial coordinates are nearly the same. Thus the Pauli exclusion principle "carves out a hole" in the position probability density of the triplet state, forcing the two electrons to be well separated in space. This is the origin [124] of "Hund's rule," which states that the ground term has the largest value of S (or the highest multiplicity $2S + 1$) of the terms consistent with the Pauli exclusion principle. If there are several such terms then the one with the largest value of L will lie lowest.

The basis of Hund's rule can be seen from a simple conceptual model. Since a high value of S indicates that the electrons all have spins in the same direction, all will have the same value of m_s. This forces the spatial quantum numbers or positions to be different, so the electrons are close in spin-space (enhancing their magnetic coupling) but distant in configuration-space (reducing their repulsive electron–electron interaction). Both of these interactions tend to increase the effective binding energy. In the two-electron case, these considerations generally lead to the result that the triplet is more tightly bound than the singlet.

It should be emphasized that Hund's rule deals only with the ordering of terms within a configuration, as specified by the exclusion principle. The ordering of levels within a term (J-dependent fine structure) requires consideration of the spin–orbit interaction, and there is no ordering of the states within a level because they are degenerate in the absence of an external field. Thus, the use of the words "ground state" to designate the lowest energy level in a complex atom can be misleading. If the lowest fine structure level of the ground term has a nonzero value for J there is not a single ground state, but rather a manifold of ground states.

As an example of Hund's rule, consider the carbon atom, which has a ground configuration $1s^2 2s^2 2p^2$. The open p^2 subshell permits the possible values $L = 0, 1, 2$ and $S = 0, 1$, but since the two electrons are equivalent, only those terms for which $L + S =$ even are permitted by the exclusion principle. The ground configuration thus consists of the three terms ^3P, ^1D, and ^1S, and since the ^3P has the highest value of S, Hund's rule predicts that it is the ground term. Extending the considerations beyond Hund's rule, it will be shown in Section 5.2.4 that the fine structure levels of the ^3P term will have "normal" ordering, with $^3P_0$ lowest, followed by $^3P_1$ and $^3P_2$. This contrasts with the oxygen atom, which has a p^4 ground configuration, thus containing two electron holes instead of two electrons. This causes the fine structure ordering of the ground term of the oxygen atom to be inverted from that of the carbon atom (and thus the $^3P_2$ ground level is properly a ground-state manifold).

5.2.3 Electrostatic corrections: the Slater parameters

For the electrostatic electron–electron interaction the first-order perturbation correction is

$$\Delta E = \left\langle \Psi \left| \frac{Ke^2}{|\mathbf{r}_1 - \mathbf{r}_2|} \right| \Psi \right\rangle, \tag{5.16}$$

where Ψ is the unperturbed wave function. Using the orthogonality of the spin wave functions

$$\Delta E = \frac{1}{2} \int d\mathbf{r}_1 \int d\mathbf{r}_2 [u_{n_1}^*(\mathbf{r}_1) u_{n_2}^*(\mathbf{r}_2) \pm u_{n_1}^*(\mathbf{r}_2) u_{n_2}^*(\mathbf{r}_1)]$$
$$\times \frac{Ke^2}{|\mathbf{r}_1 - \mathbf{r}_2|} [u_{n_1}(\mathbf{r}_1) u_{n_2}(\mathbf{r}_2) \pm u_{n_1}(\mathbf{r}_2) u_{n_2}(\mathbf{r}_1)]. \tag{5.17}$$

Expanding products and collecting like terms

$$\Delta E = \int d\mathbf{r}_1 \int d\mathbf{r}_2 u_{n_1}^*(\mathbf{r}_1) u_{n_2}^*(\mathbf{r}_2) \frac{Ke^2}{|\mathbf{r}_1 - \mathbf{r}_2|} u_{n_1}(\mathbf{r}_1) u_{n_2}(\mathbf{r}_2)$$

$$\pm \int d\mathbf{r}_1 \int d\mathbf{r}_2 u_{n_1}^*(\mathbf{r}_1) u_{n_2}^*(\mathbf{r}_2) \frac{Ke^2}{|\mathbf{r}_1 - \mathbf{r}_2|} u_{n_1}(\mathbf{r}_2) u_{n_2}(\mathbf{r}_1). \tag{5.18}$$

The noncentral nature of the interelectron distance can be made central by a Legendre polynomial expansion

$$\frac{1}{|\mathbf{r}_1 - \mathbf{r}_2|} = \sum_{k=0}^{\infty} \frac{r_<^k}{r_>^{k+1}} P_k(\cos\theta). \tag{5.19}$$

Here $r_<$ and $r_>$ are the lesser and greater, respectively, of the distances r_1 and r_2 of the electrons from the nucleus, and θ is the angle between the two radius vectors. The Legendre polynomial $P_k(\cos\theta)$ is combined with the spherical harmonics $Y_\ell^{m_l}(\theta, \phi)$ and $Y_{\ell'}^{m_l'}(\theta', \phi')$ in the wave functions. Through the use of the spherical harmonic reduction theorem [30], the expansion effects a separation not only of the radial factors from the angular factors, but also a separation of the angular variables of the individual electrons [30]. With a knowledge of the numerical factors that result from the angular integrations, it is possible to characterize the spectrum entirely in terms of the radial integrals. The direct and exchange radial integrals can be written as

$$\Delta E = \sum_{k=0}^{\infty} (F^k \pm G^k), \tag{5.20}$$

where F^k and G^k are the "Slater integrals" [177]. Note that in this widely-used notation, the k indices are superscripts and not powers (like the superscript m_l in the spherical harmonic in Eq. 5.9). These quantities can be written in terms of the radial wave functions alone as

$$F^k \equiv Ke^2 \int dr_1 r_1^2 \int dr_2 r_2^2 |R_{n_1, \ell_1}(r_1) R_{n_2, \ell_2}(r_2)|^2 \frac{r_<^k}{r_>^{k+1}}, \tag{5.21}$$

and

$$G^k \equiv Ke^2 \int dr_1 r_1^2 \int dr_2 r_2^2 R_{n_1, \ell_1}(r_1) R_{n_2, \ell_2}(r_1) R_{n_1, \ell_1}(r_2) R_{n_2, \ell_2}(r_2) \frac{r_<^k}{r_>^{k+1}}. \tag{5.22}$$

Both these quantities are positive definite [30]. For F^k this is clear, since the integrand is everywhere positive. Although it is not immediately obvious from its definition, it can be proven [167] that G^k is also positive. The angular integrals of the matrix elements of the spherical harmonics require that ℓ_1, ℓ_2 and k satisfy triangle relationships and that their sum be even. Thus, this expansion has only a small number of nonvanishing terms, and only a small number of radial integrals are needed.

For $k = 0$, G^0 vanishes and F^0 is independent of ℓ for a given n. Thus F^0 has the same ℓ-degeneracy as the potential energy, so it corresponds to a correction for repulsive electron screening to the unperturbed single-particle potential energies (which included only the attractive central-field interaction). Since F^0 is spherically symmetric, it is directly interpretable as a charge-screening parameter.

For higher orders in k, these Slater integrals couple the states of differing ℓ. Thus they can be used to remove the degeneracies between the singlet and triplet energies that occur because of the neglect of electron–electron interactions in the unperturbed potential. Using degenerate perturbation theory, the diagonalization of the energy matrix provides equations that specify the singlet–triplet splittings as a function of the Slater integrals. Through these equations, the Slater integrals can be used as empirical Slater parameters that are specified from the experimental energy-level data. If their isoelectronic behavior is regular and slowly varying, the empirical Slater parameters can be used to interpolate, extrapolate, and critically evaluate the database.

5.2.4 Spin–orbit interaction

In Eq. 2.95, it was shown from semiclassical arguments that the spin–orbit interaction energy for a single electron is given by

$$\Delta E = (g_e - 1) \frac{\hbar^2}{2m^2c^2} K Z e^2 \left\langle \frac{\mathbf{l} \cdot \mathbf{s}}{r^3} \right\rangle. \tag{5.23}$$

In the quantum mechanical formulation the same result is obtained for this case, but the more general expression is (assuming $g_e \approx 2$)

$$\Delta E = \frac{\hbar^2}{2m^2c^2} \left\langle \frac{1}{r} \frac{dV}{dr} \mathbf{l} \cdot \mathbf{s} \right\rangle \tag{5.24}$$

where $V(r)$ is the single-particle central-field potential, so the spin–orbit parameters $\xi(r)$ in Eq. 5.7 are of the form

$$\xi(r) \equiv \frac{\hbar^2}{2m^2c^2} \left\langle \frac{1}{r} \frac{dV}{dr} \right\rangle. \tag{5.25}$$

For single-electron states

$$\begin{aligned}
\mathbf{l} \cdot \mathbf{s} &= [j(j+1) - \ell(\ell+1) - 3/4]/2 \\
&= \ell/2 & (j = \ell + 1/2) \\
&= -(\ell+1)/2 & (j = \ell - 1/2).
\end{aligned} \tag{5.26}$$

Thus, the spin–orbit interaction increases the binding energy of the "jacknife" states ($j = \ell - \frac{1}{2}$) and decreases the binding energy of the "stretch" states ($j = \ell + \frac{1}{2}$). The adjective "jacknife" indicates that one vector folds back on the other in an antiparallel manner, whereas the adjective "stretch" indicates that one vector extends the other in a parallel manner. An analogy can be made to bar magnets laid side-by-side. If they are in the antiparallel jacknife configuration, the north and south poles are juxtaposed, creating an attraction. If they are in the parallel stretch mode, the two north poles and the two south poles are juxtaposed, creating a repulsion.

The splitting between these two levels is proportional to $2\ell + 1$. In practical applications, the offset of the $j = \ell - \frac{1}{2}$ level below the centroid

$$\xi_0 \equiv -\langle \xi \rangle (\ell + 1)/2 \tag{5.27}$$

is incorporated into the unperturbed energy so that the spin–orbit perturbation is characterized by a spin–orbit parameter ζ_ℓ (not to be confused with the use of the symbol ζ elsewhere
to denote the effective central charge), defined as

$$\zeta_\ell \equiv \langle \xi \rangle (2\ell + 1). \tag{5.28}$$

Here, the perturbation will couple levels within the configuration in various ways depending
on the values of j and ℓ, and degenerate perturbation theory will yield equations that
connect ζ_ℓ to the individual levels. Thus, like the Slater integrals, the spin–orbit parameter
can be empirically specified from measured energy-level data and used quantitatively to
characterize the spectra. (Thus, since the Slater integrals and the spin–orbit energies will be
used jointly to characterize spectroscopic data, the term "Slater parameters" will sometimes
be used generically to specify both types of quantities.)

    This description has been made for a single electron, but it could equally well have been
made for a subshell that lacks one electron from being closed. The electron "hole" can then
be treated exactly as though it were an electron, except for the fact that the value of ζ_ℓ has
the opposite sign from that of the corresponding electron state. This is because the vector
sum of the angular momenta of all save one of the electrons in a subshell is the negative of
the value for the electron that would close the shell. Because of the negative value of ζ_ℓ, the
ordering of the fine structure levels within a term arising from a hole state will be inverted
from that of the corresponding electron state.

5.2.5 Example: nsn'p and nsn'p^5 configurations

The usefulness of the characterization

$$E = E_0 + \sum_k (F^k + G^k) + \sum_i \zeta_i \tag{5.29}$$

can be clearly illustrated by specific examples. The approach can be used to describe
configurations that possess two valence electrons (or an electron and a hole) outside closed
shells or subshells. Let us first consider a case in which one of the electrons is in an $\ell = 0$ (s)
state, such as an nsn'p configuration in an alkaline-earthlike system or an nsn'p^5 ($n' < n$)
configuration in an inert-gaslike system. In this case there are four levels: $^3\mathrm{P}^\mathrm{o}_0$, $^3\mathrm{P}^\mathrm{o}_1$, $^3\mathrm{P}^\mathrm{o}_2$, and
$^1\mathrm{P}^\mathrm{o}_1$. It can be shown [29, 30] that for these configurations there are nonvanishing values
only for F^0 and G^1, and the s-electron has a vanishing spin–orbit splitting. Thus the four
energy levels of the configuration can be characterized by only three parameters, E_0 (which
contains F^0), G^1, and ζ_p.

    As has been pointed out by Cowan [30], the theoretical expressions for the average
energy of the atom involve the quantities F^k and G^k multiplied by rational fractions. In
the spectroscopic literature a set of parameters $F_k = F^k/D_k$ and $G_k = G^k/D_k$ has been
introduced, where the constants D_k are least common denominators in expressions for the
average energies. The F^k and G^k quantities have the advantage of generality, since the values
introduced for D_k are different for F^k and G^k, and differ among electron configurations.
In contrast, the F_k and G_k quantities have the advantages of yielding integer coefficients in
the energy level expressions, and utilizing subscripts that are not confused with powers. For

Table 5.2. Matrix elements for an sp configuration.

	$^3P_0^o$	$^3P_1^o$	$^3P_2^o$	$^1P_1^o$
$^3P_0^o$	$E_0 - G_1 - \zeta_p$	0	0	0
$^3P_1^o$	0	$E_0 - G_1 - \zeta_p/2$	0	$\zeta_p/\sqrt{2}$
$^3P_2^o$	0	0	$E_0 - G_1 + \zeta_p/2$	0
$^1P_1^o$	0	$\zeta_p/\sqrt{2}$	0	$E_0 + G_1$

all cases $D_0 = 1$. For an sp configuration $D_1 = 3$ for G^1, and all F^k vanish for $k > 0$. For a p^2 configuration, $D_2 = 25$ for F^2, F^k vanishes for all k-values except $k = 0$ and 2, and all values of G^k vanish. To conform to the usual spectroscopic notation, $G_1(sp) = G^1(sp)/3$ and $F_2(p^2) = F^2(p^2)/25$ will be used in the discussion that follows.

Table 5.2 shows the matrix elements of these various quantities [29]. In the unperturbed case these four levels are degenerate in energy (all have energy E_0), but in the eigenvector representation obtained by diagonalizing this matrix, the energies become

$$^3P_0^o = E_0 - G_1 - \zeta_p$$
$$^3P_1^o = E_0 - \zeta_p/4 - \Delta_1$$
$$^3P_2^o = E_0 - G_1 + \zeta_p/2$$
$$^1P_1^o = E_0 - \zeta_p/4 + \Delta_1, \tag{5.30}$$

where

$$\Delta_1 \equiv \sqrt{(G_1 + \zeta_p/4)^2 + \zeta_p^2/2}. \tag{5.31}$$

Here the subscript p denotes the value $l = 1$ of the p-electron.

Edlén devised a useful type of plot [87] to relate these quantities to measured data and to characterize the degree of intermediate coupling between the LS and jj limits. With small differences, the plot can be used for either the sp or the sp^5 configurations.

The quantity G_1 is positive for either of these two configurations, but the value of ζ_p is positive for the sp electron configuration and negative for the sp^5 hole configuration. Taking this into account, Edlén defined the dimensionless coordinates

$$x \equiv \frac{3|\zeta_p|/4}{G_1 + 3|\zeta_p|/4} \tag{5.32}$$

and

$$y \equiv \frac{E - E_0 + \zeta_p/4}{G_1 + 3|\zeta_p|/4}. \tag{5.33}$$

The x coordinate and the scaling factor for the y coordinate are defined in terms of the *magnitude* of the spin–orbit integral so as to exclude the inherent sign of ζ_p (positive for sp, negative for sp^5). This is done so that the abscissa will have the range $0 \le x \le 1$, and so that the ordinate will be bounded with a nonvanishing denominator. Note that the quantity ζ_p in the numerator of y is *not* a magnitude. It is only the scale factor in the denominator, and not the zero point in the numerator, that differs between the two cases.

Using these coordinate definitions, it is possible to present these intermediate coupling equations in a very compact form. Thus, after some algebra, the energy levels of Eqs. 5.30 and 5.31 can be written in terms of four values for $y(^{2S+1}P_J)$ as

$$y(^3P_0) = -1$$
$$y(^3P_1) = -\sqrt{1 - 4x(1-x)/3}$$
$$y(^3P_2) = -1 + 2x$$
$$y(^1P_1) = \sqrt{1 - 4x(1-x)/3}. \tag{5.34}$$

For an sp^5 configuration, $\zeta_p \to -\zeta_p$ in Table 5.2 and in Eqs. 5.30 and 5.31 (the vacancy state causes ζ_p to be negative). Again, after some algebra (this time taking into account the differing definitions of x and y regarding the magnitude of ζ_p), the values of y for this configuration are

$$y(^3P_2) = -1$$
$$y(^3P_1) = -\sqrt{1 - 8x(1-x)/3}$$
$$y(^3P_0) = -1 + 2x$$
$$y(^1P_1) = \sqrt{1 - 8x(1-x)/3}. \tag{5.35}$$

It is then possible to add empirical data to these generic plots. For each specific element in the isoelectronic sequence, the measured energy levels $^3P_0, ^3P_1, ^3P_2, ^1P_1$ can be used to calculate one value of x and four values of y. These can then be related to the measured energy levels for either of the two cases by the following relationships. The ordinate for a given element in the sequence is

$$x = \frac{^3P_2 - ^3P_0}{^1P_1 + ^3P_1 - 2\,^3P_{\text{lowest}}}, \tag{5.36}$$

and the abscissa for each of the energy levels is

$$y(^{2s+1}P_J) = \frac{2\,^{2s+1}P_J - ^1P_1 - ^3P_1}{^1P_1 + ^3P_1 - 2\,^3P_{\text{lowest}}}, \tag{5.37}$$

where $^3P_{\text{lowest}} = ^3P_0$ for the sp case, and $^3P_{\text{lowest}} = ^3P_2$ for the sp^5 case.

Examples of these plots are given in Fig. 5.1, and can provide insights into the relationships that will be developed below. Values of x and y can be deduced from measured energy-level data ($^3P_0, ^3P_1, ^3P_2, ^1P_1$) using Eqs. 5.32 and 5.33 (with $^{2s+1}P_J$ sequentially set equal to each of these energy-level values). The measured values of y and x can then be added to the plot to graphically display the extent of the intermediate coupling.

In the limit $\zeta_p \ll G_1$ (approaching the pure LS-coupling limit on the diagram), $\Delta_1 \approx G_1 + \zeta_p/4$, and the levels are given by

$$^3P_0^o = E_0 - G_1 - \zeta_p$$
$$^3P_1^o \approx E_0 - G_1 - \zeta_p/2$$
$$^3P_2^o = E_0 - G_1 + \zeta_p/2$$

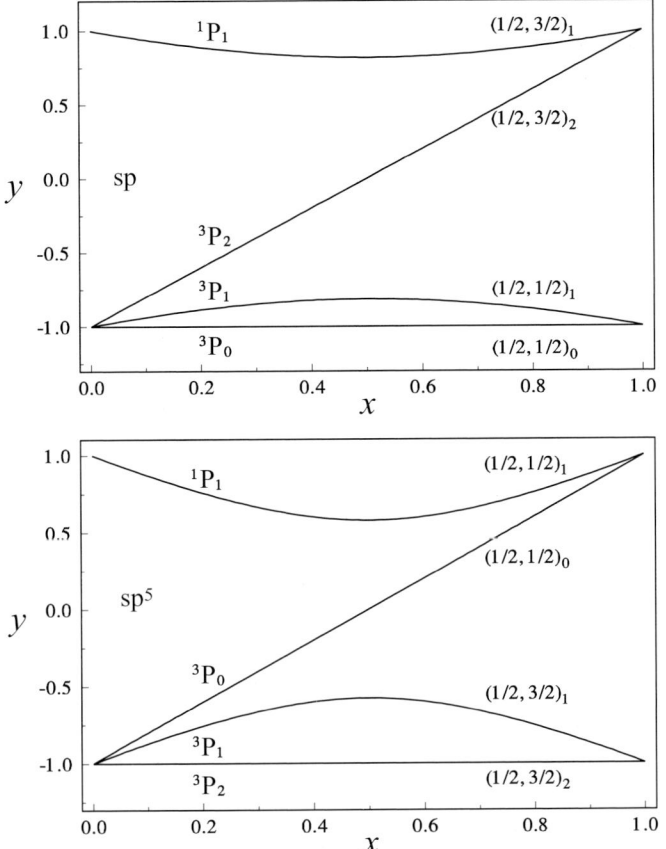

Fig. 5.1. Transition from LS to jj coupling for the sp and sp^5 configurations. (After Ref. [87].)

$$^1\mathrm{P}_1^\mathrm{o} \approx E_0 + G_1 \tag{5.38}$$

corresponding to well-separated singlet and triplet terms with triplet fine structure splittings obeying (as will be shown below) the Landé interval rule (Eq. 4.36).

In the limit $\zeta_\mathrm{p} \gg G_1$ (approaching pure jj coupling), refactorization and binomial expansion of the square root yields $\Delta_1 \approx 3\zeta_\mathrm{p}/4 + G_1/3$, and the levels separate into two widely spaced pairs given by

$$
\begin{aligned}
^3\mathrm{P}_0^\mathrm{o} &= E_0 - \zeta_\mathrm{p} - G_1 = (1/2, 1/2)_0 \\
^3\mathrm{P}_1^\mathrm{o} &\approx E_0 - \zeta_\mathrm{p} - G_1/3 = (1/2, 1/2)_1 \\
^3\mathrm{P}_2^\mathrm{o} &= E_0 + \zeta_\mathrm{p}/2 - G_1 = (1/2, 3/2)_2 \\
^1\mathrm{P}_1^\mathrm{o} &\approx E_0 + \zeta_\mathrm{p}/2 + G_1/3 = (1/2, 3/2)_1.
\end{aligned}
\tag{5.39}
$$

Here we have introduced the jj-coupling notation $(j_1, j_2)_J$. In this limit, the (j_1, j_2) manifolds are ordered as the single-particle energies. Real physical cases always lie between the LS and jj limits, so both notations are only nominal.

In the case of intermediate coupling, these relationships permit the Landé interval rule to be examined by forming the ratio

$$\frac{{}^3P_2^o - {}^3P_1^o}{{}^3P_1^o - {}^3P_0^o} = \frac{-G_1 + 3\zeta_p/4 + \Delta_1}{G_1 + 3\zeta_p/4 - \Delta_1}. \tag{5.40}$$

Near the LS-coupling limit where $\Delta_1 \approx G_1 + \zeta_p/4$ this gives

$$\frac{{}^3P_2^o - {}^3P_1^o}{{}^3P_1^o - {}^3P_0^o} \approx \frac{-G_1 + 3\zeta_p/4 + G_1 + \zeta_p/4}{G_1 + 3\zeta_p/4 - G_1 - \zeta_p/4} \approx 2 \tag{5.41}$$

just as expected. Near the jj-coupling limit where $\Delta_1 \approx 3\zeta_p/4 + G_1/3$, this becomes

$$\frac{{}^3P_2^o - {}^3P_1^o}{{}^3P_1^o - {}^3P_0^o} \approx \frac{-G_1 + 3\zeta_p/4 + 3\zeta_p/4 + G_1/3}{G_1 + 3\zeta_p/4 - 3\zeta_p/4 - G_1/3} \approx \frac{9\zeta_p}{4G_1}. \tag{5.42}$$

This is a very large value, indicative of the fact that the $J = 2$ level now lies near the nominal singlet level (see Fig. 5.1), very far above the $J = 0$ and 1 levels of the nominal triplet.

These equations for the $nsnp$ configuration can also be applied to the $nsn'p^5$ configuration by noting that the vacancy state causes ζ_p to be negative. This leads to an inverted fine structure in the LS representation and a reordering of the jj designations.

Hydrogenlike calculations for $G_1(nsnp)$

A specific calculation for $G_1(2s2p)$ using hydrogenlike wave functions is instructive. Setting $k = 1$ in Eq. 5.22 and rewriting the $r_<$ and $r_>$ formulation as twice the integral with $r_1 < r_2$, the desired integral is

$$G_1(2s2p) = \frac{2Ry}{3} \int_0^\infty dr_1 r_1^2 R_{2s}(r_1) R_{2p}(r_1) \int_0^{r_1} dr_2 r_2^2 \frac{r_2}{r_1^3} R_{2s}(r_2) R_{2p}(r_2). \tag{5.43}$$

Denoting $x_i = Zr_i/a_0$, the hydrogenlike wave functions are given [30] by

$$R_{2s}(r) = (Z/2a_0)^{3/2} 2(1 - x/2) e^{-x/2}$$
$$R_{2p}(r) = (Z/2a_0)^{3/2} x e^{-x/2}/\sqrt{3}. \tag{5.44}$$

The integral can be performed in two steps, writing it as

$$G_1(2s2p) = \frac{Ry\, Z}{2^3 3^2} \int_0^\infty dx_1 \left(x_1 - \frac{1}{2}x_1^2 \right) I_2(x_1), \tag{5.45}$$

where

$$I_2(x_1) \equiv \int_0^{x_1} dx_2 \left(x_2^4 - \frac{1}{2}x_2^5 \right) e^{-(x_1 + x_2)}. \tag{5.46}$$

The inner integral consists of incomplete gamma functions, which are given, for $n > -1$ and $0 \le x \le \infty$, by

$$\Gamma_{x_1}(n + 1) \equiv \int_0^{x_1} dx_2 x_2^n e^{-x_2} = n! \left[1 - e^{-x_1} \sum_{k=0}^{n} \frac{x_1^k}{k!} \right], \tag{5.47}$$

Table 5.3. Slater exchange energies G_1^H computed from exact hydrogenic wave functions, expressed as powers of primes. Decimal equivalents are compared with empirical fits to data by Edlén. (After Ref. [45].)

Config.	Seq.	$G_1^H/Ry\,Z$	Decimal equiv.	$G_1(\text{Edlén})/Ry\,Z$
2s2p	Be	$2^{-8}\,3\;5$	0.058 594	0.0586
3s3p	Mg	$2^{-8}\,3^{-2}\,5\;13$	0.028 212	0.0271
4s4p	Zn	$2^{-18}\,3^2\,5^2\;19$	0.016 308	0.0176
5s5p	Cd	$2^{-14}\,3^2\,5^{-2}\;13\;37$	0.010 569	0.0101
6s6p	Hg	$2^{-12}\,3^{-2}\,5^2\,7\;797$	0.007 390	0.0085

yielding

$$I_2(x_1) = 36\left[\left(1 + x_1 + \frac{x_1^2}{2!} + \frac{x_1^3}{3!} + \frac{x_1^4}{4!}\right)e^{-x_1} - 1\right] + \frac{1}{2}x_1^5 e^{-x_1}. \tag{5.48}$$

Inserting this into the outer integral yields

$$G_1(2s2p) = \frac{15}{256}\,Ry\,Z. \tag{5.49}$$

Hydrogenic calculations such as this have been performed [45] for higher homologues of the $nsnp$ configurations using a symbolic algebra package. To facilitate comparisons of hydrogenlike values with the corresponding quantities in complex atoms, the hydrogenlike values will henceforth be denoted as $G_1^H(nsnp)$. The values are presented both as powers of primes and as decimal equivalents in Table 5.3. The fact that the Slater integrals all scale linearly with Z will be used later in making an isoelectronic screening parametrization of these quantities.

It is interesting to note that these linearities were applied to measured data as early as 1963. Edlén [87] computed $G_1^H(2s2p)$ theoretically for 2s2p (as done above), and also made a semiempirical screening parametrization of values of $G_1(2s2p)$ deduced from measured data for the Be sequence (by methods discussed in Section 5.3). When he extrapolated the linear trend of the screening parametrization to infinite Z, Edlén noted that it yielded the hydrogenic value ($15Ry/256$) to within experimental uncertainties. He then extended his analysis to obtain similar screening parametrizations of the available database for 3s3p, 4s4p, 5s5p, and 6s6p in the Mg, Zn, Cd, and Hg sequences. The empirical values obtained by Edlén can now be compared with theoretical calculations for the corresponding hydrogenic values. This is done in Table 5.3, and indicates close agreement.

5.2.6 Example: the np^2 and np^4 configuration

Another interesting example involves configurations of the form np^2 where there are two equivalent electrons. The np^4 configuration is very similar, consisting of two vacancies which, with small modifications, can be deduced from the two-electron case.

Table 5.4. Matrix elements for a p^2 configuration.

	$^3P_0$	$^3P_1$	$^3P_2$	$^1D_2$	$^1S_0$
$^3P_0$	$E_0 - 5F_2 - \zeta_{pp}$	0	0	0	$-\sqrt{2}\zeta_{pp}$
$^3P_1$	0	$E_0 - 5F_2 - \zeta_{pp}/2$	0	0	0
$^3P_2$	0	0	$E_0 - 5F_2 + \zeta_{pp}/2$	$\zeta_{pp}/\sqrt{2}$	0
$^1D_2$	0	0	$\zeta_{pp}/\sqrt{2}$	$E_0 + F_2$	0
$^1S_0$	$-\sqrt{2}\zeta_{pp}$	0	0	0	$E_0 + 10F_2$

The presence of spatially equivalent electrons introduces restrictions on the number of terms possible. When two or more electrons have the same values for n and ℓ, this limits the values for m_ℓ and m_s that will lead to an antisymmetric space wave function for a triplet and a symmetric space wave function for a singlet. As discussed earlier, for two equivalent electrons $n\ell^2$ the quantity $L + S$ must be even, hence an ns^2 configuration contains a $^1S_0$ term but no $^3S_1$. For an np^2 configuration, $\mathbf{S} = 1/2 + 1/2 = 0, 1$, and $\mathbf{L} = 1 + 1 = 0, 1, 2$. Taking this into account, the singlet polyad contains $L = 0, 2$ terms, but not $L = 1$, and the triplet polyad contains $L = 1$ terms, but not $L = 0, 2$.

Thus an np^2 configuration contains the five levels $^3P_0$, $^3P_1$, $^3P_2$, $^1D_2$, and $^1S_0$. Adding up the degeneracies of two $J = 0$, one $J = 1$, and two $J = 2$ levels, there are 15 states. Since the electrons are equivalent, there is no spatial exchange and G^k is not defined and only the direct integrals remain. Of these, only the F^0 and F^2 direct integrals are nonvanishing. Since the electrons are identical, they both yield the same spin–orbit integral. Thus, the five levels can be characterized in terms of the three quantities E_0, $F_2 \equiv F^2/25$, and ζ_{pp}. Here the subscript pp denotes the values $l = 1$ for the two p-electrons. The matrix of these quantities is shown in Table 5.4.

Diagonalizing the 2×2 submatrices in this Hamiltonian yields the relationships

$$^3P_0 = E_0 + 5F_2/2 - \zeta_{pp}/2 - \Delta_0$$
$$^3P_1 = E_0 - 5F_2 - \zeta_{pp}/2$$
$$^3P_2 = E_0 - 2F_2 + \zeta_{pp}/4 - \Delta_2$$
$$^1D_2 = E_0 - 2F_2 + \zeta_{pp}/4 + \Delta_2$$
$$^1S_0 = E_0 + 5F_2/2 - \zeta_{pp}/2 + \Delta_0, \tag{5.50}$$

where

$$\Delta_0 = \sqrt{(15F_2/2 + \zeta_{pp}/2)^2 + 2\zeta_{pp}^2}$$
$$\Delta_2 = \sqrt{(3F_2 - \zeta_{pp}/4)^2 + \zeta_{pp}^2/2}. \tag{5.51}$$

Edlén [87] also devised a plot to display the degree of intermediate coupling in these configurations. Here the dimensionless coupling parameter is defined as

$$x \equiv \frac{\zeta_{pp}}{5F_2 + \zeta_{pp}} \qquad (0 \le x \le 1), \tag{5.52}$$

and the dimensionless measure of the relative separation is

$$y \equiv \frac{2E - 2E_0 - 5F_2 + \zeta_{pp}}{15F_2 + 3\zeta_{pp}}. \tag{5.53}$$

In terms of these quantities the energy levels of a p^2 configuration can be written

$$y\left({}^3P_0\right) = -\sqrt{1 - 4x(1-x)/3}.$$
$$y\left({}^3P_1\right) = -1 + x$$
$$y\left({}^3P_2\right) = -\frac{3}{5} + \frac{11}{10}x - \frac{2}{5}\sqrt{1 - \frac{17}{6}x + \frac{163}{48}x^2}$$
$$y\left({}^1D_2\right) = -\frac{3}{5} + \frac{11}{10}x + \frac{2}{5}\sqrt{1 - \frac{17}{6}x + \frac{163}{48}x^2}$$
$$y\left({}^1S_0\right) = \sqrt{1 - 4x(1-x)/3}, \tag{5.54}$$

whereas for a p^4 configuration this becomes

$$y\left({}^3P_2\right) = -\frac{3}{5} + \frac{1}{10}x - \frac{2}{5}\sqrt{1 - \frac{7}{6}x + \frac{83}{48}x^2}$$
$$y\left({}^3P_1\right) = -1 + x$$
$$y\left({}^3P_0\right) = -\sqrt{1 - 8x(1-x)/3}$$
$$y\left({}^1D_2\right) = -\frac{3}{5} + \frac{1}{10}x + \frac{2}{5}\sqrt{1 - \frac{7}{6}x + \frac{83}{48}x^2}$$
$$y\left({}^1S_0\right) = \sqrt{1 - 8x(1-x)/3} \tag{5.55}$$

Examples of these plots are given in Fig. 5.2, and can again provide insight into the relationships to be developed below.

It is again instructive to examine this in limiting cases.

In the limit $\zeta_{pp} \ll F_2$ (approaching pure LS coupling), $\Delta_0 \approx 15F_2/2 + \zeta_{pp}/2$ and $\Delta_2 \approx 3F_2 - \zeta_p/4$, so the levels are given by

$$^3P_0 \approx E_0 - 5F_2 - \zeta_{pp}$$
$$^3P_1 = E_0 - 5F_2 - \zeta_{pp}/2$$
$$^3P_2 \approx E_0 - 5F_2 + \zeta_{pp}/2$$
$$^1D_2 \approx E_0 + F_2$$
$$^1S_0 \approx E_0 + 10F_2, \tag{5.56}$$

and the singlet and triplet polyads are well separated, with the triplet obeying the Landé interval rule (Eq. 4.36).

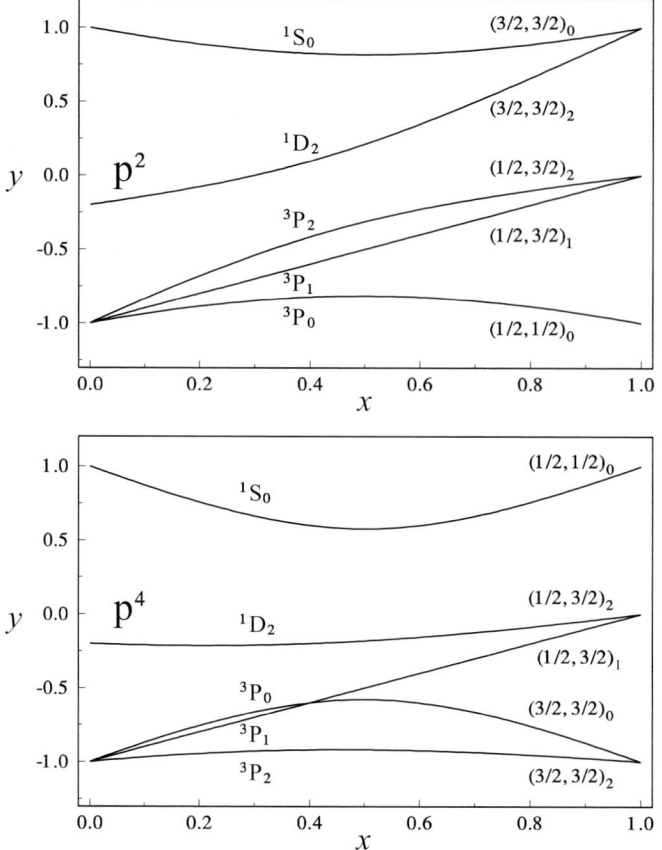

Fig. 5.2. Transition from LS to jj coupling for the p^2 and p^4 configurations. (After Ref. [87].)

In the limit $\zeta_{pp} \gg F_2$ (approaching pure jj coupling), refactorization and binomial expansion of the square root yields $\Delta_0 \approx 3\zeta_p/2 + 5F_1/2$ and $\Delta_2 \approx 3\zeta_{pp}/4 - F_2$. In this case the levels separate into three groups

$$
\begin{aligned}
{}^3P_0 &\approx E_0 - 3\zeta_{pp} = (1/2, 1/2)_0 \\
{}^3P_1 &= E_0 - \zeta_{pp}/2 - 5F_2 = (1/2, 3/2)_1 \\
{}^3P_2 &\approx E_0 - \zeta_{pp}/2 - F_2 = (1/2, 3/2)_2 \\
{}^1D_2 &\approx E_0 + \zeta_{pp} - 3F_2 = (3/2, 3/2)_2 \\
{}^1S_0 &\approx E_0 + \zeta_{pp} + 5F_2 = (3/2, 3/2)_0,
\end{aligned}
\tag{5.57}
$$

Here again, the more nearly appropriate jj notation is also included.

The equations developed above have been done for the np^2 configuration. In the case of the np^4, the hole states cause the value of ζ_{pp} to be negative, and lead to an inverted fine structure in the LS representation, and a reordering of the jj designations.

5.3 Screening parametrizations of Slater parameters

We have seen that the Balmer energy scales with nuclear charge as Z^2, whereas the Slater energies scale as Z. This is because both quantities involve integrals over the quantity $1/r$ which scales as Z, but the Balmer energy couples the nucleus of charge Z to an electron, whereas the Slater energy couples one electron to another. Since the centroid energy E_0 inseparably combines the Balmer energies and the monopole Slater energy F^0, it contains an admixture of linear and quadratic Z dependences and is not well adapted to a screening parameterization. However, the splittings of the levels about E_0 provide a set of relationships that involve the Slater and spin–orbit parameters that can be solved to isolate their characteristic charge-scaling factors.

5.3.1 $nsn'p$ configurations

The number of levels in a configuration generally exceeds the number of parameters that characterize it, so the system is overdetermined, and the goodness-of-fit can be tested. In the case of an $nsn'p$ configuration there are three independent separations among the four levels ${}^3P^o_0$, ${}^3P^o_1$, ${}^3P^o_2$, ${}^1P^o_1$ that are characterized by two parameters G_1, ζ_p. One scheme by which these quantities can be determined from the measured energy levels is given by

$$G_1(nsnp) = \left({}^1P^o_1 + {}^3P^o_1\right)/2 - \left({}^3P^o_2 + {}^3P^o_0\right)/2 \tag{5.58}$$

$$\zeta_p = 2\left({}^3P^o_2 - {}^3P^o_0\right)/3. \tag{5.59}$$

(This was the method that was used by Edlén to obtain the effective values for G_1 in Table 5.3. Notice that the quantities in these formulae involve only the off-diagonal elements of ζ_p since the quantity Δ_1 was eliminated by summing the $J = 1$ levels. Thus a second independent determination of ζ_p can be made from the quantity Δ_1 using

$$\zeta'_p = \sqrt{\left({}^1P^o_1 - {}^3P^o_1\right)^2/4 - (G_1 + \zeta_p/4)^2}\sqrt{2}. \tag{5.60}$$

The agreement between ζ'_p and ζ_p can be used as a measure of goodness of fit. This is the approach that was used by Edlén in obtaining the empirical estimates of $G_1^H(nsnp)$ in Table 5.3.

If the goodness of fit is not sufficiently close, both ζ'_p and ζ_p can be used so as to extend the number of fitting parameters to match the number of independent splittings, thus assuring that the characterization exactly matches the measured data.

This latter procedure has some advantages. It has been suggested [194] by Wolfe that this procedure can include the effects of spin–other-orbit interaction in the characterization. Since the spin–other-orbit interaction is known to add to the spin–orbit interaction for the triplet–triplet *diagonal*, and to subtract from it for the singlet–triplet *off-diagonal*, treating these two quantities as free and independent parameters can include this effect. The same effective parameterization based on a different phenomenological origin was independently suggested [132] by King and VanVleck, who argued that the physical radial wave functions (in contrast to the Schrödinger radial wave functions) may be slightly different for singlet and triplet states within the same configuration, and hence yield different on-diagonal and off-diagonal spin–orbit integrals. Since both physical mechanisms lead to the same parametrization, both can be effectively accounted for by this formalism.

Once the quantities G_1, ζ_p (and if desired, ζ_p') have been extracted from measured data along the isoelectronic sequence, these data can be mapped into screening parameters by use of the relationships

$$G_1(nsnp) = G_1^{\mathrm{H}}(nsnp)R_Z(Z - S_G) \tag{5.61}$$

$$\zeta_p = \frac{R_Z\alpha^2(Z - S_\zeta)^4}{3n^3}[1 + \mathcal{O}(\alpha^2 Z^2)] \tag{5.62}$$

where G_1^{H} is either a computed hydrogenic value or a fitted value.

The empirical values of the screening parameters S_G and S_ζ have been found to exhibit apparent linearities when plotted vs the reciprocal screened charge. An example for the 4s4p configuration in the Zn isoelectronic sequence is shown in Fig. 5.3. These parameters can thus be fitted to the expressions

$$S_G = a_G + b_G/(Z - S_G) \tag{5.63}$$

$$S_\zeta = a_\zeta + b_\zeta/(Z - S_\zeta). \tag{5.64}$$

Fits to these functions can then be used to interpolate, extrapolate, and critically evaluate data.

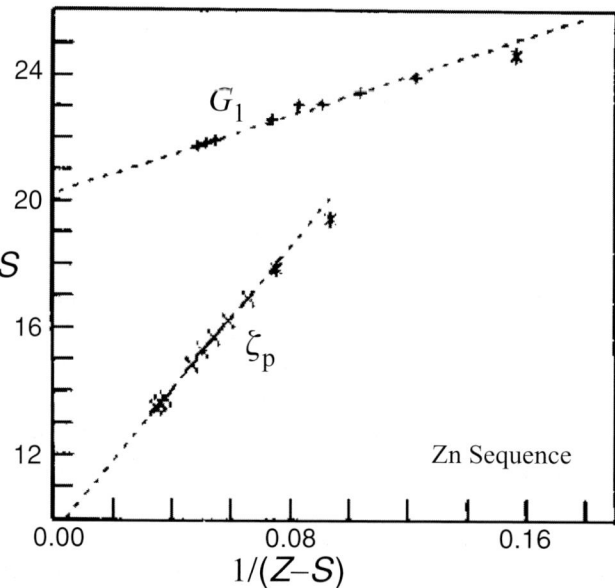

Fig. 5.3. Screening parametrization of the 4s4p configuration in the Zn isoelectronic sequence. (From Ref. [44].)

5.3.2 np^2 configurations

For np^2 configurations there is a twofold overdetermination, since there are four independent level splittings about the centroid energy of the five levels $^3P_0$, $^3P_1$, $^3P_2$, $^1D_2$, $^1S_0$, and still only two parameters, F_2 and ζ_{pp}. As in the $nsnp$ case, we can determine these parameters from the data without involving the off-diagonal (multiplicity mixing) values of ζ_{pp} by considering the singlet–triplet averages for $J = 0$ and 2, from which the Δ_0 and Δ_2 cancel. In this case the measured energy levels yield

$$\zeta_{pp} = 2\left[5\left(^1D_2 + {}^3P_2\right) - 2\left(^1S_0 + {}^3P_0\right) - 6{}^3P_1\right]/15 \tag{5.65}$$

$$F_2 = \left[\left(^1S_0 + {}^3P_0\right) - 2{}^3P_1\right]/15. \tag{5.66}$$

The off-diagonal values ζ'_{pp} and ζ''_{pp} can then be computed

$$\zeta'_{pp} = \sqrt{\left(^1S_0 - {}^3P_0\right)^2/4 - (15F_2/2 + \zeta_{pp}/2)^2}\Big/\sqrt{2} \tag{5.67}$$

$$\zeta''_{pp} = \sqrt{\left(^1D_2 - {}^3P_2\right)^2/4 - (3F_2 - \zeta_{pp}/4)^2}\sqrt{2}. \tag{5.68}$$

If ζ_{pp}, ζ'_{pp}, and ζ''_{pp} all agree to within experimental uncertainties, then the goodness-of-fit criteria are satisfied and the level splittings can be confidently characterized by ζ_{pp} and F_2. If there are discrepancies between two or among all three of the extracted spin–orbit parameters, then the number of fitting parameters can be extended to accurately reproduce the data. Here again, the measured data can be mapped into screening parameters using

$$F_2(np^2) = F_2^H(np^2)R_Z(Z - S_F) \tag{5.69}$$

$$\zeta_{pp} = \frac{R_Z\alpha^2(Z - S_\zeta)^4}{3n^3}[1 + \mathcal{O}(\alpha^2 Z^2)] \tag{5.70}$$

where F_2^H is either a computed hydrogenic value or a fitted parameter.

5.4 Systems with three or more valence electrons

In systems with many open-shell electrons the relationships become more complicated, but they can nonetheless be used to characterize spectroscopic data [68]. A relatively simple example is given by the np^3 configuration. Here the Pauli exclusion principle rules out a number of terms, and the configuration contains only the levels $^2D^o_{5/2}$, $^4S^o_{3/2}$, $^2D^o_{3/2}$, $^2P^o_{3/2}$, $^2P^o_{1/2}$. Their energy separations from the centroid E_0 are characterized by F_2 and ζ. The fact that the p^3 subshell is half-filled causes the spin–orbit integral to vanish for diagonal matrix elements. Thus, near the conditions of pure LS coupling the spin–orbit interaction produces no first-order fine structure splitting. The matrix elements of this configuration are shown in Table 5.5.

In the cases of the $nsn'p$ and np^2 configurations, no more than two levels possessed the same value of J, but in np^3 there are three levels with $J = 3/2$. Thus, whereas the former yielded expressions for the measured eigenvalues through the diagonalization of 2×2 matrices (which could each be characterized by a single mixing angle), this case will

Table 5.5. Matrix elements for a p^3 configuration.

	$^2D^o_{5/2}$	$^4S^o_{3/2}$	$^2D^o_{3/2}$	$^2P^o_{3/2}$	$^2P^o_{1/2}$
$^2D^o_{5/2}$	E_0	0	0	0	0
$^4S^o_{3/2}$	0	$E_0 - 9F_2$	0	ζ	0
$^2D^o_{3/2}$	0	0	E_0	$\sqrt{5}\zeta/2$	0
$^2P^o_{3/2}$	0	ζ	$\sqrt{5}\zeta/2$	$E_0 + 6F_2$	0
$^2P^o_{1/2}$	0	0	0	0	$E_0 + 6F_2$

require the diagonalization of a 3×3 matrix (which requires an array of amplitudes for its characterization). If theoretical values for the Slater and spin–orbit integrals are the starting point, then the matrix diagonalization must be performed to obtain the predicted values for the energy eigenvalues. In the semiempirical formalism, the measured eigenvalues are the starting point, and the Slater and spin–orbit parameters can be deduced quite directly by the formation of simple sums and products.

A general $N \times N$ symmetric matrix M_{ij} can be diagonalized by a transformation T_{ij}, where

$$\sum_{n=1}^{N}\sum_{m=1}^{N} T_{in} M_{mn} T_{nj}^{-1} = E_i \delta_{ij}. \tag{5.71}$$

The eigenvalues E_i are given by the N roots of λ in the equation

$$\det(M_{ij} - \lambda \delta_{ij}) = 0, \tag{5.72}$$

which yields the secular equation

$$\sum_{n=0}^{N} a_n \lambda^n = 0, \tag{5.73}$$

where $a_N = 1$. Although the explicit solution for E_i in terms of the a_n for $N \geq 3$ can be very complicated and cumbersome, the inverse solution is quite simple. Since a polynomial of degree N which has real coefficients has N roots, Eq. 5.73 has the equivalent form

$$(\lambda - E_1)(\lambda - E_2)\cdots(\lambda - E_N) = 0, \tag{5.74}$$

so the a_n coefficients can be written directly as sums and products of the experimentally determined real eigenvalues E_i given by

$$a_{N-1} = -\sum_{i=1}^{N} E_i,$$

$$a_{N-2} = \sum_{i=1}^{N}\sum_{j>i}^{N} E_i E_j,$$

$$a_{N-3} = -\sum_{i=1}^{N}\sum_{j>i}^{N}\sum_{k>j}^{N} E_i E_j E_k,$$

$$\vdots$$

$$a_0 = (-1)^N \prod_{i=1}^{N} E_i. \tag{5.75}$$

These equations permit the a_n coefficients to be determined from the experimental energies. Although there are N^2 matrix elements M_{ij} and only N coefficients a_n, the Hamiltonian matrix elements are interrelated by the number p, of Slater parameters, and it is possible to construct the M_{ij} from the a_n for a specific J value if $p \leq N$. Additional relationships among the Slater parameters can be obtained from other values of J within the configuration. If p is less than the total number of levels in the configuration, the Slater parameters are overdetermined, which provides tests of the single-configuration assumption.

For the matrix in Table 5.5 the quantity F_2 can be specified from the diagonal elements to obtain

$$F_2 = \left(^2P^o_{1/2} - {}^2D^o_{5/2} \right)/6. \tag{5.76}$$

If the energies are defined relative to that of the unmixed $^2P_{1/2}$ level ($E_0 + 6F_2$), the secular equation for the 3×3 matrix of $J = 3/2$ levels is

$$0 = \lambda^3 + 21F_2\lambda^2 + [90(F_2)^2 - 9\zeta^2/4]\lambda - 99F_2\zeta^2/4. \tag{5.77}$$

The coefficients of the powers of λ can then be obtained from sums and products of the measured eigenenergies

$$21F_2 = -{}^4S^o_{3/2} - {}^2D^o_{3/2} - {}^2P^o_{3/2}, \tag{5.78}$$

$$90(F_2)^2 - 9\zeta^2/4 = \left(^4S^o_{3/2}\right)\left(^2D^o_{3/2}\right) + \left(^2D^o_{3/2}\right)\left(^2P^o_{3/2}\right) + \left(^2P^o_{3/2}\right)\left(^2S^o_{3/2}\right), \tag{5.79}$$

and

$$-99F_2\zeta^2/4 = \left(^4S^o_{3/2}\right)\left(^2D^o_{3/2}\right)\left(^2P^o_{3/2}\right). \tag{5.80}$$

These, together with Eq. 5.76, comprise four equations. Since there are only two Slater parameters, the system is overdetermined. Thus the system can be parametrized either by making a least-squares adjustment of the quantities F_2 and ζ, or by defining two additional free parameters. However, since the half-filled p-shell causes the spin–orbit integral to vanish on the diagonal, only one additional parameter is gained by freeing the two off-diagonal spin–orbit parameters from each other. If we denote these as ζ_{SP} for the S–P overlap and ζ_{DP} for the D–P overlap, then Eq. 5.77 becomes

$$0 = \lambda^3 + 21F_2\lambda^2 + \left[90(F_2)^2 - \left(5\zeta_{DP}^2 + 4\zeta_{SP}^2\right)/4\right]\lambda - 3F_2\left(25\zeta_{DP}^2 + 8\zeta_{SP}^2\right)/4. \tag{5.81}$$

The eigenvector amplitudes T_{ij} for the transformation from the LS to the eigenvectors $|\psi_i\rangle$ can be obtained from

$$\left(\langle\psi_i|{}^4\mathrm{S}^{\mathrm o}_{3/2}\rangle, \langle\psi_i|{}^2\mathrm{D}^{\mathrm o}_{3/2}\rangle, \langle\psi_i|{}^2\mathrm{P}^{\mathrm o}_{3/2}\rangle\right) = \frac{(b_{\mathrm S}, b_{\mathrm D}, 1)}{\sqrt{1 + b_{\mathrm S}^2 + b_{\mathrm D}^2}} \tag{5.82}$$

where

$$b_{\mathrm S}(E_i) \equiv \frac{\zeta}{15F_2 + E_i}; \qquad b_{\mathrm D}(E_i) \equiv \frac{\sqrt 5 \zeta/2}{6F_2 + E_i}. \tag{5.83}$$

Here i denotes the ith energy eigenvalue. (The normalization condition automatically provides the quantity $b_{\mathrm P} = 1/\sqrt{1 + b_{\mathrm S}^2 + b_{\mathrm D}^2}$.) From this transformation, the eigenvectors can be written

$$\left|{}^2\mathrm{D}^{\mathrm o\,\prime}_{5/2}\right\rangle = \left|{}^2\mathrm{D}^{\mathrm o}_{5/2}\right\rangle$$

$$\left|{}^4\mathrm{S}^{\mathrm o\,\prime}_{3/2}\right\rangle = T_{\mathrm{SS}}\left|{}^4\mathrm{S}^{\mathrm o}_{3/2}\right\rangle + T_{\mathrm{SD}}\left|{}^2\mathrm{D}^{\mathrm o}_{3/2}\right\rangle + T_{\mathrm{SP}}\left|{}^2\mathrm{P}^{\mathrm o}_{3/2}\right\rangle$$

$$\left|{}^2\mathrm{D}^{\mathrm o\,\prime}_{3/2}\right\rangle = T_{\mathrm{DS}}\left|{}^4\mathrm{S}^{\mathrm o}_{3/2}\right\rangle + T_{\mathrm{DD}}\left|{}^2\mathrm{D}^{\mathrm o}_{3/2}\right\rangle + T_{\mathrm{DP}}\left|{}^2\mathrm{P}^{\mathrm o}_{3/2}\right\rangle$$

$$\left|{}^2\mathrm{P}^{\mathrm o\,\prime}_{3/2}\right\rangle = T_{\mathrm{PS}}\left|{}^4\mathrm{S}^{\mathrm o}_{3/2}\right\rangle + T_{\mathrm{PD}}\left|{}^2\mathrm{D}^{\mathrm o}_{3/2}\right\rangle + T_{\mathrm{PP}}\left|{}^2\mathrm{P}^{\mathrm o}_{3/2}\right\rangle$$

$$\left|{}^2\mathrm{P}^{\mathrm o\,\prime}_{1/2}\right\rangle = \left|{}^2\mathrm{P}^{\mathrm o}_{1/2}\right\rangle. \tag{5.84}$$

These mixing amplitudes can then be used to characterize the system. This is an extension from one singlet–triplet mixing angle that described the intermediate coupling between two levels to a full set of Euler angles that describe the intermediate coupling among three levels.

These methods will be applied to sample calculations in Chapter 8.

5.5 Antisymmetrization of a multielectron system

For a system of N noninteracting electrons with single-particle wave functions $u_{n_i}(\mathbf r_i)$, a normalized wave function that is antisymmetric under the interchange of any two electron identities can be constructed by use of the Slater determinant [177]

$$\psi_A = \frac{1}{\sqrt{N!}}
\begin{vmatrix}
u_{n_1}(\mathbf r_1) & u_{n_1}(\mathbf r_2) & \cdots & u_{n_1}(\mathbf r_N) \\
u_{n_2}(\mathbf r_1) & u_{n_2}(\mathbf r_2) & \cdots & u_{n_2}(\mathbf r_N) \\
\vdots & \vdots & \cdots & \vdots \\
u_{n_N}(\mathbf r_1) & u_{nN}(\mathbf r_2) & \cdots & u_{n_N}(\mathbf r_N)
\end{vmatrix}. \tag{5.85}$$

This formulation indicates the magnitude of the task of obtaining a comprehensive description of a complex atom, and the advantages to the separation of spin and space so as to make the radial formulation in terms of equivalent electrons.

The wave function ψ_A has $N!$ terms, each a product of N single-particle wave functions. Thus, to obtain the antisymmetrized wavefunction of all of the electrons in the element neodymium ($Z = 60$) it would be necessary to write out $60!$ sets of product wavefunctions, a number which exceeds the estimated number of protons in the known universe (10^{80}).

6

Electric dipole radiation

Lux aeterna luceat eis.
Let there be light. Take the rest of the week off.

Atomic energy levels deduced from optical spectra comprise one of the most precisely known sets of physical measurements that exist. However, the precision of the determinations of the relative oscillator strengths of these spectral lines from the relative intensities of spectral lines is much less precise. Fortunately, time-dependent methods for the study of the dynamics of the emission process exist (and are being developed) that permit the transition probability rates, oscillator strengths, and reaction rates to be determined with ever-increasing precision.

In most elementary quantum mechanics textbooks, the section on the emission of radiation is the least satisfactory section of the book. Whereas the development of relationships among various spatial overlap integrals between state vectors for various operators (such as the electrostatic dipole moment) can be formulated in a very elegant *ab initio* manner, the connection of these matrix elements to the time dependence of the system often seems driven by *a posteriori* assumptions that are extended beyond their justifiable range of applicability. As Fermi observed [96] in stating his *Golden Rule*, "the transition probability and energy perturbation can be calculated with the help of perturbation theory (*i.e.*, there is no better way known)." The Weisskopf–Wigner approximation [191] offers a scheme for making precise calculations, but does not provide the rigor that has characterized so many other areas of quantum mechanics. Merzbacher has stated it well in his observation [155] that "the fact remains that the exponential decay law, for which we have so much empirical support in radioactive processes, is not a rigorous consequence of quantum mechanics, but the result of somewhat delicate approximations." An analysis of the existing evidence [72] indicates that, while measurements confirming the validity of the exponential decay law extend over a wide range of times (10^{-13} to 10^{17} seconds, 10^{-4} to 45 half-lives), the accuracies of these tests leave much to be desired. The degree to which the exponential hypothesis is valid, on a level of parts per thousand, for decay times commensurate with the meanlife, has not been established either theoretically or experimentally.

For these reasons, no attempt will be made here to develop expressions for the transition probability rates of the atom from fundamental theory. The form of these quantities has been made plausible by the semiclassical approach of Wilhelm Wien described in Section 2.5, and will be adopted without further discussion.

6.1 Hierarchies in transition arrays

For the spontaneous emission of an electric dipole (E1) photon, the transition probability rate from an upper state u to a lower state l is

$$A_{ul} = \frac{4(2\pi)^3 K e^2}{3\hbar \lambda_{ul}^3} |\langle u|\mathbf{r}|l\rangle|^2 = \left[\frac{1265.38}{\lambda_{ul}(\text{\AA})}\right]^3 \left|\left\langle u \left|\frac{\mathbf{r}}{a_0}\right| l\right\rangle\right|^2. \tag{6.1}$$

For reasons dating back to Einstein's 1917 paper [92] in which the A and B parameters were first defined, the quantity A_{ij} has been called the "spontaneous transition probability." This is clearly a misnomer, since it is a rate and not a probability (it does not range from zero to one, and has dimensions s^{-1}). This use of this term has persisted both because of tradition and the fact that term "transition rates" is often used to indicate nonradiative transitions such as Auger and autoionization processes. Thus, many authors are choosing to use the words "transition probability rates" to designate the quantity A_{ij}. That choice of wording will be adopted here.

There is a simple relationship between emission and absorption that can be understood in terms of the radiation emitted by a classical simple harmonic oscillator $x(t) = x_0 \cos(\omega t)$. If the energy loss per cycle is small compared to the total energy of such a system, the radiation rate can be obtained from the Larmor formula for the energy loss from an accelerated charge (Eq. 2.149)

$$-\left\langle\frac{dE}{dt}\right\rangle = \frac{2Ke^2}{3c^3}\langle(\omega^2 x_0 \cos(\omega t))^2\rangle = \frac{Ke^2\omega^4 x_0^2}{3c^3}. \tag{6.2}$$

Combining this with the expression for the total energy

$$E = \frac{1}{2}m\omega^2 x_0^2, \tag{6.3}$$

the simple harmonic oscillator lifetime can be written

$$\frac{1}{\tau_{\text{SHO}}} \equiv -\frac{1}{E}\frac{dE}{dt} = \frac{2Ke^2\omega^2}{3mc^3} = \frac{2(2\pi)^2 Ke^2}{3mc\lambda^2}. \tag{6.4}$$

The emission transition probability rate is formulated in terms of a dimensionless "oscillator strength" f_{lu} (characterizing three mutually perpendicular classical simple harmonic oscillators)

$$A_{ul} \equiv \frac{1}{\tau_{\text{SHO}}}(3f_{lu}). \tag{6.5}$$

The time-reversed connections between emission and absorption between two energy levels that were described [92] by Einstein have been quantitatively formulated in terms of a transition moment that connects the two levels.

The absorption oscillator strength f_{lu} for a transition from a lower state l to an upper state u is thus defined from the equation

$$f_{lu} = \frac{4\pi mc}{3\hbar\lambda} |\langle u|\mathbf{r}|l\rangle|^2 = \left[\frac{303.75}{\lambda(\text{Å})}\right] \left|\left\langle u \left|\frac{\mathbf{r}}{a_0}\right| l\right\rangle\right|^2. \tag{6.6}$$

These quantities are defined as state-to-state component transitions, and their definitions must be broadened to describe level-to-level lines, and to examine the implications for term-to-term multiplets, polyad-to-polyad supermultiplets, and configuration-to-configuration transition arrays.

6.1.1 Components

For the state-to-state components, the transition probability rate can be written

$$A(\gamma'L'J'M_J', \gamma LJM_J) = \left[\frac{1265.38}{\lambda(\text{Å})}\right]^3 |\langle \gamma LJM_J|\mathbf{r}/a_0|\gamma'L'J'M_J'\rangle|^2. \tag{6.7}$$

In LS coupling this has the standard selection rules $\Delta S = 0$ (the electrostatic interaction does not affect the spin), $\Delta L = \pm 1$ (the parity of the atom changes), $\Delta J = 0, \pm 1$ with no $J = 0$ to $J = 0$ (the triangle relationship of vectors $\mathbf{J}$ and $\mathbf{1}$ to give $\mathbf{J}'$), and $\Delta M_J = 0, \pm 1$ (corresponding to the polarization of the emitted radiation).

Since spin-changing transitions are forbidden in electric dipole transitions, the quantum number S is suppressed here. The M_J and M_J' quantum numbers are projections with respect to an arbitrarily chosen quantization axis (specified, for example, by an anisotropic excitation mechanism). Thus the radiation into all 4π steradians must be independent of the M_J values for a field-free atom. However, the angular distribution and polarization of the radiation differs for the $\Delta M_J = 0$ (π radiation) and $\Delta M_J = \pm 1$ ($\sigma^{\pm}$ radiation). This can easily be understood from a simple model.

For a classically radiating oscillating dipole aligned with the z-axis at the origin of a spherical coordinate system, the plane of polarization (the direction of the electric field) lies along the direction of the ϑ unit vector, and has an angular distribution proportional to $\sin^2\vartheta$ (as described in the proverb "under the candle it is darkest"). As shown in Fig. 6.1, a general three-dimensional aligned system can be characterized by three mutually perpendicular dipoles of intensities $I_\sigma/2$, $I_\sigma/2$ and I_π. Here z is the quantization axis of the excitation process. The dipoles along the x- and y-axes correspond to the $\sigma^{\pm}$ radiation and the dipole along the z-axis corresponds to the π radiation.

The origin of this notation lies in the polarization of the Zeeman components that are observed in the presence of a magnetic field, as could be impressed here along the z-axis. Relative to a plane formed by the z-axis and the propagation vector of the photon (the y–z plane in Fig. 6.1), the polarization of the $\Delta m = 0$ radiation will be parallel (π) to the plane, and that of the $\Delta m = \pm 1$ will be perpendicular ($\sigma^{\pm}$, for senkrecht) to the plane. It is possible to discriminate among these components using a linear polarizer.

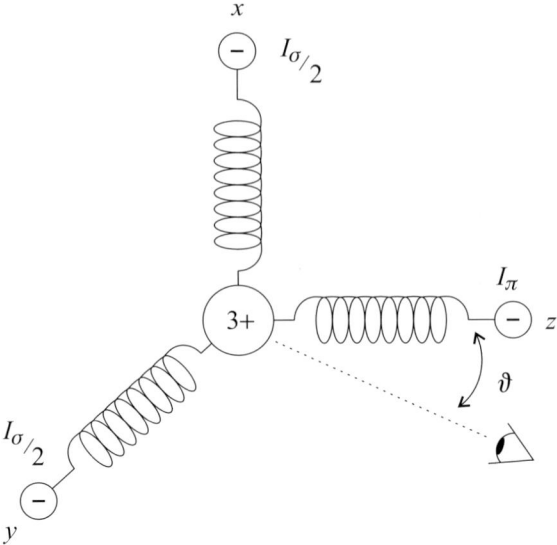

Fig. 6.1. Model for an aligned source of radiation.

Let the radiation be viewed at an angle ϑ to the z-axis. If the plane of the polarizer is aligned with the y–z plane, then no radiation from the dipole along the x-axis will enter because its radiation is polarized perpendicular to that plane. Since the angle with the y-axis is $(90° - \vartheta)$, the intensity entering the solid angle of the detector will be

$$\frac{dI_\parallel}{d\Omega} = \frac{1}{4\pi}[I_\pi \sin^2\vartheta + (I_\sigma/2)\sin^2(90° - \vartheta)]$$

$$= \frac{1}{4\pi}[I_\pi - (I_\pi - I_\sigma/2)\cos^2\vartheta]. \tag{6.8}$$

If the polarizer is rotated $90°$ so that it is aligned with the x-axis, it will similarly admit no light from the dipoles along either the y- or z-axis, and will view the dipole on the x-axis at a right angle, thus

$$\frac{dI_\perp}{d\Omega} = \frac{1}{4\pi}[I_\sigma/2]. \tag{6.9}$$

The total light intensity observed in this solid angle with no polarization selection is the sum of these two contributions

$$\frac{dI}{d\Omega} = \frac{dI_\parallel}{d\Omega} + \frac{dI_\perp}{d\Omega} = \frac{1}{4\pi}[(I_\pi + I_\sigma/2) - (I_\pi - I_\sigma/2)\cos^2\vartheta]$$

$$= \frac{1}{4\pi}(I_\pi + I_\sigma/2)[1 - P\cos^2\vartheta] \tag{6.10}$$

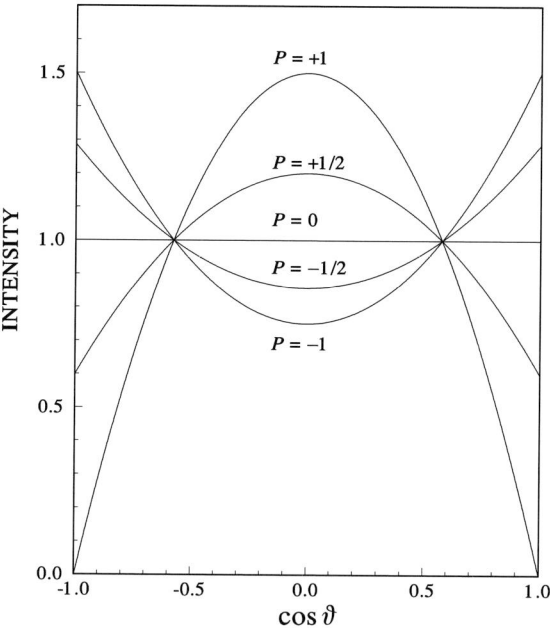

Fig. 6.2. Angular distribution of dipole radiation. (After Ref. [64].)

where P is the sample polarization

$$P \equiv \frac{I_\pi - I_\sigma/2}{I_\pi + I_\sigma/2}. \qquad (6.11)$$

Integrating over all 4π steradians

$$\int^{4\pi} d\Omega \frac{dI}{d\Omega} = (I_\pi + I_\sigma/2)(1 - P/3)$$

$$= \frac{2}{3}(I_\pi + I_\sigma). \qquad (6.12)$$

Thus, in terms of the total emitted radiation, the angular distribution can be written

$$\frac{dI}{d\Omega} = \frac{I}{4\pi}\left[\frac{1 - P\cos^2\vartheta}{1 - P/3}\right]. \qquad (6.13)$$

A plot of this distribution is shown in Fig. 6.2. Since detection is usually made within a very limited solid angle, differences in angular distributions can affect the detected relative intensities if the sample is polarized. Notice that at the "magic angle" $\vartheta = \pm 54.7°$ ($\cos^2\vartheta = \frac{1}{3}$) the curves cross. At this angle the dependence on P cancels in the numerator and denominator, hence the intensity (if detected with no instrumental polarization) is proportional to the total radiation emitted into all 4π steradians irrespective of P.

6.1.2 Lines

For a line intensity, it is necessary to weight each component by the population of the initial state and sum over the final states

$$I(\gamma L J, \gamma' L' J') = \sum_{M_J} N(\gamma L J M_J) \left[\frac{1265.38}{\lambda(\text{Å})}\right]^2 \sum_{M'_J} \left|\left\langle \gamma L J M_J \left|\frac{\mathbf{r}}{a_0}\right| \gamma' L' J' M'_J \right\rangle\right|^2. \quad (6.14)$$

For compactness of notation the quantum number S has been suppressed here, since these transitions obey the $\Delta S = 0$ selection rule.

The dipole transition element has the property that a sum over M'_J values yields a result that is independent of M_J. This provides the definition for the line-strength factor

$$S(\gamma L J, \gamma' L' J') \equiv \sum_{M_J} \sum_{M'_J} |\langle \gamma L J M_J | \mathbf{r}/a_0 | \gamma' L' J' M'_J \rangle|^2$$

$$= (2J + 1) \sum_{M'_J} |\langle \gamma L J M_J | \mathbf{r}/a_0 | \gamma' L' J' M'_J \rangle|^2$$

$$= (2J' + 1) \sum_{M_J} |\langle \gamma L J M_J | \mathbf{r}/a_0 | \gamma' L' J' M'_J \rangle|^2. \quad (6.15)$$

In the Schrödinger approximation, the radial wave functions are independent of the angular momentum quantum numbers J, L and S. Therefore the line-strength factor is the same for all members of the configuration-to-configuration transition array that obey the selection rules imposed by the separable angular portion of the transition integral. Both the emission transition probability rates A_{ul} (Eq. 6.1) and the absorption oscillator strengths f_{lu} (Eq. 6.6) can be computed from the line-strength factor $S_{ul} = S_{lu}$ (Eq. 6.15) with appropriate corrections for the upper or lower level degeneracies using

$$(2J + 1)A(\gamma L J, \gamma' L' J') = \left[\frac{1265.38}{\lambda(\text{Å})}\right]^3 S(\gamma L J, \gamma' L' J') \quad (6.16)$$

$$(2J' + 1)f(\gamma' L' J', \gamma L J) = \left[\frac{303.756}{\lambda(\text{Å})}\right] S(\gamma L J, \gamma' L' J') \quad (6.17)$$

$$(2J + 1)A(\gamma L J, \gamma' L' J') = \left[\frac{2582.68}{\lambda(\text{Å})}\right]^2 (2J' + 1)f(\gamma' L' J', \gamma L J) \quad (6.18)$$

(the number in brackets in Eq. 6.18 was obtained from those defined in Eqs. 6.1 and 6.6 with adjustments for the powers involved). In these terms the emitted intensity can be written as

$$I(\gamma L J, \gamma' L' J') = N(\gamma L J)A(\gamma L J, \gamma' L' J'), \quad (6.19)$$

where

$$N(\gamma L J) \equiv \sum_{M_J} N(\gamma L J M_J). \quad (6.20)$$

The reciprocal of the meanlife of the level γLJ is the sum of the various branches of the decay

$$\frac{1}{\tau(\gamma LJ)} = \sum_{\gamma'L'J'} A(\gamma LJ, \gamma'L'J'), \tag{6.21}$$

and the branching fraction BF from an upper level to a specific channel of its decay is given by

$$BF(\gamma LJ, \gamma'L'J') = \tau(\gamma LJ)A(\gamma LJ, \gamma'L'J'). \tag{6.22}$$

6.1.3 Multiplet values

In the nonrelativistic Schrödinger approximation the radial and spin portions of the wave function are separate, causing the dipole transition integral to be independent of spin. Thus all members of the multiplet have the same line-strength factor in such a calculation, but their transition probability rates and oscillator strengths will differ among the individual lines if the energy levels have been subjected to perturbation corrections. The fact that these perturbation corrections are made only in the energy levels, and not in the wave functions themselves, can introduce errors in the specification of the line-strength factors, particularly if there is a significant amount of cancellation in the dipole transition integral. For this reason, multiplet values for lifetimes, transition probability rates, and oscillator strengths that are often reported in theoretical studies must be treated with some care.

In time-dependent processes, the concept of a multiplet lifetime has little utility, since the sum of two exponential functions cannot be recast as a single exponential. In a situation of continuous excitation, the intensity (in photons/s) of an unresolved multiplet can be defined as

$$I(\gamma, \gamma') \equiv \sum_{J} \sum_{J'} \eta(\lambda_{JJ'}) N(\gamma J) A(\gamma J, \gamma'J'), \tag{6.23}$$

where $\eta(\lambda_{JJ'})$ is the relative detection efficiency for the various wavelengths that comprise the unresolved multiplet. If there are no significant anisotropies in the excitation process, then the individual fine structure levels can be considered to be populated according to their statistical weights

$$N(\gamma J) \approx (2J + 1)N(\gamma), \tag{6.24}$$

where $N(\gamma)$ is the individual population of each of the upper states. If, in addition, the detection efficiency is relatively flat in the region studied, the population and detection efficiency can be pulled through the sum

$$I(\gamma, \gamma') \approx \eta N(\gamma) \sum_{J} \sum_{J'} (2J + 1)A(\gamma J, \gamma'J'). \tag{6.25}$$

In such a case a weighted multiplet value for the transition rate can be useful. A similar expression exists for absorption, wherein the multiplet value is proportional to the quantity

$$\sum_{J'}\sum_{J}(2J'+1)f(\gamma'J',\gamma J).\tag{6.26}$$

However, the validity of both of these expressions requires isotropic excitation and uniform wavelength detection efficiency, and they must be used with care.

6.1.4 Supermultiplet and transition-array values

Since, in the nonrelativistic Schödinger approximation the line-strength factor is the same for all lines in the transition array, it can be theoretically applied to the supermultiplet to connect polyads of the same multiplicity in the two configurations (provided they possess levels of the appropriate J values and parity). Since the energy separations of terms of differing L in the polyads are usually much larger than the J splitting within a term, effective supermultiplet averages of the transition probability rates and oscillator strengths would not generally be of utility.

In pure LS coupling, there would be no transitions between polyads of differing multiplicity, but all real systems exhibit some degree of intermediate coupling, in which case the physical states contain a mixture of multiplicities (singlet–triplet, doublet–quartet, etc., mixing). In such a case the polyads mix, and the full transition array involves lines between all levels that are not forbidden by energy, total angular momentum, or parity considerations. In such cases the configuration value for the line-strength factor would also pertain to the individual amplitudes of the line intensities.

6.1.5 Semiempirical evaluation of line-strength factors

While the assumption that is built into the Schrödinger approximation that the line strength has a common value for the entire transition array can lead to inaccuracies in *ab initio* theoretical calculations, it can provide a valuable means for quantitatively characterizing the system if applied to measured lifetime or branching-fraction data. The line-strength factor can be used as an empirical quantity to be deduced from the measured data. It will be shown later how measured lifetimes, branching fractions, transition wavelengths, and energy-level splittings can be combined to obtain empirical values for the line-strength factor, which can then be compared within a multiplet, supermultiplet, or transition array, and followed isoelectronically.

6.2 *Ab initio* calculations

The line strength can be broken down into three factors in the following manner

$$S_{\gamma J,\gamma'J'} = D_{\text{mult}}^2 D_{\text{line}}^2 P^2 \tag{6.27}$$

where D_{mult} is the multiplet factor, D_{line} is the line factor, and P is the radial transition integral. After performing the angular integrations in Eq. 6.15, these can be written in the symbolic form

$$D^2_{mult} = (2L + 1)(2L' + 1)\ell_> \begin{Bmatrix} \ell' & L' & L_c \\ \ell & L & 1 \end{Bmatrix}^2, \tag{6.28}$$

where L and L' are the total orbital angular momenta of the upper level and lower levels respectively, $\ell_>$ is the greater of the upper and lower single-particle angular momenta of the jumping electron ℓ and ℓ', L_c is the orbital angular momentum of the core, and the bracketed symbol is the Racah 6-j symbol [30] discussed earlier in Section 4.6. Similarly,

$$D^2_{line} = (2J + 1)(2J' + 1) \begin{Bmatrix} L & S & J \\ J' & 1 & L' \end{Bmatrix}^2. \tag{6.29}$$

Here J and J' are the total angular momentum of the upper and lower levels, and S is the spin angular momentum of both the upper and lower levels. The radial transition matrix is

$$P = \frac{1}{a_0} \int_0^\infty dr r^2 R_{n\ell}(r) r R_{n'\ell'}(r). \tag{6.30}$$

As an example we shall make calculations for the hydrogenlike (one-electron ion of nuclear charge Z) Lyman-alpha 1s–2p transition. Here $\ell = L = L_c = 0, \ell' = L' = \ell_> = 1$, $S = \frac{1}{2}$, so

$$D^2_{mult} = (3)(3)(1) \begin{Bmatrix} 1 & 1 & 0 \\ 0 & 0 & 1 \end{Bmatrix}^2 = 1, \tag{6.31}$$

and there are two lines, corresponding to $J = \frac{1}{2}$ and $J' = \frac{1}{2}, \frac{3}{2}$

$$D^2_{line}\left(\frac{1}{2}, \frac{1}{2}\right) = (2)(2) \begin{Bmatrix} 0 & \frac{1}{2} & \frac{1}{2} \\ \frac{1}{2} & 1 & 1 \end{Bmatrix}^2 = \frac{2}{3} \tag{6.32}$$

$$D^2_{line}\left(\frac{1}{2}, \frac{3}{2}\right) = (2)(4) \begin{Bmatrix} 0 & \frac{1}{2} & \frac{1}{2} \\ \frac{3}{2} & 1 & 1 \end{Bmatrix}^2 = \frac{4}{3}. \tag{6.33}$$

The radial wave functions are

$$R_{1s}(r) = \left[\frac{Z}{a_0}\right]^{3/2} 2 \exp(-Zr/a_0) \tag{6.34}$$

$$R_{2p}(r) = \left[\frac{Z}{2a_0}\right]^{3/2} \frac{2}{\sqrt{3}} \frac{Zr}{2a_0} \exp(-Zr/2a_0). \tag{6.35}$$

Defining $x \equiv 3Zr/2a_0$, the integral becomes

$$P = \frac{a_0}{\sqrt{6}Z} \left(\frac{2}{3}\right)^5 \int_0^\infty dx x^4 e^{-x} = \frac{a_0}{Z} \sqrt{\frac{2^{15}}{3^9}}. \tag{6.36}$$

Collecting the factors

$$Z^2 S_{1/2,1/2} = 2^{16}/3^{10}; \quad Z^2 S_{1/2,3/2} = 2^{17}/3^{10}. \tag{6.37}$$

The transition wavelength is obtained from the Balmer formula and the DeBroglie relationship (here in SI units)

$$\frac{2\pi \hbar c}{\lambda} = \frac{1}{2}mc^2(\alpha Z)^2 \left[1 - \frac{1}{2^2} \right] \tag{6.38}$$

which, noting that $\alpha a_0 = \hbar/mc$, becomes

$$Z^2\lambda = 16\pi a_0/3\alpha. \tag{6.39}$$

The expressions for the transition probabilities

$$(2J+1)A_{J,J'} = \left(\frac{4\alpha c a_0^2}{3} \right) \left(\frac{2\pi}{\lambda} \right)^3 S_{J,J'} \tag{6.40}$$

thus yield

$$A_{1/2,1/2} = A_{3/2,1/2} = \left(\frac{2}{3} \right)^8 \frac{c(\alpha Z)^4}{a_0} = \frac{Z^4}{1.5953}(\text{ns}^{-1}). \tag{6.41}$$

Here, and in other numerical results quoted in this section, the infinite mass value for a_0 is used. For specific members of the hydrogenlike isoelectronic sequence, the appropriate reduced mass corrections should be made.

Similarly, the expressions for the oscillator strengths

$$(2J'+1)f_{J',J} = \left(\frac{2a_0}{3\alpha} \right) \left(\frac{2\pi}{\lambda} \right) S_{J,J'} \tag{6.42}$$

yield

$$f_{1/2,1/2} = \frac{1}{2}f_{1/2,3/2} = \frac{2^{13}}{3^{10}} = \frac{1}{3}(0.4162). \tag{6.43}$$

These computations can be extended to prescribe any transition in hydrogen. For example, an arbitrary member of the Lyman sequence can be obtained [14] from

$$A_{np,1s} = \frac{2^7}{3^2} \frac{n}{(n^2-1)^2} \left(\frac{n-1}{n+1} \right)^{2n} \frac{(\alpha Z)^4 c}{a_0}. \tag{6.44}$$

A similar expression was presented in Eq. 2.160 for the circular orbit yrast transitions $(n, \ell = n-1 \rightarrow n-1, \ell = n-2)$.

For the $n=3$ levels, the decay involves the blend of three independent but degenerate lines comprising the Balmer H-alpha (the $n=3 \rightarrow n=2$) manifold, with a second channel to the 3p via Lyman-β (3p $\rightarrow$ 1s)

$$A_{3s,2p} = \frac{2^8 3}{5^9} \frac{(\alpha Z)^4 c}{a_0} = 0.006\ 317 Z^4\ \text{ns}^{-1}$$

$$A_{3p,2s} = \frac{2^{13}}{3\ 5^9} \frac{(\alpha Z)^4 c}{a_0} = 0.022\ 461 Z^4\ \text{ns}^{-1}$$

$$A_{3p,1s} = \frac{1}{2^5 3} \frac{(\alpha Z)^4 c}{a_0} = 0.167\ 344 Z^4\ \text{ns}^{-1}$$

$$A_{3d,2p} = \frac{2^{16} 3}{5^{11}} \frac{(\alpha Z)^4 c}{a_0} = 0.064\ 686 Z^4\ \text{ns}^{-1}. \tag{6.45}$$

(Again, the infinite mass value for a_0 has been used.) As indicated in Eq. 6.21, this corresponds to the lifetimes

$$\tau_{3s} Z^4 = (0.006\ 317)^{-1} = 158\ \text{ns}$$
$$\tau_{3p} Z^4 = (0.022\ 461 + 0.167\ 344)^{-1} = 5.27\ \text{ns}$$
$$\tau_{3d} Z^4 = (0.064\ 686)^{-1} = 15.46\ \text{ns}. \tag{6.46}$$

From Eq. 6.22, the branching fractions of the 3p decay are

$$BF_{3p,2s} = 5.27(0.0225) = 12\%; \qquad BF_{3p,1s} = 5.27(0.1673) = 88\%. \tag{6.47}$$

6.3 Commutation relations involving the E1 transition moment

There are a number of relationships involving the electric dipole moment that can be developed from familiar quantum mechanical commutation equations. By applying the basic commutation relationships $[p_x, x] = \hbar/i$ and $[AB, C] = A[B, C] - [C, A]B$ to a Hamiltonian $H = p^2/2m + V$, it can be shown that

$$[H, x] = -\frac{\hbar^2}{m} \frac{\partial}{\partial x} \tag{6.48}$$

$$[H, [H, x]] = \frac{\hbar^2}{m} \frac{\partial V}{\partial x} \tag{6.49}$$

$$[[H, x], x] = -\frac{\hbar^2}{m}. \tag{6.50}$$

The standard expression for the E1 transition moment $\langle a\,|r|\,b\rangle$ is known as the "length form." In analogy to the classical simple harmonic oscillator, for which the magnitudes of the average displacement, speed, and acceleration are all related, there is also a "velocity form" and an "acceleration form" for this transition moment [25].

6.3.1 Velocity form of the transition moment

An expression involving the transition integral can be formed from the first commutator relation, Eq. 6.48

$$-\frac{\hbar^2}{m}\left\langle a\left|\frac{\partial}{\partial x}\right|b\right\rangle = \langle a|Hx - xH|b\rangle = (E_a - E_b)\langle a|x|b\rangle. \tag{6.51}$$

Here we have used the hermiticity of the Hamiltonian to obtain the eigenvalues. This can be rewritten as

$$\left\langle a\left|\frac{p_x}{m}\right|b\right\rangle = -i\omega_{ab}\langle a|x|b\rangle. \tag{6.52}$$

This is the analogue of the classical simple harmonic oscillator relationship $\langle dx/dt\rangle = -\omega\langle x\rangle$, hence it is called [25] the velocity form (as opposed to the standard length form) of the electric dipole moment. If the quantum mechanical wave functions were exactly correct, the values deduced from the length form and velocity form would be identical. Since the unperturbed wave functions are not physically exact, this alternative method is sometimes used as a test of the reliability of the calculation. Unfortunately, agreement between length form and velocity form does not guarantee their correctness, and the discrepancy between the two forms does not necessarily bracket the correct result. The nature of the Hamiltonian formulation optimizes the specification of the total energy of the system, which derives primarily from the infinite-range Coulomb potential. Thus, the wave functions so obtained tend to be much more accurate in the large-r region. In the small-r region, many perturbations that affect the dynamics of the system are not included in the wave function. The velocity form of the transition integral emphasizes the inner r regions much more than does the length form, and their comparison can probe the accuracy of the wave functions.

6.3.2 Acceleration form of the transition moment

Forming a transition integral from the second commutation relation, Eq. 6.49, yields

$$\frac{\hbar^2}{m}\left\langle a \left| \frac{\partial V}{\partial x} \right| b \right\rangle = \langle a|HHx - 2HxH + xHH|b\rangle$$

$$= (E_a - E_b)^2 \langle a\,|x|\,b\rangle \tag{6.53}$$

From the classical relationship $\mathbf{F} = m\mathbf{a} = -\nabla V$, this can be written

$$\left\langle a \left| \frac{1}{m}\frac{dp_x}{dt} \right| b \right\rangle = \omega_{ab}^2 \langle a\,|x|\,b\rangle . \tag{6.54}$$

Here again we have made use of hermiticity to obtain the eigenvalues, and factored these into a squared binomial. This is the analogue of the classical [25] simple harmonic oscillator relationship $\langle dx^2/dt^2\rangle = -\omega^2\langle x\rangle$, hence it is called the acceleration form of the electric dipole moment. This formulation weights the inner portion of the wave function even more strongly than either the length form or the velocity form, and is of limited value in characterizing this moment in complex atoms. However, it does have useful applications in characterizing the polarization of the core for transitions between nonpenetrating orbitals.

For $V(r) = k\zeta e^2/r$, this yields the relationship

$$\langle a\,|\mathbf{r}|\,b\rangle = \frac{4\zeta Ry^2 a_0^3}{(E_a - E_b)^2}\left\langle a \left| \frac{\mathbf{r}}{r^3} \right| b \right\rangle . \tag{6.55}$$

that will be used in Section 6.5.

6.3.3 The Thomas–Reiche–Kuhn sum rule

Forming an expectation value from the third commutator relation, Eq. 6.50, one obtains

$$-\frac{\hbar^2}{m}\langle a\,|\,a\rangle = \langle a\,|Hxx - 2xHx + xxH|a\rangle . \tag{6.56}$$

Using the hermiticity of the Hamiltonian, and the completeness of the basis states

$$\frac{\hbar^2}{m} = 2E_a \sum_i \langle a\,|x|\,i\rangle\langle i\,|x|\,a\rangle - 2\sum_i\sum_j \langle a\,|x|\,i\rangle\langle i\,|H|\,j\rangle\langle j\,|x|\,a\rangle . \tag{6.57}$$

Since H is diagonal in this representation, this becomes

$$1 = \frac{2m}{3\hbar^2} \sum_i (E_a - E_i) |\langle a \,|\mathbf{r}|\, i \rangle|^2, \tag{6.58}$$

where we have also used the isotropy of space

$$|\langle i \,|x|\, j \rangle|^2 = |\langle i \,|y|\, j \rangle|^2 = |\langle i \,|z|\, j \rangle|^2 = |\langle i \,|\mathbf{r}|\, j \rangle|^2 /3. \tag{6.59}$$

Writing the quantum numbers in more specific detail as $|n\rangle \to |\gamma J M_J\rangle$ and $|i\rangle \to |\gamma' J' M'_J\rangle$, the sum over M'_J eliminates the dependence upon M_J which we can average over by introducing a factor of its degeneracy $(2J + 1)$

$$1 = \frac{2ma_0^2}{3\hbar^2}(2J + 1) \sum_{\gamma' J'} (E_{\gamma J} - E_{\gamma' J'}) \sum_{M_J M'_J} |\langle \gamma J M_J \,|\mathbf{r}/a_0|\, \gamma' J' M'_J\rangle|^2, \tag{6.60}$$

which can be recognized as

$$1 = \sum_{\gamma' J'} f_{\gamma J, \gamma' J'}. \tag{6.61}$$

This is known as the Thomas–Reiche–Kuhn sum rule [137, 169]. Since the completeness condition requires that both discrete and continuum states be included, the sum indicated here symbolizes both a sum over the discrete states and an integral over the continuum. If more than one active electron is involved, the factor unity on the left-hand side of the equality is replaced by the number of active electrons. This relationship makes clear the statement that a strong transition can "steal" oscillator strength from other transitions. Since there are unitarity considerations imposed by the number of active electrons, the existence of a very strong transition from a given level limits the oscillator strength available to other channels.

6.4 Quadratic Stark effect and atomic polarizability

6.4.1 Polarizabilities and oscillator strengths

When an electron in an atom is placed in a weak electric field E in, e.g., the x-direction, it is subjected to a perturbing potential

$$\Delta H = e\mathsf{E}x. \tag{6.62}$$

The shift in the energy of an eigenstate $|\gamma J M_J\rangle$ is given to second-order perturbation theory by

$$\Delta E_{\gamma J M_J} = e\mathsf{E}\langle \gamma J M_J |x| \gamma J M_J\rangle$$
$$+ e^2\mathsf{E}^2 \sum_{\gamma' J' M'_J} \frac{\langle \gamma J M_J |x| \gamma' J' M'_J\rangle \langle \gamma' J' M'_J |x| \gamma J M_J\rangle}{E_{\gamma J} - E_{\gamma' J'}}. \tag{6.63}$$

Since the unperturbed Hamiltonian is invariant under spatial inversions, the linear term vanishes unless there are degenerate states of opposite parity. (The degeneracy of the magnetic substates is irrelevant since they have the same parity). Averaging this over M_J substates, taking into account the y- and z-components by multiplying by a factor of 3, and noting that $Ry \equiv Ke^2/2a_0$, this becomes

$$\Delta E_{\gamma J} = \frac{1}{(2J+1)} \sum_{M_J} \Delta E_{\gamma J M_J}$$

$$= \mathsf{E}^2 \frac{2Ry\, a_0^3}{3(2J+1)K} \sum_{\gamma'J} \sum_{M_J M_J'} \frac{|\langle \gamma J M_J | \mathbf{r}/a_0 | \gamma' J' M_J' \rangle|^2}{E_{\gamma J} - E_{\gamma'J'}}. \tag{6.64}$$

The M-summed matrix element can be written in terms of the oscillator strength using the relationships from Section 6.1.2

$$(2J+1)f_{\gamma'J',\gamma J} = \frac{1}{3Ry}(E_{\gamma J} - E_{\gamma'J'}) \sum_{M_J M_J'} |\langle \gamma J M_J | \mathbf{r}/a_0 | \gamma' J' M_J' \rangle|^2. \tag{6.65}$$

Combining these two equations, and comparing this to the classical formulation of the energy of an induced dipole

$$\Delta E_{\gamma J} = \frac{1}{2}\alpha_d' \mathsf{E}^2 = \mathsf{E}^2 \left(\frac{2Ry^2 a_0^3}{K} \right) \sum_{\gamma'J'} \frac{f_{\gamma'J',\gamma J}}{(E_{\gamma J} - E_{\gamma'J'})^2} \tag{6.66}$$

an expression for the dipole polarizability α_d' is obtained (the prime denotes that the definition is in SI units). If, according to standard practices in atomic physics, α_d is measured in units of a_0^3/K (where K is the electrostatic Coulomb constant),

$$\alpha_d \equiv \frac{K\alpha_d'}{a_0^3} = \sum_{\gamma'J'} \frac{4Ry^2 f_{\gamma'J',\gamma J}}{(E_{\gamma J} - E_{\gamma'J'})^2} = 4Ry^2 \sum_{\gamma'J'} f_{\gamma'J',\gamma J}\, \lambda_{\gamma J, \gamma'J'}^2. \tag{6.67}$$

This a very useful expression, since it relates two quantities, the polarizability and the oscillator strengths, both of which can be measured. Thus one can be computed from the other, or measurements of both can be compared for consistency.

6.4.2 Determination of dipole polarizabilities from lifetime measurements

The dipole polarizability of an atom or ion describes the response, in lowest order in the field strength, of the electron cloud to an external electric field. This atomic structure property plays an important role in many processes that motivate its theoretical and experimental determination. It is used to specify core effects in the spectroscopy of high Rydberg states, to account for electron correlation in model potential and transition moment calculations, in atomic scattering processes, and in the formulation of van der Waals forces, refractive indices, ion mobility in gases, diamagnetic susceptibilities, etc. Until recently, experimental determinations were made directly, by observation of one or more of the quantities described above. However, as a result of advances in the field of atomic meanlife measurements, it has become possible to make use of Eq. 6.67 to make very accurate indirect determinations. This method utilizes high-precision lifetime measurements and a knowledge of the fortuitous cancellations that occur in certain types of systems.

A substantial database now exists for high-precision lifetime measurements of the lowest resonance transitions in alkali-metallike atoms and ions. Primarily through the use of selective excitation using lasers, measurements accurate to within 1% are now available for ns–np transitions in these systems. For Li, Na, Mg$^+$, Ag, Cd$^+$, and Cs the lowest resonance transitions are unbranched, so lifetime values can be converted directly to absorption oscillator strengths. For Ca$^+$, Cu, Sr$^+$, and Ba$^+$ the transitions have weak alternative branches to ^{2}D terms, and can also be converted to f-values if accurate theoretical estimates of these small branching fractions are available. The f-values for these intrashell ns–np transitions are characteristically large and close to unity, which implies (through f sum rules in the single active electron approximation) that extrashell ns–n'p f-values are small. This tendency is especially pronounced for the singly charged ions in several of the sequences, because fortuitous cancellations (discussed later in Section 6.6) affect the ns–n'p transition moments for all intershell ($n' > n$) cases. Theoretical estimates of these higher-order contributions to the sum can be made to provide error estimates on the determination.

The method can be made clearer by writing the relationships between the polarizability and the lifetime more explicitly. The ground-state dipole polarizability can expressed in the form

$$\alpha_d(n) = 4Ry^2 \left[\sum_{n'} \frac{f_{nn'}}{(E_{n'} - E_n)^2} + \int_0^\infty \frac{\mathrm{d}f/\mathrm{d}E}{(E - E_n)^2} \mathrm{d}E \right], \tag{6.68}$$

where $f_{nn'}$ is the absorption oscillator strength for a transition from an ns ^{2}S$_{1/2}$ ground state to an n'p ^{2}P$_{J'}$ excited state and $E_{n'} - E_n \equiv E_{nn'}$ is the corresponding excitation energy. The summation notation includes an implicit sum over $J' = \frac{1}{2}, \frac{3}{2}$, and so does the integral over continuum states. If the sum is dominated by the $n' = n$ term, a lower limit α_{d0} for the dipole polarizability can be deduced directly from the measured lifetime of the ns–np transition through its oscillator strength and transition energy

$$\alpha_{d0} \equiv f_{nn'}(2Ry/E_{nn'})^2. \tag{6.69}$$

Thus if the $n' = n$ term dominates, the uncertainties in the determination can be accurately estimated by propagating the tolerances in the lifetime measurement, and combining these with theoretical estimates of the magnitudes and uncertainties in the specification of the $n' > n$ transitions. Moreover, the contributions due to higher-order terms in the summation could also be examined in the context of the Thomas–Reiche–Kuhn sum rule

$$N = \sum_{n'} f_{nn'} + \int_0^\infty \frac{\mathrm{d}f}{\mathrm{d}E} \mathrm{d}E, \tag{6.70}$$

where N is the number of active electrons (unity in the single-particle approximation). Some departure from the $N = 1$ value and the single-particle approximation is expected in the theoretical estimates if the calculations explicitly include core polarization effects in the transition matrix elements. Because n represents the ground state, no transitions involve emission, and all f-values are positive. Since $E_{nn'}$ increases with increasing n', the value

Table 6.1. Determination of α_d for the Mg^+ ion from measured
f-values. The 3p levels were deduced from measured lifetime data, and
the contributions for $n > 3$ levels and the continuum were estimated by
quantum defect methods. (After Ref. [181].)

n'	J'	$f_{nn'}$	$E_{nn'}$	$f_{nn'}(2Ry/E_{nn'})^2$
3	1/2	0.304(2)	35 669	11.52(9)
	3/2	0.610(6)	35 761	22.99(24)
4	1/2	4.365×10^{-4}	80 620	3.240×10^{-3}
	3/2	7.704×10^{-4}	80 650	5.713×10^{-3}
5	1/2	7.080×10^{-4}	97 455	3.594×10^{-3}
	3/2	1.341×10^{-3}	97 469	6.806×10^{-3}
6	1/2	4.789×10^{-4}	105 622	2.069×10^{-3}
	3/2	9.165×10^{-4}	105 630	3.958×10^{-3}
7	1/2	3.122×10^{-4}	110 204	1.238×10^{-3}
	3/2	5.998×10^{-4}	110 208	2.379×10^{-3}
8	1/2	2.100×10^{-4}	113 030	7.918×10^{-4}
	3/2	4.043×10^{-4}	113 033	1.524×10^{-3}
9	1/2	1.467×10^{-4}	114 897	5.352×10^{-4}
	3/2	2.827×10^{-4}	114 899	1.032×10^{-3}
10–∞	1/2	5.632×10^{-4}		1.938×10^{-3}
	3/2	1.088×10^{-3}		3.744×10^{-3}
contin.	1/2	5.119×10^{-2}		3.127×10^{-2}
	3/2	1.014×10^{-1}		6.137×10^{-2}
	f sum	1.075	α_d	34.62(26)

of α_d can be bracketed

$$\alpha_{d0} \leq \alpha_d \leq [\alpha_{d0} + (N - f_{nn'})(2Ry/E_{n,n+1})^2]. \tag{6.71}$$

Theoretical estimates of the contributions of higher-n transitions can also be made using energy-level data and the semiempirical Coulomb approximation.

Table 6.1 displays an application of this approach to the Mg^+ ion. The oscillator strengths for the $3s\,^2S_{1/2}$–$3p\,^2P_{1/2}$ and $3s\,^2S_{1/2}$–$3p\,^2P_{3/2}$ transitions are available from high-precision lifetime measurements (cf. Ref. [181]), which yielded values $\tau(^2P_{1/2}) = 3.854 \pm 0.030$ ns and $\tau(^2P_{3/2}) = 3.810 \pm 0.040$ ns. To buttress the determination, Coulomb approximation calculations were made up to $n' = 20$, using the measured energy levels and the Ritz parametrizations of the np levels. Notice in Table 6.1 that the calculated oscillator strengths for $4 \leq n' \leq 20$ make a negligible contribution to α_d, and (in the single active electron approximation) 85% of the oscillator strength is contained in the 3s–3p transitions.

6.5 Core polarization contributions to the E1 transition moment

Up to this point the development has treated the transition moment as a property of a single jumping electron in the presence of a passive core of running electrons. While this approximation is a useful starting point for some types of calculations, the full formulation of the transition moment involves overlap integrals over all of the electrons and all of the states in the atom, before and after the emission or absorption process, including correlations. If such a fully inclusive calculation were made, the core electrons would automatically be included, and there would be no reason to characterize their polarization. For complex atoms, such all-inclusive calculations often introduce too many degrees of freedom to be theoretically tractable. Moreover, for the predictive semiempirical systematization of measured data, the parametrized single-electron model offers many advantages. Thus the core polarization model, which was discussed in Section 3.4 in the context of high Rydberg states, is also useful in characterizing transition moments [13, 114].

In earlier discussions we have assumed that the dipole moment d arises entirely from dipole oscillations of the orbital electron, which can be classically formulated as

$$\mathbf{d} = \mathbf{d}_{\text{max}} \cos \omega t. \tag{6.72}$$

Thus, although the average value of the dipole moment vanishes, the average of the square of the moment is nonvanishing

$$\langle \mathbf{d} \rangle = 0; \qquad \langle \mathbf{d}^2 \rangle = \mathbf{d}_{\text{max}}^2 / 2. \tag{6.73}$$

In an N_e-electron atom, the dipole moment is given by

$$\mathbf{d} = - \sum_{i=1}^{N_e} e \mathbf{r}_i, \tag{6.74}$$

but in the core polarization model the jumping electron is singled out, and the running electrons are treated as a correlated, deformable core of charge. Since the core and jumping electrons repel each other, the core-plus-nucleus and the jumping-electron-plus-nucleus form oscillating dipoles $\mathbf{d}_c$ and $\mathbf{d}_e$, respectively. Assuming that the displacement of the core electrons from the nucleus is small compared to the orbital radius of the jumping electron, the electron dipole moment is

$$\mathbf{d}_e = -e\mathbf{r}, \tag{6.75}$$

and the core dipole moment is given by the core polarizability α'_d (with the prime indicating SI units) and the electric field produced by the orbital electron near the nucleus as

$$\mathbf{d}_c = \alpha'_d \frac{K e \mathbf{r}}{r^3}. \tag{6.76}$$

The total dipole moment of the system is then [13, 114]

$$\mathbf{d} = -e\mathbf{r} \left[1 - \frac{K \alpha'_d}{r^3} \right]. \tag{6.77}$$

This is an asymptotic expression for large r, and clearly breaks down near or inside some effective core radius r_c. Numerous attempts have been made to correct this expression for core penetration, using expressions of the form

$$\mathbf{d} = -e\mathbf{r}\left[1 - \frac{K\alpha_d' F(r)}{r^3}\right],$$ (6.78)

where $F(r)$ is a small-r cutoff function, such as

$$F(r) \approx 1 - \exp[-(r/r_c)^m]$$ (6.79)

or

$$F(r) \approx r^4/\left(r^2 + r_c^2\right)^2.$$ (6.80)

For high Rydberg states, where the core penetration corrections can be neglected, the identity obtained earlier in Eq. 6.55 for the acceleration form of the transition element is useful.

For a single-electron transition in a quasi-hydrogenic atom the dipole moment is given by

$$\langle n\ell\,|\mathbf{d}|\,n'\ell'\rangle = -\langle n\ell\,|e\mathbf{r}|\,n'\ell'\rangle + \alpha_d'\left\langle n\ell\left|F(r)\frac{Ke\mathbf{r}}{r^3}\right|n'\ell'\right\rangle.$$ (6.81)

If $F(r) \cong 1$, we can use the identity obtained in Eq. 6.55 for the acceleration form of the dipole moment for a potential $V(r) = K\zeta e^2/r$ to obtain

$$\left\langle n\ell\left|\frac{\mathbf{r}}{r^3}\right|n'\ell'\right\rangle = \frac{\langle n\ell\,|\mathbf{r}|\,n'\ell'\rangle}{4\zeta a_0^3 Ry^2\lambda_{n\ell,n'\ell'}^2}.$$ (6.82)

In this nonpenetrating limit (with the substitution $\alpha_d \equiv \alpha_d' K/a_Z^3$)

$$\left\langle n\ell\left|\frac{\mathbf{d}}{ea_0}\right|n'\ell'\right\rangle - -\left\langle n\ell\left|\frac{\mathbf{r}}{a_0}\right|n'\ell'\right\rangle\left[1 - \frac{\alpha_d}{4\zeta Ry^2\lambda_{n\ell,n'\ell'}^2}\right],$$ (6.83)

and the core polarization can be taken into account through a simple correction for the charge of the core and the wavelength of the transition.

This expression can be used to obtain predictions for transition probability rates between high-n and high-ℓ levels in complex atoms using hydrogenic values for the radial integrals. If we denote the transition probability rates and wavelengths in hydrogen by $A_{ij}(\mathrm{H})$ and $\lambda_{ij}(\mathrm{H})$ and the corresponding values for the high Rydberg atom of core charge ζ by $A_{ij}(\zeta)$ and $\lambda_{ij}(\zeta)$, the two systems are (from Eq. 6.1) related by

$$\frac{A_{ij}(\zeta)}{A_{ij}(\mathrm{H})} = \frac{1}{\zeta^2}\left[\frac{\lambda_{ij}(\mathrm{H})}{\lambda_{ij}(\zeta)}\right]^3\left[1 - \frac{\alpha_d}{\zeta[2Ry\,\lambda_{ij}(\zeta)]^2}\right]^2.$$ (6.84)

Corrections of this type are made for transitions between high-n and high-ℓ levels in the Mg-like ion Cl VI in Table 6.2. Here the unbranched transitions from the yrast (most nearly circular orbit) [112] 7i-level, and the branched transitions from the yrare (more nearly circular) 7h- and 7g-levels are studied. Empirical estimates of the polarizabilities are available [38] which yield (in reduced-mass atomic units as defined above and in Eq. 3.38) $\alpha_d = 1.62$

Table 6.2. Corrections for contributions from the dipole moment of the core to the branching fractions (*BF*) and lifetimes (τ) of high Rydberg levels in Mg-like Cl VI. (After Ref. [37].)

Trans.	$|\langle i|r/a_0|j\rangle|^2$	λ(Å)	*BF*	τ(ns)	Corr.	*BF*(Corr.)	τ(ns)
6h–7i	1498.49	3404.72	1.0	1.014	0.990	1.000	1.024
6g–7h	1093.28	3261.85	0.529		0.989	0.545	
5g–7h	55.6124	1256.93	0.471	0.657	0.930	0.455	0.683
6h–7g	5.63666	3551.60	0.0006		0.991	0.0008	
6f–7g	790.608	2057.58	0.389		0.974	0.459	
5f–7g	61.2871	932.54	0.324		0.875	0.343	
4f–7g	7.28156	477.86	0.286	0.171	0.569	0.197	0.207

and $\alpha_Q = 9$. These polarizabilities were used to compute the transition wavelengths using the energy-level formulae of Section 3.4. The values for the hydrogenic transition integrals in the compilation of Green *et al.* [109] were used to deduce the line-strength factors, and combined with the wavelengths to obtain the transition probability rates.

These calculations were made both without and with the corrections for the contribution of the core to the dipole moment. These are quoted in Table 6.2 in terms of branching fractions and level lifetimes. Because the contributions to the total dipole moment from the electron and the core have opposite signs, the effect is to lengthen the lifetime slightly. Since the correction depends on the wavelengths of the exit channels, it also has an effect on the branching fractions.

In this example the corrections are small but measurable. However, such corrections can be very large in situations where there are severe cancellations in the transition integral of the type that are discussed in the next section. If cancellations in the length form of the dipole transition moment are not mirrored by the same cancellations in the acceleration form, then the core polarization contribution will be revealed by the suppression of the normally dominant contribution from the dipole moment of the orbital electron.

6.6 Cancellation

Occasionally a specific oscillator strength in a Rydberg series is anomalously small compared with other members of the same Rydberg series and with the corresponding transition in other members of the same isoelectronic sequence. This occurs whenever the radial transition integral undergoes a change of sign along an isoelectronic sequence near values of the effective quantum numbers which correspond to a physical ion. The conditions for cancellation are very sensitive to the effective quantum numbers of the participating levels, and the anomaly is usually very sharp, restricted to a single Rydberg transition and a single member of the isoelectronic sequence.

A knowledge of the occurrence of anomalously small oscillator strengths can be valuable for a number of reasons. It can be utilized in term-analysis studies, since it explains the absence of lines that would otherwise be expected to be present. Anomalously low oscillator strengths in low-lying resonance transitions are useful for astrophysical abundance determinations because they are unsaturated in absorption spectra and permit lineshape studies. Knowledge that a normally strong transition becomes insignificant for a specific ion is also valuable in atomic lifetime measurements that incorporate cascade information into the analysis.

A simple graphical exposition has been developed [58] that predicts regions of likely cancellation in a space that is defined by the effective quantum numbers of the upper and lower levels. This permits the use of energy-level data (either directly measured or interpolated or extrapolated by Ritz parametrizations of the quantum defect) to make simple yes or no statements about the likelihood of severe cancellations in oscillator strengths.

A useful pictorial context for visualizing these cancellations is provided by the Coulomb (or quantum defect) approximation. This model utilizes the fact that, at sufficiently large values of the radial coordinate r the wave function of a complex atom resembles that of a hydrogenlike atom, with the radial coordinate shifted to $r' \rightarrow r + a_0 \pi \delta / \zeta$ and a small-r cutoff employed to exclude the core region. This method permits the effective quantum number to be treated as a continuous variable, since it deduces a wave function from a specified value for the energy (as parametrized by the effective quantum number $n^* \equiv n - \delta$). In this manner the possibilities for cancellations in a transition connecting levels with quantum numbers ℓ and ℓ' can be expressed on a generic plot of the locus of the cancellation nodes as a continuous function of n_ℓ^* vs $n_{\ell'}^*$.

The basis of this graphical exposition can be seen from the formulation of Burgess and Seaton [21], who derived a simple analytic representation of the Coulomb approximation that is essentially equivalent to the numerical tables of Bates and Damgaard [8]. The Burgess and Seaton formulation has the advantage that it combines tabulated parameters with simple analytic functions, to obtain

$$(n_\ell^*)^2 f_{n\ell,n'\ell'} = A \cos^2[\pi(n_{\ell'}^* - n_\ell^* - \chi)], \qquad (6.85)$$

where A and χ are slowly varying functions of the effective quantum numbers for given values of ℓ and ℓ'. Figure 6.3 (a) shows a landscape of this periodic function for an s–p system. The locus of cancellations is specified by the condition $(n_{\ell'}^* - n_\ell^* - \chi) = k + \frac{1}{2}$, where k is an integer. Using Burgess and Seaton's expression for χ, the cancellation condition is given by

$$n_{\ell'}^* = n_\ell^* + k + \frac{1}{2} + a_{\ell\ell'} + \frac{b_{\ell\ell'}}{n_\ell^*} + \frac{c_{\ell\ell'}}{(n_\ell^*)^2} - \frac{\alpha_{\ell\ell'}}{(n_\ell^*)^2/n_\ell^2 - 1} - \frac{\beta_{\ell\ell'}}{(n_\ell^*/n_\ell)^2 - 1}, \qquad (6.86)$$

where $a_{\ell\ell'}$, $b_{\ell\ell'}$, $c_{\ell\ell'}$, $\alpha_{\ell\ell'}$, $\beta_{\ell\ell'}$ are fitted constants that are given in Table 6.3.

The formal solution to this equation would require the extraction of the roots of polynomial equations of up to sixth order. However, for values of k such that $n_{\ell'}^*$ and n_ℓ^* are not too close together (in which case the oscillator strengths and transition probabilities become small due to the wavelength dependence), the appropriate root can be obtained by choosing a value for n_ℓ^* and iterating the equation to obtain successively better approximations for $n_{\ell'}^*$. With this single physical root, the equation represents a family of curves (approaching

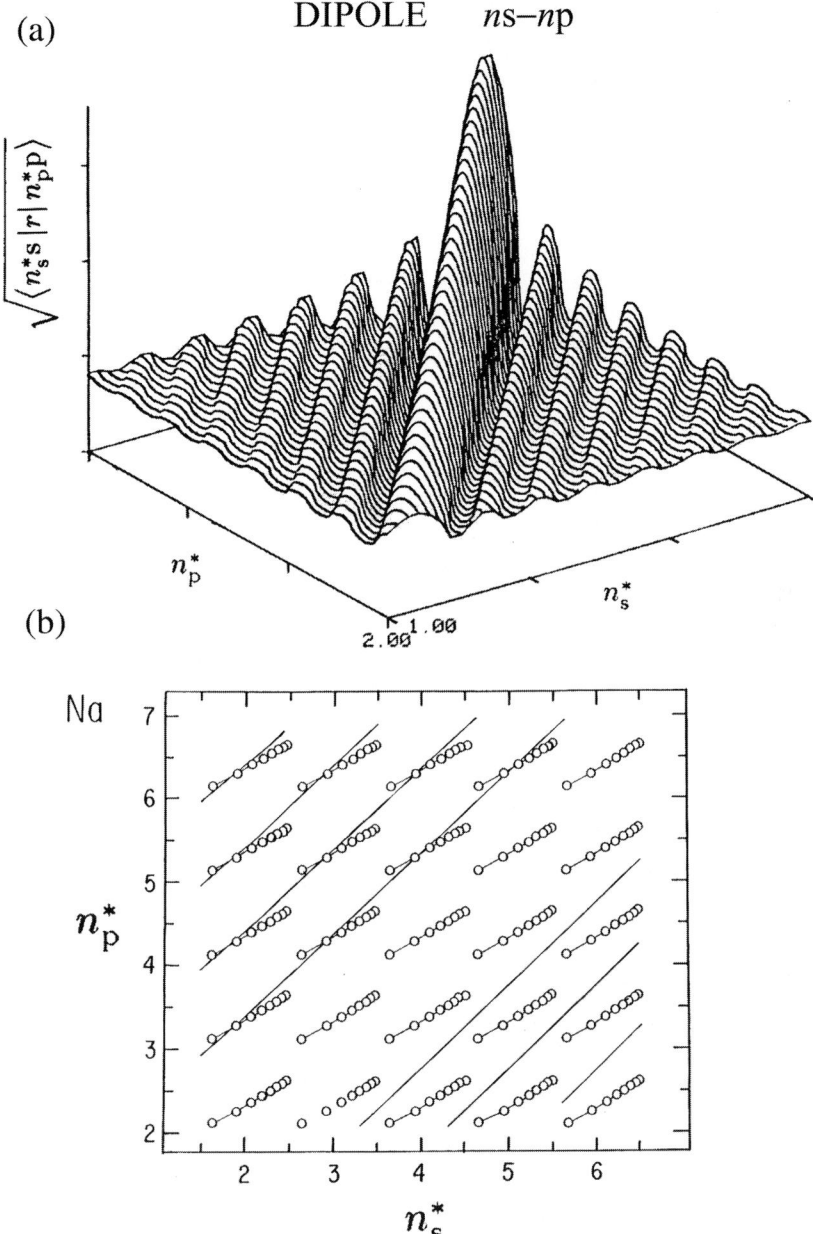

Fig. 6.3. Expositions of the cancellations in the dipole transition matrix as a function of the effective quantum numbers for the ns–$n'p$ transitions in the sodium isoelectronic sequence. (a) The square root of the dipole matrix element (with the sign preserved) is plotted as a landscape on the plane of the effective quantum numbers. (The square root was taken to place a wider range of oscillations on a similar scale of amplitudes, with the sign preserved to trace the sign changes in the regions of cancellation.) (b) A two-dimensional plot in the space of the effective quantum numbers. The cancellation nodes are indicated by the solid lines and the isoelectronic loci for the individual ns–$n'p$ transitions are denoted by circles.

Table 6.3. Coefficients for the equation for cancellation nodes.
(From Ref. [21].)

ℓ	ℓ'	$a_{\ell\ell'}$	$b_{\ell\ell'}$	$c_{\ell\ell'}$	$\alpha_{\ell\ell'}$	$\beta_{\ell\ell'}$
0	1	-0.147	$+0.2515$	-0.078	$+0.310$	0.000
1	0	-0.216	-0.171	0.000	0.000	0.000
1	2	-0.120	$+0.600$	0.000	$+0.362$	$+0.0535$
2	1	-0.247	-0.272	0.000	-0.010	-0.019
2	3	-0.117	$+1.170$	0.000	$+0.321$	$+0.106$
3	2	-0.362	$+0.599$	-2.432	-0.390	$+0.050$

parallel straight lines with a $45°$ slope for large quantum numbers) on a plot of n_ℓ^* vs $n_{\ell'}^*$. The formulation of Burgess and Seaton is approximate, and not symmetric under interchange of upper and lower level quantum numbers, so values of $n_{\ell'}^* < n_\ell^*$ must be obtained by interchanging ℓ and ℓ' and not by allowing k to take on negative values.

A comparison with experimental energy-level data can be obtained by plotting the corresponding $n_{\ell'}^*$ and n_ℓ^* values for a specific transition, following it along an isoelectronic sequence, and looking for coincidences or near coincidences between the experimental points and the lines of nodes predicted by this equation. An example is shown in Fig. 6.3(b) for the ns–n'p transitions in the Na isoelectronic sequence. The slope of the isoelectronic trajectory of the measured data depends on the degree of penetration that affects the upper and lower levels. For a level that is affected only by polarization, n_ℓ^* tends to decrease with increasing ζ, whereas for a penetrating orbital, n_ℓ^* tends to increase with increasing ζ. These tendencies can be understood by some simple classical considerations.

As one moves up in charge state along an isoelectronic sequence the core shrinks relative to the active electron, causing the penetration portion of δ to sharply diminish.

For polarization, δ first increases with increasing ζ, then passes through a maximum and ultimately decreases toward zero. This can be understood by examining the charge-scaling properties of the factors that make up the polarization contribution. The radial expectation value of the orbital electron scales quartically as the net charge of the core,

$$\langle r^{-4} \rangle \propto \zeta^4, \tag{6.87}$$

whereas the polarizability scales as the quartic reciprocal of a larger effective charge within the core

$$\alpha_d \propto 1/(\zeta + p)^4, \tag{6.88}$$

(where p is an empirical parameter introduced by Edlén [87] to account for the "screening defect" or penetration). So the polarization quantum defect scales as

$$\delta \propto \alpha_d \langle r^{-4} \rangle / R\zeta^2 \propto \zeta^2/(\zeta + p)^4. \tag{6.89}$$

Forcing the derivative with respect to ζ to vanish

$$2/\zeta - 4/(\zeta + p) = 0 \tag{6.90}$$

demonstrates that isoelectronically the polarization quantum defect initially increases, then reaches a maximum at $\zeta = p$, and ultimately decreases as ζ^{-2}. This can be phenomeno-logically stated by noting that at low ζ differential screening causes the core polariz-abilities to stiffen much faster with ζ than the valence electron draws in, whereas for large ζ the differential screening becomes negligible relative to the gross energy scaling factor ζ^2.

The relativistic energy is proportional to ζ^4; hence, when scaled to the gross energy, the corresponding portion of δ increases monotonically as ζ^2. However, its relative magnitude is small for low and moderate values of ζ.

As the quantum defect of a level increases or decreases along an isoelectronic sequence, this will cause its radial wave function to shift outward or inward relative to the core. Consider a case, for example, in which the upper level in a transition is affected only by polarization, and the lower level has significant core penetration. As ζ initially increases, the wave function of the upper level will move outward and that of the lower level will move inward. For overlap integrals that favor the external region (such as the dipole transition element) the relative differences in radial shifts of the nodes of the upper and lower wave functions causes these regular oscillations, and hence regular cancellations, in the integrand.

An application of this approach to the d–f transitions of the Cu isoelectronic sequence is shown in Fig. 6.4. A locus is drawn in the space of the effective quantum numbers of the upper and lower state for each transition, with the ions (from left to right) Cu I through Mo XIV indicated by division markers. The downward slope of the locus signifies that the d-state quantum defects arise mainly from penetration effects (since they are large and decrease

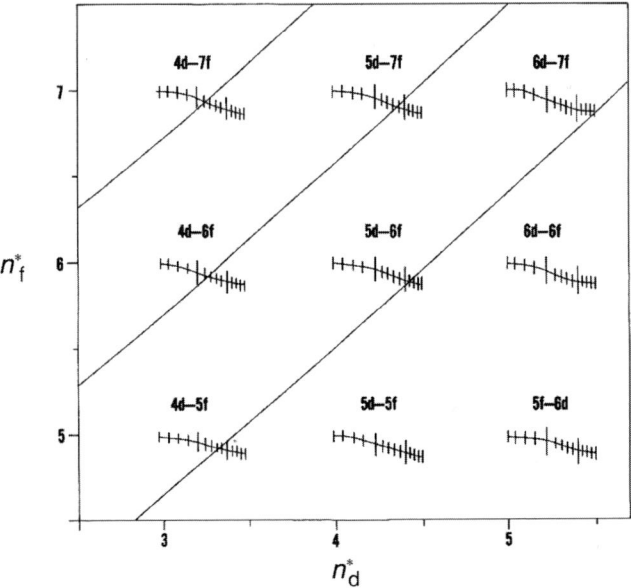

Fig. 6.4. Cancellation plot of the nd–$n'f$ transitions in the Cu isoelectronic sequence. (From Ref. [39].)

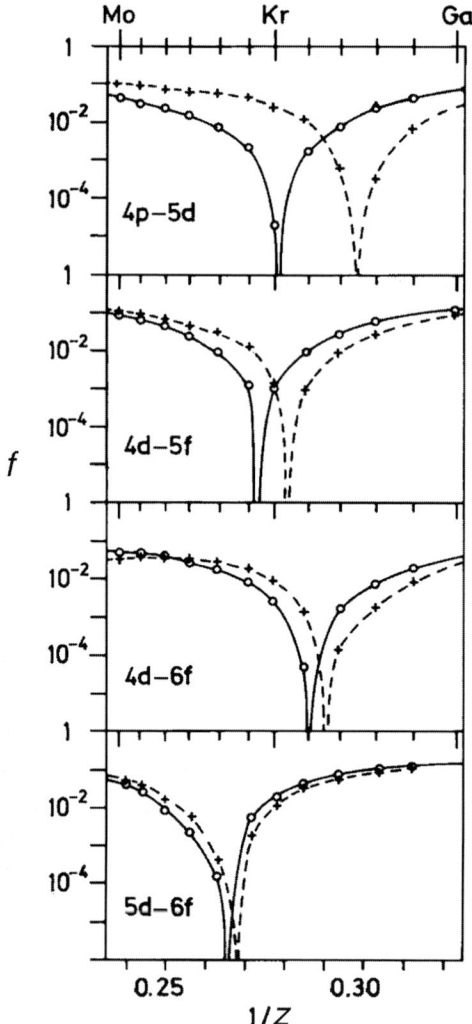

Fig. 6.5. Effect of core polarization on isoelectronic ions in the Cu sequence for which cancellation is most severe. (After Ref. [39].)

with ζ) whereas the f-state quantum defects arise mainly from polarization effects (which have not yet reached their maximum value with increasing ζ). The solid lines designate the nodes of sign change in the dipole transition integral, which correspond to cancellations in the transition probability.

Notice that for the ions of low charge state near the neutral atom limit of the sequence these integrals have the opposite sign from the hydrogenic (high-ζ) case, indicating that at low ζ there are nodes in the radial wave function that are concealed within the core. (Since hydrogenic radial wave functions have $n-\ell-1$ nodes, and the quantum defect reduces the effective principal quantum number, some nodes must collapse into the nonhydrogenic region inside the core cutoff radius.) Intersections on this plot indicate that intensity anomalies should be exhibited for 4d–5f transitions near Kr VIII, for 4d–6f transitions near Br VII and for 5d–6f transitions near Y XI.

Some of these indications have been verified by observations in a program of "disappearance spectroscopy" in which missing lines are studied [39]. By a comparative examination of relative intensities along either an isoelectronic sequence or a Rydberg series or within a fine structure multiplet, the position of exact cancellation can be experimentally pinpointed. This knowledge can be used to probe small effects (core polarization, configuration interaction, higher moments, relativistic effects, etc.), normally masked by the dominant interaction.

While the approximate position of these cancellation effects can be located in effective quantum number space by use of the Coulomb approximation, the exact positions of the cancellation nodes are very sensitive to the details of the calculation and the perturbations on the system. Thus it is very difficult to predict the exact ion in the isoelectronic sequence, or the exact fine structure level in a term, which will be most severely affected. In the vicinity of a cancellation, small perturbations can shift the exact position of the node in effective quantum number space, and cause substantial changes in the predicted oscillator strengths for the physical ions.

As a qualitative example of this type of shift, estimates have been made of the effect of the inclusion of the contribution of core polarization to the transition moment (as described in the previous section), and the results are displayed in Fig. 6.5. These studies include the 4p–5d, 4d–5f, 4d–6f and 5d–6f transitions for the Cu sequence (the cancellations in the latter three are indicated in the exposition of Fig. 6.4), with computed values shown for Se VI, Br VII, Kr VIII, and Rb IX. Basic Hartree–Fock calculations were made with the option of including core polarization corrections in the transition moment. The solid lines denote calculations with α_d set equal to zero, and the dashed lines were made with values for α_d taken from Ref. [38]. It can be seen that a shift in the position of the cancellation node by a fraction of a charge state causes changes of several orders of magnitude in the oscillator strengths.

Notice that these shifts in the positions of the cancellation nodes due to inclusion of core polarization effects in the transition moment require either that penetration be included in formulation of the moment (by the use of a cutoff function such as in Eq. 6.78), or in the wave functions (by a multielectron calculation such as the Hartree–Fock method). In the ideally nonpenetrating Coulomb approximation, the length form and the acceleration form of the transition moment would yield the same result, so the contributions from the orbital electron and the core could be factored out as in Eq. 6.84, and would undergo congruent cancellations.

7

Line strengths in two-valence-electron systems

All that is not forbidden is obligatory.

7.1 Relativistic E1 transitions

In a complex atom or ion, the only rigorous constraints that are imposed on radiative transitions between levels are those of conservation of energy, conservation of angular momentum, and conservation of parity. For electric dipole transitions, conservation of parity leads [141] to "Laporte's rule," which states that the parity of the atom must change because the E1 photon carries away one unit of parity. For a single out-of-shell electron, the parity is given (nonrelativistically) by $(-1)^{\ell}$ and the angular momentum is given by $j = \ell \pm \frac{1}{2}$. Thus it is not possible for two different levels with the same parity to also have the same total angular momentum. For systems with multiple out-of-shell electrons it is possible for two levels with the same parity to have the same total angular momentum, and the eigenvectors of these levels can (and in real cases always do) contain an admixture of other LS quantum numbers. In the simplest LS formulation (nonrelativistic E1), this mixing is neglected, and the spectrum consists of levels of noninteracting multiplicities (singlets and triplets for two valence electrons, doublets and quartets for three-valence-electron systems, etc.). If the exact LS-coupling assumption is relaxed, the individual multiplicity amplitudes in the admixtures lead to E1-allowed "intersystem" or "intercombination" (relativistic E1) transitions between the levels despite their nominal LS labels.

7.1.1 Selection rules

The fact that an E1 photon carries away one unit of angular moment and one unit of parity imposes the selection rules on the atom $\Delta J = 0, \pm 1$ (no $0 \rightarrow 0$), $\Delta M_J = 0, \pm 1$ (no $0 \rightarrow 0$), with a parity change. For a one-electron transition there is a selection rule $\Delta \ell = \pm 1$, whereas Δn is unrestricted. Since the nonrelativistic electrostatic Schrödinger picture separates space from spin, the unperturbed system is in LS coupling, which leads to the conditions $\Delta L = 0, \pm 1$ (no $0 \rightarrow 0$) and $\Delta S = 0$.

In the relativistic Dirac picture, the electron spin enters not as a magnetic correction, but as a dynamical consequence of the four-dimensional space–time formulation. The wave functions are eigenvectors of $\mathbf{J}$ but not of $\mathbf{L}$ or $\mathbf{S}$, and are therefore admixtures of the nonrelativistic LS eigenstates. Since these states have mixed multiplicity, there are intercombination transitions between the intermediate coupling amplitudes that break the $\Delta S = 0$ selection rule.

7.1.2 LS singlet–triplet mixing angles

In the configurations sp and p^2 there are at most two levels of differing multiplicity (one singlet and one triplet) with the same J, and these are coupled by the spin–orbit interaction. Since there are only two amplitudes, unitarity permits them to be characterized by a single "mixing angle." The determination of the intermediate coupling eigenfunctions and eigenenergies is accomplished through the diagonalization of a 2×2 matrix of the form

$$\mathsf{M} = \begin{pmatrix} a & c \\ c & b \end{pmatrix} \tag{7.1}$$

with a transformation

$$\mathsf{T} = \begin{pmatrix} \cos\theta & \sin\theta \\ -\sin\theta & \cos\theta \end{pmatrix}. \tag{7.2}$$

The elements of the basis transformation $\mathsf{T}^{-1}\mathsf{M}\mathsf{T}$ are

$$\begin{aligned}
\mathsf{M}'_{11} &= a\cos^2\theta + b\sin^2\theta - 2c\sin\theta\cos\theta \\
\mathsf{M}'_{12} &= (a-b)\sin\theta\cos\theta + c(\cos^2\theta - \sin^2\theta) \\
\mathsf{M}'_{21} &= (a-b)\sin\theta\cos\theta - c(\cos^2\theta - \sin^2\theta) \\
\mathsf{M}'_{22} &= a\sin^2\theta + b\cos^2\theta + 2c\sin\theta\cos\theta.
\end{aligned} \tag{7.3}$$

Using the double-angle reduction formulae $\cos 2\theta = \cos^2\theta - \sin^2\theta$ and $\sin 2\theta = 2\sin\theta\cos\theta$, these can be written

$$\mathsf{M}'_{11}, \mathsf{M}'_{22} = \frac{a+b}{2} \pm \left[\frac{a-b}{2}\cos 2\theta - c\sin 2\theta\right]$$

$$\mathsf{M}'_{12}, \mathsf{M}'_{21} = \frac{a-b}{2}\sin 2\theta \pm c\cos 2\theta. \tag{7.4}$$

To diagonalize the matrix, we force $\mathsf{M}'_{12} = \mathsf{M}'_{21} = 0$, which yields

$$\cot 2\theta = \pm\frac{b-a}{2c} \tag{7.5}$$

and

$$\mathsf{M}'_{11}, \mathsf{M}'_{22} = \frac{a+b}{2} \mp \sqrt{\left(\frac{a-b}{2}\right)^2 + c^2}. \tag{7.6}$$

This will be applied below to the sp and p^2 systems individually.

7.1.3 *LS* singlet–triplet mixing of the sp configuration

The development in Section 5.2.5 and Table 5.2 showed that for the $^3P_1$ and $^1P_1$ levels for an $nsnp$ configuration the 2×2 matrix has the elements

$$a = E_0 - G_1 - \zeta_p/2$$
$$b = E_0 + G_1$$
$$c = \zeta_p'/\sqrt{2}. \tag{7.7}$$

Here we have allowed the diagonal and off-diagonal values of ζ_p and ζ_p' to vary freely. This is to include additional interactions in the formulation, and also to remove the overdetermination, thus insuring that the fitted parameters specify the measured energy separations exactly. In terms of these quantities, the mixing angle is given by

$$\cot(2\theta_1) = \pm\frac{2G_1 + \zeta_p/2}{\sqrt{2}\zeta_p'}. \tag{7.8}$$

This yields the eigenvectors

$$\left|^3P_0^{o'}\right\rangle = \left|^3P_0^o\right\rangle$$
$$\left|^3P_1^{o'}\right\rangle = \cos\theta_1\left|^3P_1^o\right\rangle - \sin\theta_1\left|^1P_1^o\right\rangle$$
$$\left|^3P_2^{o'}\right\rangle = \left|^3P_2^o\right\rangle$$
$$\left|^1P_1^{o'}\right\rangle = \sin\theta_1\left|^3P_1^o\right\rangle + \cos\theta_1\left|^1P_1^o\right\rangle, \tag{7.9}$$

where the primes indicate that the LS symbols are only nominal, and describe states in intermediate coupling. In terms of these eigenvectors, the transition elements to a ground term $ns^2\,^1S_0$ are given by

$$\left\langle^1S_0|\mathbf{r}|^3P_1^{o'}\right\rangle = \sin\theta_1\left\langle^1S_0|\mathbf{r}|^1P_1^o\right\rangle$$
$$\left\langle^1S_0|\mathbf{r}|^1P_1^{o'}\right\rangle = \cos\theta_1\left\langle^1S_0|\mathbf{r}|^1P_1^o\right\rangle, \tag{7.10}$$

and the line-strength factors for the resonance (singlet–singlet) and intercombination (singlet–triplet) transitions differ only by factors of the singlet–triplet mixing angles.

In this case both the singlet and triplet $J = 1$ levels are threefold degenerate, hence the line strengths can be deduced from the measured lifetimes using the relationship

$$S_{u\ell} = \left[\frac{\lambda_{u\ell}(\text{Å})}{1265.38}\right]^3\frac{3}{\tau_u(ns)}. \tag{7.11}$$

If we define the measured values for the line strengths of the resonance and intercombination lines

$$S(\text{Res}) \equiv \sum_{M_J M_J'}\left|\left\langle^1S_0|\mathbf{r}|^1P_1^{o'}\right\rangle\right|^2$$
$$S(\text{Int}) \equiv \sum_{M_J M_J'}\left|\left\langle^1S_0|\mathbf{r}|^3P_1^{o'}\right\rangle\right|^2, \tag{7.12}$$

then both of these quantities can be expressed in terms of an effective reduced value for the line-strength factor

$$S_r \equiv \sum_{M_J M_J'} |\langle {}^1S_0|\mathbf{r}|{}^1P_1^o \rangle|^2. \tag{7.13}$$

A separate evaluation of this quantity can be obtained from each of the two transitions

$$S_r(\text{Res}) = S(\text{Res})/\cos^2\theta_1$$
$$S_r(\text{Int}) = S(\text{Int})/\sin^2\theta_1. \tag{7.14}$$

When the reduced line-strength factor is deduced from the measured lifetime data in this manner, it has been found that the isoelectronic databases for many isoelectronic sequences can be systematized by linear representations. The mixing-angle data can often be accurately represented by

$$\cot(2\theta_1) \cong \cot(2\Theta_1) + \frac{C_1}{(Z - C_2)^p} \tag{7.15}$$

where Θ_1 is the value of the LS-coupling mixing angle in the limit of pure jj coupling, and C_1, C_2 and p are empirical fitting constants.

Similarly, consistent with the results described in Section 3.7.1 for single-valence-electron systems, the reduced line-strength factors can often be accurately represented by

$$Z^2 S_r \cong S_H + \frac{C_3}{(Z - C_4)}. \tag{7.16}$$

Again here S_H is the hydrogenic value (obtained from Eq. 3.82, with an additional factor of two to account for the equivalent electrons in the ns^2 ground configuration), and C_3 and C_4 are empirical fitting constants. Small differences between the values for $S_r(\text{Res})$ and $S_r(\text{Int})$ serve to incorporate empirical deviations from the single-configuration nonrelativistic model into the characterization, and thus enhance the reliability of this predictive model.

Application: the Cd sequence

An application of this approach to the Cd isoelectronic sequence is shown in Fig. 7.1 [73].

The values of $\cot(2\theta_1)$ deduced from the base of measured data are represented by circles in the top panel of Fig. 7.1, and plotted vs $1/(Z - 46.47)$. In this case the power-law fit yields $p = 1$, indicating a linear dependence on the reciprocal screened charge. In this exposition it can be seen that the data form a straight line that approaches a high-Z limit (denoted by a solid diamond) of $\cot(2\Theta_1) = \sqrt{2}/4$. This corresponds to the jj limit $\cos^2\Theta_1 = \frac{1}{3}$, $\sin^2\Theta_1 = \frac{2}{3}$.

These mixing angles were used together with the wavelength factors to convert the measured lifetime data to reduced line-strength factors that are plotted vs $1/(Z - 45.73)$ in the bottom panel of Fig. 7.1. This exposition shows that each of the quantities $Z^2 S_r(\text{Res})$ and $Z^2 S_r(\text{Int})$ exhibits a linear variation that approaches the hydrogenic limit $S_H = 2700$ at high Z.

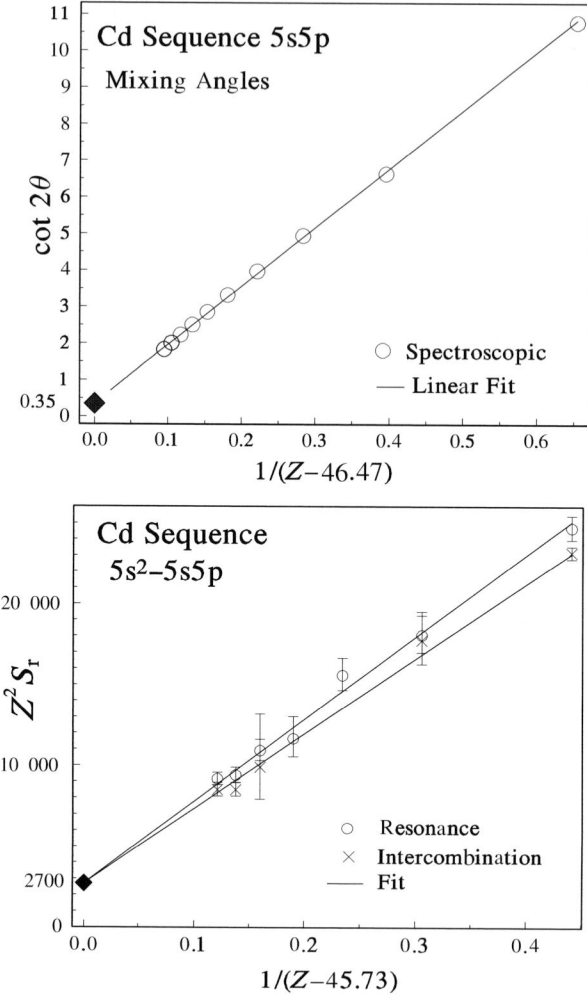

Fig. 7.1. Isoelectronic plots of the mixing angles and line-strength factors for the $5s^2$–$5s5p$ transitions in the Cd isoelectronic sequence. (After Ref. [73].)

Fits to the measured data yield the explicit relationships

$$\cot(2\theta_1) = 0.298 + 16.628/(Z - 46.47)$$
$$Z^2 S_r(\text{Res}) = 2700[1 + 18.78/(Z - 45.73)]\cos^2\theta_1$$
$$Z^2 S_r(\text{Int}) = 2700[1 + 17.14/(Z - 45.73)]\sin^2\theta_1 \qquad (7.17)$$

which predict the line-strength factors for any ion in the Cd isoelectronic sequence to within the experimental accuracies of the existing database. Moreover, as additional data become available, these fitting constants can be sharpened to improve the predictive power.

7.1.4 Application in $\Delta n = 1$ transitions

Many experimental studies exist for the intrashell line strengths of the $ns^2\ ^1S_0$–$nsnp\ ^{1,3}P_1$ resonance and intercombination lines in the spectra of ions of the alkaline-earthlike isoelectronic sequences. For the $\Delta n = 0$ transitions in these two-valence-electron systems, experimental determination of line strengths can be made directly from lifetime measurements (since $nsnp$ is the first excited configuration, and there is only the single decay branch to the ground state). This is in contrast to the situation for the intershell transitions of the form $ns^2\ ^1S_0$–$nsn'p\ ^{1,3}P_1$ with $n' \neq n$, for which other decay branches (e.g., to $nsn's$ and np^2 levels) compete. Thus, although a base of experimental lifetime data exists for these transitions, it is not possible to convert these data to oscillator strengths and line strengths, since virtually no branching ratio information is available for multiply charged ions. However, examples do exist for which unbranched $\Delta n = 1$ transitions occur, as do cases in which a fine structure dependence on the branching permits specification of a branching fraction.

As one example, consider the $2p^6\ ^1S_0$–$2p^5 3s\ ^{1,3}P_1$ transitions in the Ne sequence. Here the first excited state is extrashell to the ground state, giving rise to unbranched $\Delta n = 1$ resonance and intercombination transitions. Thus the measured lifetimes of these levels can be used directly in the mixing-angle reduction approach [70].

The top panel in Fig. 7.2 provides an exposition of the linearity of the isoelectronic variation of the quantity $\cot(2\theta_1)$ when plotted vs a screened charge. (Here the effective screening is negative, with the effective central charge exceeding the number of core electrons, sometime referred to as "exchange core polarization.") Excluding neutral neon, the linear trend of the points deduced from measured spectroscopic energy-level data approaches the jj limit of $-2\sqrt{2}$ at high Z. Notice that, unlike the case of the $nsnp$ levels, for the $np^5(n+1)s$ levels the quantity $\cot(2\theta_1)$ passes through zero in achieving the jj limit. Thus the singlet and triplet undergo an avoided crossing in their energies and a true crossing in their eigenvectors, which must be taken into account when the low-Z and high-Z characteristics are compared.

The bottom panel in Fig. 7.2 shows the isoelectronic linearity of the line-strength factors of the resonance and intercombination transitions, and their approach to the hydrogenic value [70]

$$Z^2 S_{2p^6, 2p^5 3s} = 3Z^2 S_{2s,3p} = 2^{16} 3^8 / 5^{12} = 1.761\ 205 \ldots \tag{7.18}$$

for high Z.

The intershell nature of these $\Delta n = 1$ inert-gaslike transitions causes them to have substantially shorter wavelengths (and hence shorter lifetimes) than those that characterize the corresponding unbranched $\Delta n = 0$ ns^2–$nsnp$ transitions in alkaline-earthlike systems. These shorter lifetimes severely limit the members of the sequence that are accessible to measurement by time-of-flight methods (but are still too long-lived to be accessible to lifetime determination from natural line width). This strongly motivates the use of these data-based extrapolation methods for their prediction.

These transitions are particularly interesting because it is possible to create a population inversion between the $2p^5 3s$ and $2p^5 3p$ levels that can produce extreme Ultraviolet XUV laser amplification. Modeling studies have demonstrated that amplification is optimal near

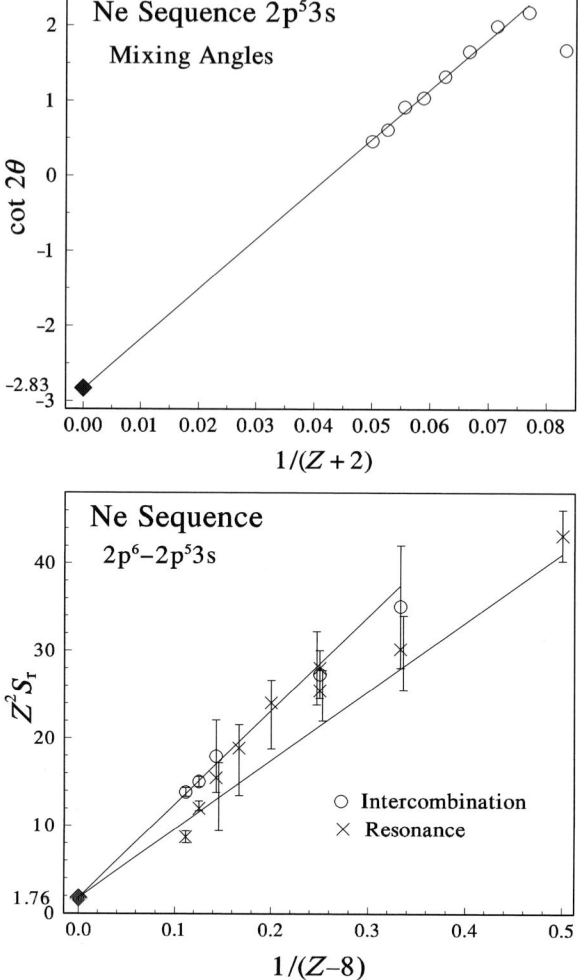

Fig. 7.2. Isoelectronic plots for the mixing angles and reduced line-strength factors for the $2p^6$–$2p^5$3s transitions in the Ne sequence. (After Ref. [70].)

$Z = 36$, leading to tests at $Z = 34$ and $Z = 39$ in Se and Y plasmas (cf. Ref. [70]). Although the lifetimes of these levels are too short for time-of-flight measurements in the vicinity of $Z = 36$, the mixing-angle formulation and the high-Z asymptote permit interpolation of trends if precision results at lower Z are available.

Another interesting example concerns the $2s^2\ ^1S_0$–$2s3p\ ^{1,3}P_1$ transitions in the Be isoelectronic sequence [70]. Here, differential lifetime measurements in the $2s3s\ ^3S_1$–$2s3p\ ^3P_{2,1,0}$ decay channels reveal the branch to ground as an extra channel available only to the $^3P_1$ level. Moreover, the $^1P_1$ level decays dominantly to the ground state, and other branching can be estimated theoretically without significant error.

The intercombination transition probability data are available for this particular system because of a specialized experimental technique that studies the fine structure variation of the lifetimes of the ^3P_J levels for $J = 0, 1, 2$, as measured in the allowed branch 2s3s $^3S_1$–2s3p ^3P_J. The $J = 0, 2$ lifetimes are essentially equal, but the $J = 1$ lifetime is shortened by the additional relativistic E1 channel to the ground state that is available to it (but ruled out for $J = 0, 2$ by angular momentum considerations). Thus by the use of differential lifetime measurements, the 2s^2 $^1S_0$–2s3p $^3P_1$ transition probability is obtained. A description of the level system, the experimental procedures, and the results will be presented in Section 10.3.1.

For the 2s3p $^1P_1$ level, the decay is branched among the 2s^2 $^1S_0$, 2s3s $^1S_0$, 2p^2 $^1D_2$ and 2p^2 $^1S_0$ levels. However, for multiply charged ions the total branching is small, and the use of theoretical estimates is possible. Available theoretical estimates indicate that the branching fraction of the ground state transition 2s^2 $^1S_0$–2s3p $^1P_1$ is 56% for Be I, 69% for B II, and smoothly varies from 87% in C III to 96% in Fe XXIII. Thus the uncertainty introduced by the use of the theoretical branching fractions is probably smaller than the uncertainties in the lifetime measurements.

A plot of the reduced line-strength factors for the 2s^2 $^1S_0$–2s2p $^{1,3}P_1$ transitions in this sequence is given in Fig. 7.3 Here again the high-Z trend is toward the hydrogenic limit [70]

$$Z^2 S_{2s^2,2s3p} = 3Z^2 S_{2p,3s} = 2^{21} 3^7 / 5^{12} = 18.786\,18\,\ldots \tag{7.19}$$

for high Z.

7.1.5 Extension to sp–sd transitions

For the s^2–sp transitions described in the previous section, the lower configuration was considered to be a ground-state singlet. Thus, there was no branching either within the multiplet or to other configurations. If this analysis is extended along the chain of transitions

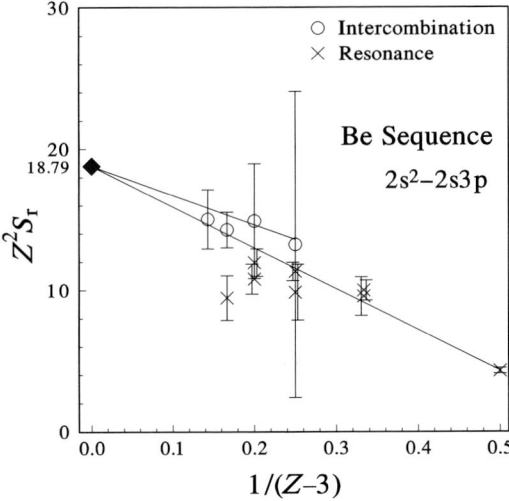

Fig. 7.3. Isoelectronic plot of the reduced line-strength factors for the 2s^2–2s3p transitions in the Be sequence. (After Ref. [70].)

to the sp–sd multiplet there is still no branching to other configurations, but there is interconfigurational branching within the multiplet. If intermediate coupling is present, there is
also branching in the supermultiplets of the singlet and triplet polyads. In the Schrödinger
picture the radial matrix element is an overall constant for the supermultiplet, so a mixing-
angle formulation can be used to determine these interconfigurational branching fractions.
If lifetime measurements are available, these branching fractions can be used to specify the
transition probability rates and the oscillator strengths.

The mixing angle formulation of an sp level can be extended to describe an $s\ell$ level by
noting that the matrix elements to be diagonalized are given by

$$
\begin{aligned}
a &= E_0 - G_\ell - \zeta_\ell/2 \\
b &= E_0 + G_\ell \\
c &= \sqrt{\ell(\ell+1)}\zeta'_\ell/2.
\end{aligned}
\tag{7.20}
$$

For an nd level $\ell = 2$ and mixing between the $^1D_2$ and $^3D_2$ levels is prescribed by

$$
\cot(2\theta_2) = \frac{2G_2 + \zeta_d/2}{\sqrt{6}\zeta'_d}.
\tag{7.21}
$$

The physical levels are specified by

$$
\begin{aligned}
^3D_3 &= E_0 - G_2 + \zeta_d \\
^1D'_2,\ ^3D'_2 &= E_0 - \zeta_d/4 \pm \sqrt{(G_2 + \zeta_d/4)^2 + 3(\zeta'_d)^2/2} \\
^3D_1 &= E_0 - G_2 - 3\zeta_d/2.
\end{aligned}
\tag{7.22}
$$

The Slater and spin–order parameters, G_2 and ζ_d respectively, can be obtained as a one-to-
one mapping of the measured intervals between these energy levels using

$$
\begin{aligned}
G_2 &= \left(^1D'_2 + {}^3D'_2 - {}^3D_3 - {}^3D_1\right)/2 \\
\zeta_d &= 2\left(^3D_3 - {}^3D_1\right)/5 \\
\zeta'_d &= \left[\left(^1D'_2 - {}^3D'_2\right)^2 - (2G_2 + \zeta_d/2)^2\right]^{1/2}/\sqrt{6}.
\end{aligned}
\tag{7.23}
$$

Here the formalism will be applied to supermultiplets connecting sp and sd configurations.
The eigenvectors of the mixed sd terms are

$$
\begin{aligned}
\left|^3D'_1\right\rangle &= \left|^3D_1\right\rangle \\
\left|^3D'_2\right\rangle &= \cos\theta_2\left|^3D_2\right\rangle - \sin\theta_2\left|^1D_2\right\rangle \\
\left|^3D'_3\right\rangle &= \left|^3D_3\right\rangle \\
\left|^1D'_2\right\rangle &= \sin\theta_2\left|^3D_2\right\rangle + \cos\theta_2\left|^1D_2\right\rangle,
\end{aligned}
\tag{7.24}
$$

and those of the mixed sp terms were given in Eqs. 7.9. The transition moments among these
wave functions can be evaluated in terms of the mixing angles and the transition moments

in pure spin–orbit coupling, which are given [30] by

$$\left\langle^{2S+1}L_J|r|^{2S+1}L'_{J'}\right\rangle = (-)^{L+S+J'+1}\sqrt{(2J+1)(2J'+1)}\begin{Bmatrix} L & S & J \\ J' & 1 & L' \end{Bmatrix}\mathcal{M}, \quad (7.25)$$

where $\mathcal{M}$ is the interconfigurational radial matrix element. In the nonrelativistic Schrödinger approximation $\mathcal{M}$ is the same for all members of the supermultiplet. In calculations of emission branching fractions $\mathcal{M}$ is often assumed to be the same for all transitions from the same upper level, whereas in calculations of absorption branching fractions it is similarly assumed to be the same for all transitions from the same lower level. In cases in which the transition moment is not strongly influenced by cancellation effects or CI (configuration interaction), its variation over the multiplet is often negligible.

The transition moments can be formed for this system using $L = 1$, $L' = 2$, and $\mathcal{M} = \langle\text{sp}|r|\text{sd}\rangle$. These equations yield, for the upper level $^3D_1$

$$\left\langle^3P_0^o|\mathbf{r}|^3D_1\right\rangle = \frac{1}{\sqrt{3}}\langle\text{sp}|r|\text{sd}\rangle$$

$$\left\langle^3P_1^{o'}|\mathbf{r}|^3D_1\right\rangle = \frac{1}{2}\cos\theta_1\langle\text{sp}|r|\text{sd}\rangle$$

$$\left\langle^3P_2^o|\mathbf{r}|^3D_1\right\rangle = \frac{1}{\sqrt{60}}\langle\text{sp}|r|\text{sd}\rangle$$

$$\left\langle^1P_1^{o'}|\mathbf{r}|^3D_1\right\rangle = \frac{1}{2}\sin\theta_1\langle\text{sp}|r|\text{sd}\rangle, \quad (7.26)$$

for the upper level $^3D_2'$

$$\left\langle^3P_1^{o'}|\mathbf{r}|^3D_2'\right\rangle = \left[\frac{\sqrt{3}}{2}\cos\theta_1\cos\theta_2 + \sin\theta_1\sin\theta_2\right]\langle\text{sp}|r|\text{sd}\rangle$$

$$\left\langle^3P_2^o|\mathbf{r}|^3D_2'\right\rangle = \frac{1}{2}\cos\theta_2\langle\text{sp}|r|\text{sd}\rangle$$

$$\left\langle^1P_1^{o'}|\mathbf{r}|^3D_2'\right\rangle = \left[\frac{\sqrt{3}}{2}\sin\theta_1\cos\theta_2 - \cos\theta_1\sin\theta_2\right]\langle\text{sp}|r|\text{sd}\rangle, \quad (7.27)$$

for the upper level $^3D_3$

$$\left\langle^3P_2^o|\mathbf{r}|^3D_3\right\rangle = \sqrt{\frac{7}{5}}\langle\text{sp}|r|\text{sd}\rangle, \quad (7.28)$$

and for the upper level $^1D_2'$

$$\left\langle^3P_1^{o'}|\mathbf{r}|^1D_2'\right\rangle = \left[\frac{\sqrt{3}}{2}\cos\theta_1\sin\theta_2 - \sin\theta_1\cos\theta_2\right]\langle\text{sp}|r|\text{sd}\rangle$$

$$\left\langle^3P_2^o|\mathbf{r}|^1D_2'\right\rangle = \frac{1}{2}\sin\theta_2\langle\text{sp}|r|\text{sd}\rangle$$

$$\left\langle^1P_1^{o'}|\mathbf{r}|^1D_2'\right\rangle = \left[\frac{\sqrt{3}}{2}\sin\theta_1\sin\theta_2 + \cos\theta_1\cos\theta_2\right]\langle\text{sp}|r|\text{sd}\rangle. \quad (7.29)$$

Table 7.1. Fitted parameters obtained from energy-level data.

$nsnp$	Ga II	In II	$nsnd$	Ga II	In II
G_1	11 999.00	9140.41	G_2	6165.28	5831.42
ζ_p	921.49	2368.00	ζ_d	26.91	87.84
ζ_p'	806.07	2075.20	ζ_d'	72.46	119.84
θ_1	1.427	4.287	θ_2	0.412	0.718

The transition probabilities can then be written as

$$g_u A_{ul}(\text{ns}^{-1}) = \left[\frac{1265.38}{\lambda_{u\ell}(\text{Å})}\right]^3 |\langle u|r|l\rangle|^2 . \qquad (7.30)$$

The relative transition probabilities for the supermultiplet fractions R_{ul} can be defined

$$R_{ul} \equiv A_{ul} \Big/ \sum_{u'l'} A_{u'l'} . \qquad (7.31)$$

Under the assumption that $\langle np|r|nd\rangle$ is the same for all transitions in the supermultiplet it cancels in the ratio of Eq. 7.31. In this approximation it is possible to predict from R_{ul} both the ratios of the lifetimes of the various upper levels and the branching fractions for the various decay branches of each upper level. If, for a given upper level u, the sum C_u over all lower levels is formed, this quantity should be proportional to the reciprocal lifetime

$$C_u = \sum_\ell R_{ul} \propto 1/\tau_u . \qquad (7.32)$$

By comparing the triplet-to-triplet and singlet-to-triplet values of $C_u\tau_u$ it is possible to test the assumptions of this method and to make predictions where data are not available. The branching fractions can be obtained from $(BF)_{ul} \equiv R_{ul}/C_u$, and used to predict transition probabilities where lifetime measurements are available.

A study of transitions of this type has been carried out [54]. The results for the 4s4p–4s4d transitions in Ga II and the 5s5p–5s5d transitions in In II will be summarized below. The Slater parameters and mixing angles obtained are given in Table 7.1.

These mixing angles were used to compute branching fractions, assuming that $\langle sp|r|sd\rangle$ is a constant for each upper level, and used to compute transition probabilities by combining these branching fractions with measured level lifetimes, where such measurements exist. The results are given in Table 7.2.

The available values for measured lifetimes are listed in Table 7.3, and are used to predict other lifetimes in the systems.

7.1.6 *LS* singlet–triplet mixing of the p² configuration

The development in Section 5.2.6 and Table 5.3 showed that the matrix for the np^2 configuration contains two 2×2 submatrices with off-diagonal elements (as in Eq. 7.1), one for $J = 2$ and one for $J = 0$. For the $J = 2$ nominal $^3P_2$ and $^1D_2$ levels, the matrix elements

Table 7.2. Wavelengths (in air for $\lambda_{u\ell} \geq 2000$ Å), multiplet fractions $R_{u\ell}$ (in %), branching fractions $(BF)_{u\ell}$ (in %) and transition probability rates $A_{u\ell}$ (in ns^{-1}) for the two supermultiplets. The transition probability rate predictions are based on the branching fractions obtained by this formalism and the measured and predicted lifetimes given in Table 7.3. (After Ref. [54].)

Trans.	Ga II				In II			
	$\lambda_{u\ell}$(Å)	$R_{u\ell}$	$(BF)_{u\ell}$	$A_{u\ell}$	$\lambda_{u\ell}$(Å)	$R_{u\ell}$	$(BF)_{u\ell}$	$A_{u\ell}$
$^3P_0^o$–$^3D_1$	1505	15.94	56.13	0.854	1672	16.57	57.17	0.665
$^3P_1^o$–	1515	11.71	41.23	0.627	1703	11.70	40.38	0.470
$^3P_2^o$–	1537	0.75	2.63	0.040	1778	0.69	2.38	0.028
$^1P_1^o$–	2319	0.002	0.01	0.0001	2560	0.02	0.07	0.001
Sum		28.40	100	1.522		28.98	100	1.163
$^3P_1^o$–$^3D_2$	1515	21.11	75.79	1.131	1700	21.20	77.22	0.849
$^3P_2^o$–	1536	6.74	24.21	0.361	1775	6.23	22.70	0.249
$^1P_1^o$–	2317	0.002	0.01	$<10^{-4}$	2554	0.023	0.08	0.001
Sum		27.85	100	1.493		27.46	100	1.099
$^3P_2^o$–$^3D_3$	1535	27.03	100	1.448	1771	25.11	100	1.063
$^3P_1^o$–$^1D_2$	1276	0.16	0.02	0.0003	1418	0.20	1.08	0.014
$^3P_2^o$–	1291	$<10^{-3}$	$<10^{-3}$	10^{-5}	1469	0.002	0.01	0.0001
$^1P_1^o$–	1802	16.70	99.98	1.460	1967	18.25	98.91	1.285
Sum		16.72	100	1.460		18.45	100	1.299

Table 7.3. Lifetime studies of the 4s4d levels in Ga II and the 5s5p levels in In II. Experimental measurements (Exp) are combined with semiempirical branching-fraction values (SE) to predict the lifetimes of the $^3D_1$ and $^3D_3$ levels on the basis of measurements for the $^3D_2$ level. Values are in ns and experimental uncertainties are given in parentheses. (After Ref. [54].)

	Ga II		In II	
	Exp	SE	Exp	SE
$^3D_1$	–	0.66	0.86(3)	0.86
$^3D_2$	0.67(6)	–	0.91(3)	–
$^3D_3$	–	0.69	0.94(3)	1.00
$^1D_2$	0.67(4), 0.73(7)	–	0.77(3)	–

are

$$a = E_0 - 5F_2 + \zeta_{pp}/2$$
$$b = E_0 + F_2$$
$$c = \zeta'_{pp}/\sqrt{2}. \tag{7.33}$$

Again here we have allowed the diagonal and off-diagonal quantities ζ_{pp} and ζ'_{pp} to vary independently, both to include other interactions empirically, and to remove the overdetermination so that the parametrization reproduces the measured energy separations exactly. This mixing angle is given by

$$\cot(2\theta_2) = \pm \frac{12F_2 - \zeta_{pp}}{2\sqrt{2}\zeta'_{pp}}. \tag{7.34}$$

For the $J = 0$ nominal $^3P_0$ and $^1S_0$ levels, the matrix elements are

$$a = E_0 - 5F_2 - \zeta_{pp}$$
$$b = E_0 + 10F_2$$
$$c = -\sqrt{2}\zeta''_{pp}, \tag{7.35}$$

with the diagonal and off-diagonal quantities ζ_{pp} and ζ''_{pp} again allowed to vary independently as discussed above. This mixing angle is given by

$$\cot(2\theta_0) = \mp \frac{15F_2 + \zeta_{pp}}{2\sqrt{2}\zeta''_{pp}}. \tag{7.36}$$

The eigenvectors are given by

$$\left|^3P'_0\right\rangle = \cos\theta_0 \left|^3P_0\right\rangle - \sin\theta_0 \left|^1S_0\right\rangle$$
$$\left|^3P'_1\right\rangle = \left|^3P_0\right\rangle$$
$$\left|^3P'_2\right\rangle = \cos\theta_2 \left|^3P_2\right\rangle - \sin\theta_2 \left|^1D_2\right\rangle$$
$$\left|^1D'_2\right\rangle = \sin\theta_2 \left|^3P_2\right\rangle + \cos\theta_2 \left|^1D_2\right\rangle$$
$$\left|^1S'_0\right\rangle = \sin\theta_0 \left|^3P_0\right\rangle + \cos\theta_0 \left|^1S_0\right\rangle. \tag{7.37}$$

One very useful application of these results involves combining a ground p^2 configuration with an excited sp configuration to characterize the resulting transition array.

7.1.7 *LS* branching fractions for the ns^2np^2–$ns^2npn's$ transition array

In the ns^2–$nsnp$ transition array discussed in Section 7.1.3, the mixing angles deduced from measured energy-level data permitted the decay rates from two different upper levels to the same lower level to be characterized by similar effective line-strength factors, after being freed of factors arising from angular momentum considerations. An extension of this approach to a more complex example can be found in transitions of the form ns^2np^2–$ns^2npn's$, which occur in C-like, Si-like, Ge-like, Sn-like, and Pb-like atoms and ions. These homologous sequences possess many fascinating properties that can be characterized by an extension of the famous statement by Sir James Jeans [126] that "Life exists in the universe only because the carbon atom possesses certain exceptional properties." One need only replace the word "life" by the words "computer chips" and "transistors" to illustrate the unusual properties of the isoelectronic sequences of these polymer-forming elements.

These systems contain both of the types of configurations discussed in the previous sections, with the ground p^2 configuration containing the levels $^3P_0$, $^3P_1$, $^3P_2$, $^1D_2$, $^1S_0$, and the excited sp configuration containing the levels $^3P_0^o$, $^3P_1^o$, $^3P_2^o$, $^1P_1^o$. With the exception of the $^3P_0^o$ (for which the $J = 0 - 0$ transition is forbidden by conservation of angular momentum), all of the upper levels have multiply branched transitions to the lower configuration. Since, in the nonrelativistic Schrödinger picture, all of these transitions involve the same radial transition moment, the branching fractions can be specified from mixing-angle information alone, provided the Schrödinger picture and the single-configuration approximation are both valid.

The branching fractions can be specified from the effective mixing angles that have been extracted from measured energy-level data through the use of the LS-coupling values for the angular transition matrices given by Eq. 7.25 [30]. The nonvanishing values are

$$\langle ^3P_0^o|r|^3P_1\rangle = \langle ^1P_1^o|r|^1S_0\rangle = -\langle ^3P_1^o|r|^3P_0\rangle = -\sqrt{20}$$

$$2\langle ^3P_2^o|r|^3P_1\rangle = -2\langle ^3P_1^o|r|^3P_2\rangle = \langle ^1P_1^o|r|^1D_2\rangle = 10$$

$$\sqrt{5}\langle ^3P_1^o|r|^3P_1\rangle = \langle ^3P_2^o|r|^3P_2\rangle = \sqrt{75}. \tag{7.38}$$

These equations yield, for the upper level $^3P_1^{o\prime}$

$$\langle ^3P_0{}'|\mathbf{r}|^3P_1^{o\prime}\rangle = -\sqrt{20}\cos(\theta_1 + \theta_0)\,\langle p^2|r|sp\rangle$$

$$\langle ^3P_1{}'|\mathbf{r}|^3P_1^{o\prime}\rangle = \sqrt{15}\cos\theta_1\,\langle p^2|r|sp\rangle$$

$$\langle ^3P_2{}'|\mathbf{r}|^3P_1^{o\prime}\rangle = 5(2\sin\theta_1\sin\theta_2 + \cos\theta_1\cos\theta_2)\,\langle p^2|r|sp\rangle$$

$$\langle ^1D_2{}'|\mathbf{r}|^3P_1^{o\prime}\rangle = -5(2\sin\theta_1\cos\theta_2 - \cos\theta_1\sin\theta_2)\,\langle p^2|r|sp\rangle$$

$$\langle ^1S_0{}'|\mathbf{r}|^3P_1^{o\prime}\rangle = -\sqrt{20}\sin(\theta_1 + \theta_0)\,\langle p^2|r|sp\rangle, \tag{7.39}$$

for the upper level $^3P_2^{o\prime}$

$$\langle ^3P_1{}'|\mathbf{r}|^3P_2^{o\prime}\rangle = 5\,\langle p^2|r|sp\rangle$$

$$\langle ^3P_2{}'|\mathbf{r}|^3P_2^{o\prime}\rangle = \sqrt{15}\cos\theta_2\,\langle p^2|r|sp\rangle$$

$$\langle ^1D_2{}'|\mathbf{r}|^3P_2^{o\prime}\rangle = \sqrt{15}\sin\theta_2\,\langle p^2|r|sp\rangle, \tag{7.40}$$

and for the upper level $^1P_1^{o\prime}$

$$\langle ^3P_0{}'|\mathbf{r}|^1P_1^{o\prime}\rangle = -\sqrt{20}\sin(\theta_1 + \theta_0)\,\langle p^2|r|sp\rangle$$

$$\langle ^3P_1{}'|\mathbf{r}|^1P_1^{o\prime}\rangle = \sqrt{15}\sin\theta_1\,\langle p^2|r|sp\rangle$$

$$\langle ^3P_2{}'|\mathbf{r}|^1P_1^{o\prime}\rangle = 5(2\cos\theta_1\sin\theta_2 - \sin\theta_1\cos\theta_2)\,\langle p^2|r|sp\rangle$$

$$\langle ^1D_2{}'|\mathbf{r}|^1P_1^{o\prime}\rangle = -5(2\cos\theta_1\cos\theta_2 + \sin\theta_1\sin\theta_2)\,\langle p^2|r|sp\rangle$$

$$\langle ^1S_0{}'|\mathbf{r}|^1P_1^{o\prime}\rangle = -\sqrt{20}\cos(\theta_1 + \theta_0)\,\langle p^2|r|sp\rangle. \tag{7.41}$$

Table 7.4. Comparison of semiempirical (S) and measured (M) branching fractions (BF, in %) for ns^2np^2–$ns^2npn's$ transitions in neutral atoms. (After Ref. [52].)

Transition	Si I		Ge I		Sn I	
	BF(S)	(M)	BF(S)	(M)	BF(S)	(M)
$^3P'_0 \leftarrow {}^3P''_1$	33.3	33.3(3)	31.2	32.5(16)	32.3	27
$^3P_1 \leftarrow$	24.7	24.7(4)	21.2	22.1(11)	17.5	17
$^3P'_2 \leftarrow$	41.1	40.7(4)	38.1	37.1(19)	39.7	39
$^1D'_2 \leftarrow$	0.88	1.2(1)	8.8	8.1(8)	10.0	17
$^1S'_0 \leftarrow$	0.06	<0.20(6)	0.52	0.23(2)	0.5	–
$^3P_1 \leftarrow {}^3P^o_2$	25.2	24.6(3)	26.4	27.2(14)	28.3	22
$^3P'_2 \leftarrow$	74.8	74.5(3)	73.1	72.1(14)	68.5	71
$^1D'_2 \leftarrow$	0.020	0.027(4)	0.53	0.72(7)	3.2	7
$^3P'_0 \leftarrow {}^1P''_1$	0.24	0.30(2)	2.9	4.6(5)	4.2	8
$^3P_1 \leftarrow$	0.25	0.20(2)	3.3	3.6(4)	6.8	4
$^3P'_2 \leftarrow$	0.15	0.20(2)	1.0	1.68(17)	0.01	–
$^1D'_2 \leftarrow$	92.0	93.4(2)	86.2	86.1(14)	82.2	88
$^1S'_0 \leftarrow$	7.4	5.7(12)	6.6	4.0(4)	6.8	–

The degree to which the single-configuration model is valid for a given atom, ion, or isoelectronic sequence can be tested by observing how well the empirical values for the diagonal and off-diagonal elements of the effective spin-orbit parameters agree with each other (i.e., $\zeta_p \cong \zeta'_p$ and $\zeta_{pp} \cong \zeta'_{pp} \cong \zeta''_{pp}$). If these conditions are satisfied, then the branching fractions so deduced should be reliable.

These matrix elements can be converted to transition probability rates using

$$g_u A_{u\ell} = \left[\frac{1265.38}{\lambda_{u\ell}(\text{Å})} \right]^3 |\langle \ell | r | u \rangle|^2 . \qquad (7.42)$$

The branching fractions can be obtained through the appropriate sums and ratios.

Application: Si, Ge, and Sn sequences

A test of this approach applied to the neutral atoms Si I, Ge I, and Sn I has been carried out [52, 53, 55]. First, the measured values for the seven energy-level separations among the five ground term and four excited term energy levels were converted to the Slater and spin–order parameters, F_2, ζ_{pp}, ζ'_{pp}, ζ''_{pp} and G_1, ζ_p, ζ'_p. The various alternative determinations of the spin–orbit parameters agreed well, indicating that both the lower and upper configurations are relatively pure. Next, the mixing angles θ_0, θ_2 and θ_1 were deduced from the Slater parameters, and used in conjunction with the equations of Sections 7.1.3 and 7.1.6 to obtain the branching fractions predicted by these mixing angles in the context of the single-configuration model. These results are displayed in Table 7.4.

For these three neutral atoms, accurate experimental measurements of the branching fractions exist (cf. [53, 55]). These are also displayed in Table 7.4. The semiempirical results agree nearly perfectly within the accuracies of the experimental measurements.

Table 7.5. Empirical Slater and spin–orbit parameters (in cm^{-1}). (After Ref. [55].)

Z	Ion	F_2	ζ_{pp}	ζ'_{pp}	ζ''_{pp}	G_1	ζ_p	ζ'_p
50	Sn I	919	2097	1867	2305	451	2659	2635
51	Sb II	1186	3480	3312	3853	511	4091	3942
52	Te III	1392	5105	5042	5575	666	6166	6401
53	I IV	1568	6966	7010	7523	791	7858	6973
54	Xe V	1729	9049	9219	9767	2518	10 067	10 066
55	Cs VI	1871	11 410	11 693	12 192	2390	12 493	12 226

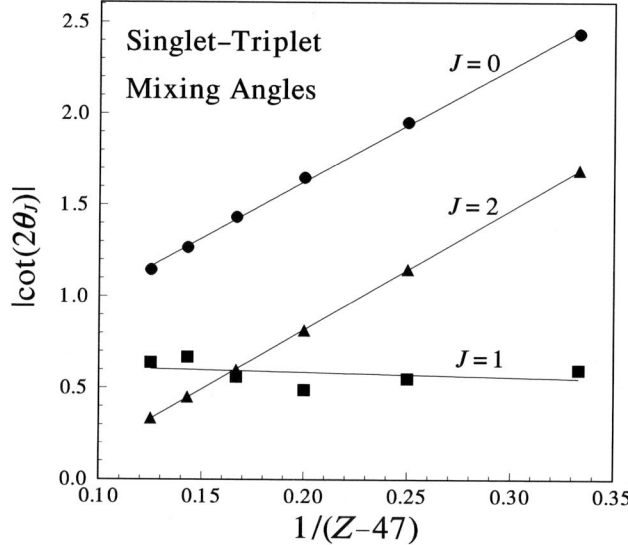

Fig. 7.4. Mixing angles for the Sn isoelectronic sequence. (After Ref. [55].)

This agreement provides confidence that similar accuracy can be attained by extending these semiempirical calculations to singly and multiply charged ions in these isoelectronic sequences. Calculations have been carried out in the Si sequence for P II–Ar V [53], in the Ge sequence for As II–Br IV [53], and in the Sn sequence for Sb II–Cs VI [55]. These results are particularly valuable, since virtually *no* branching-fraction measurements presently exist for multiply charged ions. Reliable semiempirical values for branching fractions can therefore be used to provide UV line calibration standards.

Slater and spin–orbit parameters, computed [55] from the available energy-level database for the Sn isoelectronic sequence, are presented in Table 7.5. The absence of any significant degree of CI affecting the $5s^2 5p^2$ and $5s^2 5p6s$ configurations is apparent from the agreement in the extracted values for the spin–orbit parameters, i.e., $\zeta_p \cong \zeta'_p$ and $\zeta_{pp} \cong \zeta'_{pp} \cong \zeta''_{pp}$.

The mixing angles obtained from these Slater and spin–orbit parameters are displayed in Fig. 7.4 as a plot of $|\cot(2\theta_J)|$ vs $1/(Z-47)$. Again here the power-law fit yields $p = 1$, indicating an approximately linear variation of this quantity with the reciprocal screened

Table 7.6. Wavelengths and semiempirical branching fractions (BF, in %) for the $5s^2 5p^2 - 5s^2 5p6s$ transitions in multiply charged ions of the Sn sequence. (After Ref. [55].)

Transition	Te III λ(Å)	BF	I IV λ(Å)	BF	Xe V λ(Å)	BF	Cs VI λ(Å)	BF
$^3P'_0 \leftarrow {}^3P^{o'}_1$	928.3	29.2	666.3	30.0	512.8	31.1	410.3	31.0
$^3P_1 \leftarrow$	971.2	15.0	698.0	15.3	538.6	16.1	431.9	15.6
$^3P'_2 \leftarrow$	1004.4	46.7	718.9	49.0	552.9	50.0	442.3	51.2
$^1D'_2 \leftarrow$	1106.6	8.2	784.0	5.0	600.4	2.4	479.3	1.8
$^1S'_0 \leftarrow$	1293.2	0.9	885.7	0.7	664.8	0.4	522.7	0.3
$^3P_1 \leftarrow {}^3P^o_2$	893.2	27.7	649.3	27.6	500.6	27.6	402.0	27.7
$^3P'_2 \leftarrow$	921.2	61.7	667.3	57.8	513.0	54.3	411.0	51.1
$^1D'_2 \leftarrow$	1006.4	10.7	723.0	14.6	553.5	18.1	442.7	21.2
$^3P'_0 \leftarrow {}^1P^{o'}_1$	848.9	2.6	619.7	1.5	469.2	0.7	378.5	0.6
$^3P_1 \leftarrow$	884.7	6.0	647.0	5.2	490.6	4.3	396.8	4.4
$^3P'_2 \leftarrow$	912.2	2.0	664.9	5.2	502.6	9.4	405.5	11.9
$^1D'_2 \leftarrow$	995.7	77.3	720.2	74.8	541.4	71.6	436.4	68.6
$^1S'_0 \leftarrow$	1144.2	12.1	805.1	13.3	593.3	14.1	472.1	14.6

charge. The magnitude of the cotangent is plotted because the values for $J = 2$ differ in sign from those of $J = 1$ and 2. The slopes of these quantities for the ground configuration ($J = 0$ and 2) on this plot are identical to within experimental uncertainties. For the excited state, the cotangent is nearly independent of the degree of ionization.

The branching fractions deduced from this mixing-angle data for the Sn-like multiply charged ions Te III–Cs VI are presented in Table 7.6, together with their wavelengths. The usefulness of these data can be illustrated by the consideration of a possible application.

Consider a proposal for a means to measure the branching fractions of the $3s^2 3p^3 - 3s^2 3p^2 4s$ transition array in singly ionized sulfur. The transition array consists of three multiplets: one with $\lambda = 907, 911, 913$ Å; another with $\lambda = 1053, 1056$ Å, and a third with $\lambda = 1167, 1173$ Å. Although no intensity calibration standards presently exist in this region, the Te III results in Table 7.6 provide intensity ratios from the same upper level that could provide calibration pairs. For example, transitions from the $5p6s\ ^3P^{o'}_1$ level should have the intensity ratios $I(1004\text{Å})/I(928\text{Å}) = 1.6$ and $I(1004\text{Å})/I(1107\text{Å}) = 5.7$, which connect the three wavelength regions spanned by the S II multiplet.

If the lifetimes of the upper levels of these Te III transitions could also be measured (to verify or refute the constancy of the interconfigurational transition moment that occurs in the nonrelativistic Schrödinger approximation), additional intensity ratios among the three multiplets could be established. Thus, by utilizing IC relationships obtained from spectroscopic data and a system possessing little CI, a set of calibration standards can be obtained whereby systems of greater configurational complexity can be accurately measured.

7.2 *jj*-based reformulation

To this point, the methods used in this section have been based on the nonrelativistic Schrödinger equation, using an LS-coupling angular basis set and radial wavefunctions that are independent of J. In the same single-configuration spirit, these methods can be reformulated using the relativistic Dirac equation, with a jj-coupling basis set and j-dependent radial wavefunctions. The major modification will be that this yields a more general formulation involving two j-dependent radial matrix elements, from which the nonrelativistic results can be recovered by equating these two radial matrix elements.

This formalism is quite generally applicable, and its use will be demonstrated here for the example of sp and p^2 configurations that have been treated by the Schrödinger model above in Sections 7.1.3, 7.1.6, and 7.1.7.

7.2.1 *jj* mixing-angle formulation

We denote the IC wavefunction by $\Psi_{ljl'j'J}$ and the jj basis states by $|l_j l'_{j'} J\rangle$. In order to make comparisons with results obtained in the LS basis, the mixing angle relative to the jj basis will be written as $\Theta_J - \theta_J$, where Θ_J is the jj-limit value of the mixing angle θ_J which is defined in the LS basis.

For an sp configuration there are four levels (denoted in LS notation as ${}^3P_0^o$, ${}^3P_1^o$, ${}^3P_2^o$, ${}^1P_1^o$) which can be written as

$$\Psi_{s\frac{1}{2}p\frac{1}{2}0} = \left|s\frac{1}{2}p\frac{1}{2}0\right\rangle$$
$$\Psi_{s\frac{1}{2}p\frac{1}{2}1} = \cos(\Theta_1 - \theta_1)\left|s\frac{1}{2}p\frac{1}{2}1\right\rangle - \sin(\Theta_1 - \theta_1)\left|s\frac{1}{2}p\frac{3}{2}1\right\rangle$$
$$\Psi_{s\frac{1}{2}p\frac{3}{2}2} = \left|s\frac{1}{2}p\frac{3}{2}2\right\rangle$$
$$\Psi_{s\frac{1}{2}p\frac{3}{2}1} = \sin(\Theta_1 - \theta_1)\left|s\frac{1}{2}p\frac{1}{2}1\right\rangle + \cos(\Theta_1 - \theta_1)\left|s\frac{1}{2}p\frac{3}{2}1\right\rangle. \tag{7.43}$$

For a p^2 configuration there are five levels (denoted in LS notation as ${}^3P_0$, ${}^3P_1$, ${}^3P_2$, ${}^1D_2$, ${}^1S_0$) which can be written as

$$\Psi_{p\frac{1}{2}p\frac{1}{2}0} = \cos(\Theta_0 - \theta_0)\left|p\frac{1}{2}p\frac{1}{2}0\right\rangle - \sin(\Theta_0 - \theta_0)\left|p\frac{3}{2}p\frac{3}{2}0\right\rangle$$
$$\Psi_{p\frac{1}{2}p\frac{3}{2}1} = \left|p\frac{1}{2}p\frac{3}{2}1\right\rangle$$
$$\Psi_{p\frac{1}{2}p\frac{3}{2}2} = \cos(\Theta_2 - \theta_2)\left|p\frac{1}{2}p\frac{3}{2}2\right\rangle - \sin(\Theta_2 - \theta_2)\left|p\frac{3}{2}p\frac{3}{2}2\right\rangle$$
$$\Psi_{p\frac{3}{2}p\frac{3}{2}2} = \sin(\Theta_2 - \theta_2)\left|p\frac{1}{2}p\frac{3}{2}2\right\rangle + \cos(\Theta_2 - \theta_2)\left|p\frac{3}{2}p\frac{3}{2}2\right\rangle$$
$$\Psi_{p\frac{3}{2}p\frac{3}{2}0} = \sin(\Theta_0 - \theta_0)\left|p\frac{1}{2}p\frac{1}{2}0\right\rangle + \cos(\Theta_0 - \theta_0)\left|p\frac{3}{2}p\frac{3}{2}0\right\rangle. \tag{7.44}$$

The s^2 configuration provides a convenient transition partner for the sp, and has only one level (in LS notation ${}^1S_0$)

$$\Psi_{s\frac{1}{2}s\frac{1}{2}0} = \left|s\frac{1}{2}s\frac{1}{2}0\right\rangle. \tag{7.45}$$

The one-electron radial integrals will be denoted as

$$R_{11} \equiv \langle s_{\frac{1}{2}} | r | p_{\frac{1}{2}} \rangle \tag{7.46}$$

$$R_{13} \equiv \langle s_{\frac{1}{2}} | r | p_{\frac{3}{2}} \rangle. \tag{7.47}$$

7.2.2 *jj* formulation of s²–sp transitions

Using the wavefunctions from Eqs. 7.43 and 7.45, the reduced matrix elements can be written in terms of these single-particle radial integrals using the Wigner–Eckart theorem [83, 193]

$$\langle \Psi_{s\frac{1}{2}s\frac{1}{2}0} | \mathbf{r} | \Psi_{s\frac{1}{2}p\frac{1}{2}1} \rangle = \frac{1}{\sqrt{3}} [R_{11} \cos(\Theta_1 - \theta_1) - \sqrt{2} R_{13} \sin(\Theta_1 - \theta_1)]$$

$$\langle \Psi_{s\frac{1}{2}s\frac{1}{2}0} | \mathbf{r} | \Psi_{s\frac{1}{2}p\frac{3}{2}1} \rangle = \frac{1}{\sqrt{3}} [R_{11} \sin(\Theta_1 - \theta_1) + \sqrt{2} R_{13} \cos(\Theta_1 - \theta_1)]. \tag{7.48}$$

Noting that the *jj* limit for an sp configuration is $\sin \Theta_1 = \sqrt{\frac{1}{3}}$ and $\cos \Theta_1 = \sqrt{\frac{2}{3}}$, the trigonometric addition formulae yield

$$\sin(\Theta_1 - \theta_1) = (\cos \theta_1 - \sqrt{2} \sin \theta_1)/\sqrt{3}$$

$$\cos(\Theta_1 - \theta_1) = (\sqrt{2} \cos \theta_1 + \sin \theta_1)/\sqrt{3}, \tag{7.49}$$

and the expressions for the wave functions can be rewritten as

$$\langle \Psi_{s\frac{1}{2}s\frac{1}{2}0} | \mathbf{r} | \Psi_{s\frac{1}{2}p\frac{1}{2}1} \rangle = \frac{1}{\sqrt{3}} [\sqrt{2}(R_{11} - R_{13}) \cos \theta_1 + (R_{11} + 2R_{13}) \sin \theta_1]$$

$$\langle \Psi_{s\frac{1}{2}s\frac{1}{2}0} | \mathbf{r} | \Psi_{s\frac{1}{2}p\frac{3}{2}1} \rangle = \frac{1}{\sqrt{3}} [(R_{11} + 2R_{13}) \cos \theta_1 - \sqrt{2}(R_{11} + 2R_{13}) \sin \theta_1]. \tag{7.50}$$

These can be simplified by noting that the sum of the squares of the quantities involving the radial integrals reduce to

$$2(R_{11} - R_{13})^2 + (R_{11} + 2R_{13})^2 = 3(R_{11}^2 - 2R_{13}^2). \tag{7.51}$$

These relationships allow the equations to be rewritten in the very compact form

$$\langle \Psi_{s\frac{1}{2}s\frac{1}{2}0} | \mathbf{r} | \Psi_{s\frac{1}{2}p\frac{1}{2}1} \rangle = \sqrt{\frac{R_{11}^2 + 2R_{13}^2}{3}} \sin(\theta_1 - \xi) \tag{7.52}$$

$$\langle \Psi_{s\frac{1}{2}s\frac{1}{2}0} | \mathbf{r} | \Psi_{s\frac{1}{2}p\frac{3}{2}1} \rangle = \sqrt{\frac{R_{11}^2 + 2R_{13}^2}{3}} \cos(\theta_1 - \xi), \tag{7.53}$$

where ξ is an effective mixing angle for the two radial integrals

$$\tan \xi \equiv \sqrt{2} \frac{R_{13}/R_{11} - 1}{2R_{13}/R_{11} + 1}. \tag{7.54}$$

Table 7.7. Ratios of radial matrix elements for the alkaline-earthlike sequences. Here $\Delta \equiv (R_{13}/R_{11} - 1) \times 1000$. (From Ref. [71].)

ζ	Ion	Δ	Ion	Δ	Ion	Δ	Ion	Δ	Ion	Δ
I	Be	0.037	Mg	–	Zn	−1.99	Cd	−6.74	Hg	−40.2
II	B	0.113	Al	0.18	Ga	+0.16	In	−1.30	Tl	−18.2
III	C	0.212	Si	0.41	Ge	1.00	Sn	+0.35	Pb	−12.5
IV	N	0.342	P	0.61	As	1.54	Sb	1.24	Bi	−10.3
V	O	0.503	S	0.82	Se	1.97	Te	1.78	Po	−9.20
VI	F	0.682	Cl	1.04	Br	2.37	I	2.26	At	−8.64
VII	Ne	0.888	Ar	1.29	Kr	2.75	Xe	2.66	Rn	−8.27

Notice that in the nonrelativistic LS-coupling limit where $R = R_{11} = R_{13}$, this becomes

$$\langle {}^1S_0|\mathbf{r}|{}^3P_1\rangle = R \sin\theta_1 \tag{7.55}$$

$$\langle {}^1S_0|\mathbf{r}|{}^1P_1\rangle = R \cos\theta_1 \tag{7.56}$$

which is the form of the Schrödinger model that was developed earlier in Section 7.1.3.

Here again (as occurred in, e.g., the case of the Sommerfeld fine structure formula) there is an interesting morphology between the Breit–Schrödinger and Dirac formulations. This extension of the semiempirical systematization of the intermediate coupling model combines contributions from both the LS singlet–triplet mixing of the angular wave function with the jj mixing of the radial wave functions in a common parametrization.

Theoretical calculations indicate [71] that the radial mixing angle ξ is usually quite small, and can be neglected unless there are severe cancellation effects affecting the radial transition integrals. Moreover, in cases where ξ becomes significant, it usually affects only ions near the neutral end of the sequence where θ is smallest. One example where this correction has been shown to be significant is the Be isoelectronic sequence, where the LS singlet–triplet mixing in the 2s2p configuration in the neutral ion is so small that the jj radial mixing becomes non-negligible [71].

The results of Dirac–Hartree–Fock calculations of these radial transition integrals for the homologous Be, Mg, Zn, Cd and Hg isoelectronic sequences are given in Table 7.7.

While these values increase with the number of electrons in the system, they remain relatively small for all but the Hg sequence. For the Mg, Zn, and Cd sequences, the LS singlet–triplet mixing angle is large compared to the jj radial mixing angle for all ions in the sequence. In the case of the Hg sequence, the size of these corrections has grown to a point where they are significant for the neutral and singly ionized members of the sequence. For this reason, a detailed analysis of the $6s^2$–6s6p transitions in the Hg sequence has been carried out [74].

Application: the Hg sequence

The upper panel of Fig. 7.5 displays a plot of the values (open circles) of $\cot(2\theta_1)$ obtained from the measured base of energy-level data, plotted vs $1/(Z - 78)$. Again here (as was

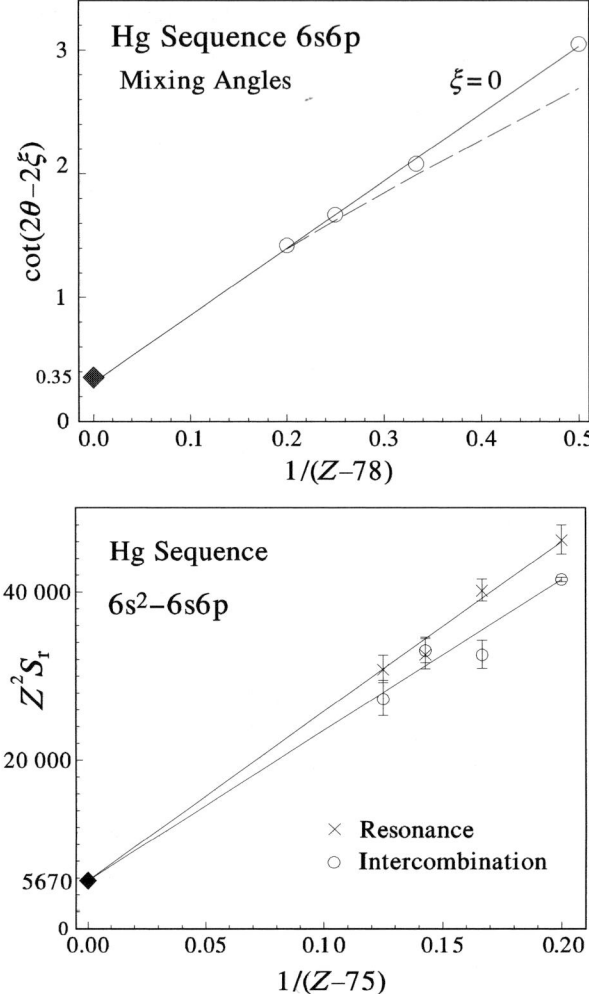

Fig. 7.5. Isoelectronic plots of the mixing angles and line-strength factors for the $6s^2$–$6s6p$ transitions in the Hg sequence. (After Ref. [74].)

observed for the Cd sequence – see Fig. 7.1) the power-law fit yields $p = 1$, so the cotangent is accurately represented as a linear function of the reciprocal charge. Also displayed as the dashed line on the same plot is the quantity $\cot(2\theta_1 - 2\xi)$, where ξ is obtained from the theoretical calculations in Table 7.7. The corrections for ξ become significant for the neutral and once-ionized members of the sequence.

The lower panel of Fig. 7.5 displays a plot of the Z-scaled effective reduced line-strength factors, obtained using the two mixing angles obtained from the singlet–triplet mixing and from the radial matrix element mixing, $(\theta_1 - \xi)$. The inclusion of the quantity ξ significantly improves the linearity of the plot for the intercombination transition near the neutral end of the sequence. In this exposition, the data for both transitions is linear to within the accuracy of the measurements, and both approach the hydrogenic limit $S_H = 5670$ at high Z.

In terms of these linearized parametrizations, the current state of knowledge of the line strengths of the $6s^2\ ^1S_1$–$6s6p\ ^{1,3}P_1$ transitions can be summarized very concisely by the simple fitted formulae

$$\cot(2\theta_1) = 0.3096 + 5.437/(Z - 78)$$

$$R_{13}/R_{11} = 0.9931 - 0.0334/(Z - 79)^{1.6}$$

$$Z^2 S_r(\text{Res}) = 5670[1 + 35.55/(Z - 75)]$$

$$Z^2 S_r(\text{Int}) = 5670[1 + 31.57/(Z - 75)]. \tag{7.57}$$

7.2.3 *jj* formulation of p²–sp transitions

Empirical incorporation of possible differences between the j-dependent Dirac wave functions is also possible for the p²–sp transition array, albeit with increased mathematical complexity over that of the result obtained for the s²–sp transition array.

Since these corrections are expected to be significant only for heavy, complex systems, we shall focus attention in this development on the Pb isoelectronic sequence. In contrast to the homologous C, Si, Ge, and Sn sequences, in the Pb sequence there is a large separation between the $6s^2 6p7s\ ^3P'_0$ and $^3P'_1$ levels ($(\frac{1}{2}, \frac{1}{2})_0$ and $(\frac{1}{2}, \frac{1}{2})_1$ in jj notation), and the higher-lying $6s^2 6p7s\ ^3P'_2$ and $^1P'_1$ levels ($(\frac{3}{2}, \frac{1}{2})_2$ and $(\frac{3}{2}, \frac{1}{2})_1$ in jj notation). Thus these two lower levels have transitions only to the ground term, whereas the two upper levels have possible transitions to other configurations. Therefore the $6s^2 6p7s\ ^3P_1$ upper level is of special interest since its branching fractions are completely specified by an analysis of this type.

We shall therefore restrict consideration here to those transitions involving $\Psi_{s\frac{1}{2}p\frac{1}{2}1}$. Using the wavefunctions of Eqs. 7.43 and 7.44, substituting $\tan\Theta_0 = \cot\Theta_2 = 1/\sqrt{2}$, and using trigonometric reduction formulae, these integrals become

$$\langle\Psi_{p\frac{1}{2}p\frac{1}{2}0}|\mathbf{r}|\Psi_{s\frac{1}{2}p\frac{1}{2}1}\rangle = -\frac{\sqrt{20}}{3}[(R_{13} + 2R_{11})\cos\theta_0\cos\theta_1 - (2R_{13} + R_{11})\sin\theta_0\sin\theta_1$$

$$+ \sqrt{2}(R_{13} - R_{11})\sin(\theta_0 - \theta_1)] \tag{7.58}$$

$$\langle\Psi_{p\frac{1}{2}p\frac{1}{2}1}|\mathbf{r}|\Psi_{s\frac{1}{2}p\frac{1}{2}1}\rangle = \frac{\sqrt{15}}{3}[(2R_{13} + R_{11})\cos\theta_1 + \sqrt{2}(R_{13} - R_{11})\sin\theta_1] \tag{7.59}$$

$$\langle\Psi_{p\frac{1}{2}p\frac{3}{2}2}|\mathbf{r}|\Psi_{s\frac{1}{2}p\frac{1}{2}1}\rangle = \frac{5}{3}[(4R_{13} + 2R_{11})\sin\theta_1\sin\theta_2 + (4R_{13} - R_{11})\cos\theta_1\cos\theta_2$$

$$- \sqrt{2}(R_{13} - R_{11})\sin(\theta_1 - \theta_2)] \tag{7.60}$$

$$\langle\Psi_{p\frac{3}{2}p\frac{3}{2}2}|\mathbf{r}|\Psi_{s\frac{1}{2}p\frac{1}{2}1}\rangle = -\frac{5}{3}[(4R_{13} + 2R_{11})\sin\theta_1\cos\theta_2 - (4R_{13} - R_{11})\cos\theta_1\sin\theta_2$$

$$+ \sqrt{2}(R_{13} - R_{11})\cos(\theta_1 - \theta_2)] \tag{7.61}$$

$$\langle\Psi_{p\frac{3}{2}p\frac{3}{2}0}|\mathbf{r}|\Psi_{s\frac{1}{2}p\frac{1}{2}1}\rangle = -\frac{\sqrt{20}}{3}[(R_{13} + 2R_{11})\sin\theta_0\cos\theta_1 + (2R_{13} + R_{11})\cos\theta_0\sin\theta_1$$

$$- \sqrt{2}(R_{13} - R_{11})\cos(\theta_0 - \theta_1)]. \tag{7.62}$$

Notice that in the nonrelativistic LS-coupling limit where $R = R_{11} = R_{13}$ these become

$$\langle {}^3P_0|r|{}^3P_1^o\rangle = -\sqrt{20}R\cos(\theta_0 + \theta_1)$$

$$\langle {}^3P_1|r|{}^3P_1^o\rangle = \sqrt{15}R\cos\theta_1$$

$$\langle {}^3P_2|r|{}^3P_1^o\rangle = 5R[2\sin\theta_1\sin\theta_2 + \cos\theta_1\cos\theta_2]$$

$$\langle {}^1D_2|r|{}^3P_1^o\rangle = -5R[2\sin\theta_1\cos\theta_2 - \cos\theta_1\sin\theta_2]$$

$$\langle {}^1S_0|r|{}^3P_1^o\rangle = -\sqrt{20}R\sin(\theta_0 + \theta_1), \tag{7.63}$$

which is the form that was used in the Schrödinger model development in Section 7.1.7.

Application: the Pb sequence

The $6s^2 6p^2$–$6s^2 6p7s$ transitions near the neutral end of the Pb isoelectronic sequence provide an example in which both the upper and lower configurations are relatively free of configuration interaction (CI), but both are significantly mixed by intermediate coupling. The lack of CI in these levels is evidenced by the agreement that is observed among the empirically extracted values for the spin–orbit energies ζ_p, ζ_p' and ζ_{pp}, ζ_{pp}', ζ_{pp}''. Another measure of the presence of CI can be obtained from comparisons of IC-predicted and measured values for the magnetic g-factors, as discussed (and computed for Pb I) in the next section.

The E1 transition moments in this sequence are significantly affected by cancellations in the radial integral. This can be seen from an application to this system of the cancellation exposition discussed in Section 6.6, which is shown in Fig. 7.6. The degree of cancellation is indicated on this plot by the proximity of the effective quantum numbers of the physical levels to the lines indicating the loci of the cancellation nodes. Figure 7.6 displays the effective quantum numbers of the transitions from the $6p^2$ ${}^3P_0$, ${}^3P_1$, ${}^3P_2$, ${}^1D_2$ and ${}^1S_1$ lower levels to the $6s6p$ ${}^3P_1^o$ upper level for Pb I and Bi II. Although the physical points do not fall directly on the nearest node (which would imply complete cancellation) their proximity to it suggests that small differences between the j-dependent wavefunctions could lead to large differences in the integrand.

This semiempirical prediction of large cancellation effects is confirmed by Dirac–Hartree–Fock calculations for the radial matrix elements, which yield the values $R_{13}/R_{11} = 1.4590$ for Pb I and $R_{13}/R_{11} = 1.4224$ for Bi II. For ions heavier than Bi II the values increase approximately as $(Z - 81)^2$, reaching 1.551 at U XI. Table 7.8 compares the results of nonrelativistic and relativistic calculations of the branching fractions for Pb I. Consistent with the deviation from unity of the quantity R_{13}/R_{11}, there are significant differences between the relativistic and nonrelativistic calculations. Branching-fraction measurements for Pb I are also available, and are compared with the semiempirical values in Table 7.8. It is clear that the measurements are in much closer agreement with the relativistically computed branching fractions.

On the basis of these results, the relativistic form of this semiempirical treatment is clearly preferable for systems as complex and heavy as the Pb sequence, and has been employed

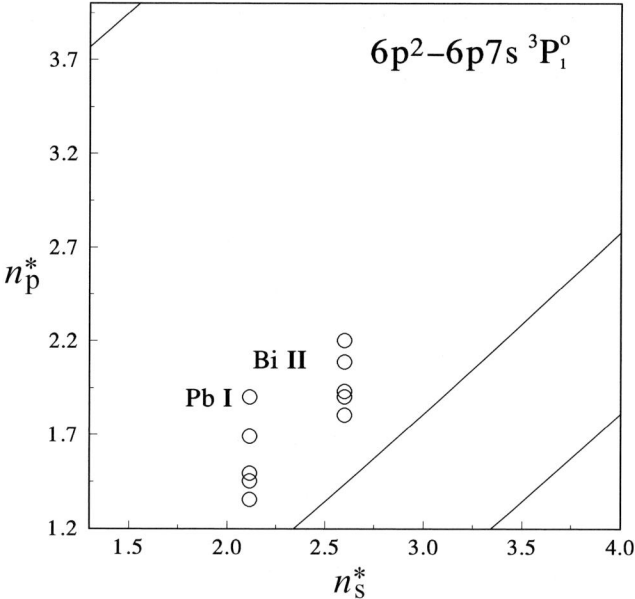

Fig. 7.6. Cancellation plot for the Pb sequence. (After Ref. [71].)

Table 7.8. Pb I branching fractions for the $^3P_1^o$ upper level in the $6s^2 6p^2 - 6s^2 6p7s$ multiplet. Wavelengths are in vacuum. (After Ref. [71].)

Transition	λ(Å)	BF(NonRel)	BF(Rel)	BF(Meas)
$^3P_0 - ^3P_1^o$	2833.89	0.489	0.310	0.324
$^3P_1 -$	3640.61	0.128	0.166	0.188
$^3P_2 -$	4058.95	0.381	0.520	0.500
$^1D_2 -$	7230.96	0.0029	0.0040	0.0005
$^1S_0 -$	17181	7×10^{-5}	3×10^{-5}	–

[71] to specify branching fractions for Bi II. Although spectroscopic data for the members of this sequence past Bi II are not available because of their nuclear instability, the slowly varying behavior exhibited by the mixing angles and radial matrix ratios should permit a reliable extrapolation to these systems.

7.3 Gyromagnetic ratios in intermediate coupling

An independent test of the applicability of the single-configuration model to the specification of line-strength factors can be obtained from a comparison of the deviation of the effective magnetic g-factor from the LS-coupling value, in cases where such data are available.

The energy of interaction of an atom with an external field B can be written as

$$E = \mu_{\mathbf{B}} \langle i | (\mathbf{L} + g_e \mathbf{S}) \cdot \mathbf{B} | k \rangle = \mu_B g_J M_J B. \tag{7.64}$$

Table 7.9. IC calculation of the effective g-factor for the
$4p^5 5s$ levels in Kr I. (After Ref. [48].)

Kr I $4p^5 5s$ Level	$\sin\theta_1 = 0.697$ $E(\text{cm}^{-1})$	g_J(Landé)	g_J(Pred)	g_J(Expt)
$^3P^o_2$	79 971.7321	3/2	1.500	1.502
$^3P^o_1$	80 916.7575	3/2	1.243	1.242
$^3P^o_0$	85 191.6075	0/0	0/0	–
$^3P^o_1$	85 846.6945	1	1.257	1.259

Table 7.10. IC calculation of the effective g-factor for the $6s^2 6p^2$
levels in Pb I. (After Ref. [48].)

Pb I $6s^2 6p^2$ Level	$\sin\theta_0 = 0.37633$ $E(\text{cm}^{-1})$	$\sin\theta_2 = 0.64248$ g_J(Landé)	g_J(Pred)	g_J(Expt)
$^3P_0$	0	0/0	0/0	–
$^3P_1$	7 819.2626	3/2	1.5000	1.501
$^3P_2$	10 650.3271	3/2	1.2936	1.279
$^3D_2$	21 457.7982	1	1.2064	1.227
$^3S_0$	29 466.8303	0/0	0/0	–

The magnetic moment operator was defined Section 4.9.1, μ_B is the Bohr magneton, and g_J is the magnetic g-factor. In the limiting case of pure LS coupling and the Dirac value for the electron moment $g_e = 2$, the g-factor assumes (for $J \neq 0$) the Landé from of Eq. 4.4.1

$$g_J = [3J(J+1) - L(L+1) + S(S+1)]/2J(J+1). \tag{7.65}$$

If there is intermediate coupling between two levels LSJ and $L'S'J$, then the physical g-factors can be written in terms of the mixing angles and the Landé g-factors as

$$g'_{LSJ} = g_{LSJ}\cos^2\theta_J + g_{L'S'J}\sin^2\theta_J$$
$$g'_{L'S'J} = g_{LSJ}\sin^2\theta_J + g_{L'S'J}\cos^2\theta_J. \tag{7.66}$$

Thus measured energy-level data can be used as discussed for the case of line strengths to determine the mixing angles, and combined with LS values obtained from Eq. 7.65 to specify physical g-factors by Eq. 7.66.

Tables 7.9 and 7.10 present applications of the method to the 4p5s configuration in Kr I and the 6s6p configuration in Pb I. Mixing angles were obtained from measured energy-level data (taken from Ref. [180] for Kr and from Ref.[196] for Pb). In both systems, the LS-coupling values for the Landé g-factors g_J were obtained from Eq. 7.66. For Kr I, the equations of Section 7.1.3 were used with the energy-level data in Table 7.9 to obtain for the Slater and spin–orbit parameters (in cm^{-1}) the values $G_1 = 800.06$, $\zeta_p = 3480$, and $\zeta_{p'} = 3485$, which yield the mixing-angle factor $\sin\theta_1 = 0.697$. For Pb I, the equations of

Section 7.1.6 were used with energy-level data in Table 7.10 to obtain for the Slater and spin–orbit parameters (again in cm^{-1}) the values $F_2 = 921.89$, $\zeta_{pp} = 7292$, $\zeta_{pp'} = 7264$, and $\zeta_{pp''} = 7525$, which yield the mixing-angle factors $\sin\theta_0 = 0.376\,33$ and $\sin\theta_2 = 0.642\,48$. The agreement between the fitted values for the two spin–orbit parameters for Kr I and the three spin–orbit parameters for Pb I provides confidence in the validity of the single-configuration approximation for these cases. The values for the LS-coupling g-factor and the emperical mixing angles were then used with Eqs. 7.67 to predict the IC values g_J (Pred). These are tabulated and compared with the experimental values g_J (Expt) for Kr I [180] and Pb I [197] in Tables 7.9 and 7.10.

These comparisons between predicted and experimental values provide a measure of the relative importance of configuration interaction and intermediate coupling effects. Deviations of a physical g-factor from its (nonzero) Landé value for levels that are unique in their J value within the configuration ($^3P_2^o$ for sp configurations, $^3P_1$ for p^2 configurations) indicate the presence of CI. Agreement with Eqs. 7.66 for levels with nonzero Landé values that can mix with another level of the same J in the same configuration ($^3P_1$ and $^1P_1$ for sp configurations; $^3P_2$ and $^1D_2$ for p^2 configurations) indicate that IC dominates over CI.

This method is particularly useful for applications to ionized atomic systems, for which little experimental g-factor data presently exist.

8

Magnetic dipole transitions

All things decay. Strive diligently.

– Last words of Buddha

A first-rate theory predicts; a second-rate theory forbids; a third-rate theory explains after the event.

– Alexsander Isaakovich Kitaigorodskii [133]

Much of the language of atomic physics was inspired by early studies of hydrogen and other atoms of low Z to moderate Z, where the dynamics are dominated by the electrostatic Coulomb interaction. Thus the n-dependent Balmer energy splittings, which are proportional to Z^2, are called the "gross energy," and the J-dependent Sommerfeld energy splittings, which are proportional to Z^4, are called the "fine structure." E1 processes, which are the primary radiative coupling between gross structure levels, are called "allowed" transitions, and M1 processes, which are the primary radiative coupling among fine structure levels, are called "forbidden" transitions. Clearly this Z-scaling causes the situation to enter a new domain for highly ionized heavy atoms, where "fine structure" can exceed "gross structure" and "forbidden" transition rates can exceed "allowed" transition rates. Isoelectronic studies can provide a fine-tuning mechanism, whereby the interactions can be studied at values of Z where they are strong to elucidate their contributions at values of Z where they are weak.

8.1 M1 transitions

In the zeroth-order nonrelativistic limit there is only electrostatics, and thus no magnetic dipole moment. In the first-order relativistic correction to the electrostatic problem, the magnetic dipole moment is given by $\boldsymbol{\mu} = \mu_B(\mathbf{L} + g_e\mathbf{S})$, and contains only angular factors. Since the Schrödinger picture separates radial and angular portions completely, the orthogonality of the radial wave functions restricts M1 processes in this approximation to occur only in transitions between levels within the same configuration. Unlike the E1 case, for M1 there is no change in parity of the atomic system. In the LS approximation these transitions would couple only fine structure levels within a term, but with intermediate coupling they also include intraconfigurational intercombination lines.

However, it is only in first order that the magnetic dipole operator involves only angular operators, and in higher relativistic orders the magnetic dipole moment acquires a radial dependence. This permits M1 transitions to occur between two different configurations of the same parity. These transitions are referred to as relativistic M1 lines.

There is a simple semiclassical picture than can convey some of the properties of E1 and M1 transitions. In terms of the single-electron Schrödinger (or Sommerfeld) model, each orbital (or orbit) is characterized by the nonrelativistic quantum numbers n and ℓ. The E1 transition moment couples two different orbitals and opens a mode of linear radial oscillation between them. By means of the electric dipole antenna thus formed, the electron radiates and moves to an orbit of lower energy. In the M1 case, the electron remains within a stable nonrelativistic orbit, but as an orbiting charge it possesses a magnetic dipole moment (and, classically, a centripetal acceleration), and thus comprises a magnetic dipole antenna. In the nonrelativistic approximation there is only one energy state of the nl-orbital, so energy cannot be radiated. However, the relativistic correction introduces electron spin, so the magnetic dipole antenna can radiate with an accompanying flip of the electron spin relative to the orbital angular momentum. For multiple electron configurations additional recouplings of the spin and angular momenta can be imagined, at least in the spirit of this simple picture.

Various models have been proposed to conceptualize the parity selection rules for radiative transitions. One model [26] utilizes the fact that the electric dipole moment is a polar vector connecting the negative and positive charges, whereas the magnetic dipole moment involves an axial vector defining a current loop. Thus, the electric dipole moment undergoes a sign change under a coordinate inversion, but the magnetic dipole moment is unchanged by a coordinate inversion. This provides a correspondence with the fact that the atomic state changes parity in an E1 transition, but there is no change in parity in an M1 transition. Another model is sometimes used that is based on conservation of angular momentum and parity. These models require that the intrinsic angular momentum and parity of the photon be specified. While these arguments work well for E1 radiation, care must be exercised in applying them to M1 and higher-order moments, which require a more detailed accounting of the properties of the photon. In contrast to massive particles, the photon has no rest frame in which a parity inversion can be performed, and its intrinsic angular momentum $J = 1$ possesses only the two helicity projections $M_J = \pm 1$, with no $M_J = 0$. The limitations of these models have been discussed by Ellis [93], who also presents vector methods for justifying the selection rules for dipole as well as quadrupole transitions.

The angular distribution of M1 radiation is the same as that of E1 radiation, but the direction of the polarization differs between E1 and M1 radiation because of the interchange of the electric and magnetic field vectors.

8.2 M1 line strengths

As shown in Section 2.4.9, the magnetic dipole moment of an atom arises as a relativistic correction to the Coulomb potential. To first order, this quantity is specified by

$$\boldsymbol{\mu} = -\frac{e}{2mc} \sum_i (\mathbf{L}_i + g_e \mathbf{S}_i), \tag{8.1}$$

where i is summed over all electrons. In the approximation of the Dirac moment of the electron ($g_e \cong 2$), this becomes (using $\hbar/mc = a_0\alpha$)

$$\boldsymbol{\mu} = -\frac{ea_0\alpha}{2} \sum_i (\mathbf{J}_i + \mathbf{S}_i)/\hbar \ . \tag{8.2}$$

Consistent with the definition of the E1 transition moment

$$\sum_i \left\langle \gamma SLJM_J \left| \frac{\mathbf{d}_i}{ea_0} \right| \gamma' S'L'J'M'_J \right\rangle = \sum_i \left\langle \gamma SLJM_J \left| \frac{\mathbf{r}_i}{a_0} \right| \gamma' S'L'J'M'_J \right\rangle, \tag{8.3}$$

the M1 transition moment is defined by

$$\sum_i \left\langle \gamma SLJM_J \left| \frac{\boldsymbol{\mu}_i}{ea_0} \right| \gamma' S'L'J'M'_J \right\rangle = \frac{\alpha}{2} \sum_i \left\langle \gamma SLJM_J \left| \frac{\mathbf{J}_i + \mathbf{S}_i}{\hbar} \right| \gamma' S'L'J'M'_J \right\rangle. \tag{8.4}$$

For a single active electron the magnetic line-strength factor is

$$S_{\mathrm{M1}} = \sum_{M_J M'_J} \left| \left\langle \gamma SLJM_J \left| \frac{\mathbf{J} + \mathbf{S}}{\hbar} \right| \gamma' S'L'J'M'_J \right\rangle \right|^2 \tag{8.5}$$

where the fine structure factor $\alpha^2/4$ has been excluded from the definition so that it can be used to explicitly characterize its status as a first-order relativistic correction. (We shall henceforth denote by L, S, J the quantum numbers in units of $\hbar$.)

Since this quantity lacks radial operators, $S_{\mathrm{M1}} = 0$ unless $\gamma = \gamma'$, $L = L'$, $S = S'$, and $J = J'$, $J' \pm 1$ (no $0 \rightarrow 0$). From simple Racah algebra it can be shown [161, 176] that

$$S_{\mathrm{M1}}(L, S, J : L, S, J \pm 1) = \frac{[(L + S + 1)^2 - J_>^2][J_>^2 - (L - S)^2]}{4J_>} |\langle n^* \mid n^{*\prime} \rangle|^2, \tag{8.6}$$

where $J_>$ is the greater of J and J', and

$$S_{\mathrm{M1}}(L, S, J : L, S, J) = \frac{(2J + 1)}{4J(J + 1)} [S(S + 1) - L(L + 1) + 3J(J + 1)]^2 |\langle n^* \mid n^{*\prime} \rangle|^2. \tag{8.7}$$

(The second expression is equivalent to $J(J + 1)(2J + 1)[g_J(SL)]^2$, where $g_J(SL)$ is the Landé g-factor.) The quantity $\langle n^* \mid n^{*\prime} \rangle$ is the "monopole moment" of the radial transition element. In the Schrödinger picture it is unity because all levels have the same radial wave function. It can be computed, for example, by the quantum defect or Coulomb approximation approach, where wave functions are empirically generated from energy-level data. Since we shall have occasion to combine these expressions coherently, it should be noted that for both Eqs. 8.6 and 8.7, the transition integrals had a positive phase before they were squared to obtain the line-strength factor. Table 8.1 provides values for the M1 line-strength factors for some of the simpler configurational examples.

Table 8.1. S_{M1} for simple configurations. Given are nonvanishing values for transitions between levels as well as self-transitions for J values shared by multiple levels. (Alternative p-electron hole configurations that lead to the same results are indicated in parentheses.)

Configuration	Transition	L	S	$J_>$	ΔJ	S_{M1}
p (p^5)	$^2P^o_{1/2}-^2P^o_{3/2}$	1	1/2	3/2	1	4/3
sp (sp^5)	$^3P^o_1-^3P^o_2$	1	1	2	1	5/2
	$^3P^o_0-^3P^o_1$	1	1	1	1	2
	$^3P^o_1-^3P^o_1$	1	1	1	0	27/2
	$^1P^o_1-^1P^o_1$	1	0	1	0	6
p^2 (p^4)	$^3P_1-^3P_2$	1	1	2	1	5/2
	$^3P_0-^3P_1$	1	1	1	1	2
	$^3P_0-^3P_0$	1	1	0	0	0
	$^1S_0-^1S_0$	0	0	0	0	0
	$^1D_2-^1D_2$	2	0	2	0	80
	$^3P_2-^3P_2$	1	1	2	0	135/2
p^3	$^2P^o_{1/2}-^2P^o_{1/2}$	1	1/2	1/2	0	2/3
	$^2P^o_{1/2}-^2P^o_{3/2}$	1	1/2	3/2	1	4/3
	$^2P^o_{3/2}-^2P^o_{3/2}$	1	1/2	3/2	0	80/3
	$^2D^o_{3/2}-^2D^o_{3/2}$	2	1/2	3/2	0	48/5
	$^2D^o_{3/2}-^2D^o_{5/2}$	2	1/2	5/2	1	12/5
	$^2D^o_{5/2}-^2D^o_{5/2}$	2	1/2	5/2	0	378/5
	$^4S^o_{3/2}-^4S^o_{3/2}$	0	3/2	3/2	0	60

The M1 transition probability has the same form as the E1 case except for the factor involving the fine structure constant

$$(2J_i + 1)A_{ij}(\mathrm{ns}^{-1}) = \frac{\alpha^2}{4}\left[\frac{1265.38}{\lambda_{ij}(\text{Å})}\right]^3 S_{M1}(i, j)$$

$$= \left[\frac{29.990}{\lambda_{ij}(\text{Å})}\right]^3 S_{M1}(i, j) . \tag{8.8}$$

In the same way that the $\Delta S = 0$ selection rule was broken for the E1 transitions if LS coupling is not exact, the intermediate coupling effects also cause intraconfigurational M1 intercombination lines to occur.

In ground configurations that possess fine structure, only the lowest level is the true ground level, and it is only a "ground state" if it has no magnetic degeneracy. Since there are no interconfigurational channels for decay of the excited fine structure levels, their relaxation occurs through M1 or E2 transitions, with M1 usually the stronger. Figure 8.1 shows energy-level (Grotrian) diagrams for ground configurations of the form ns^2np^x ($x = 1 - 5$) with the M1 transitions indicated. These types of configurations will be examined (with examples) below.

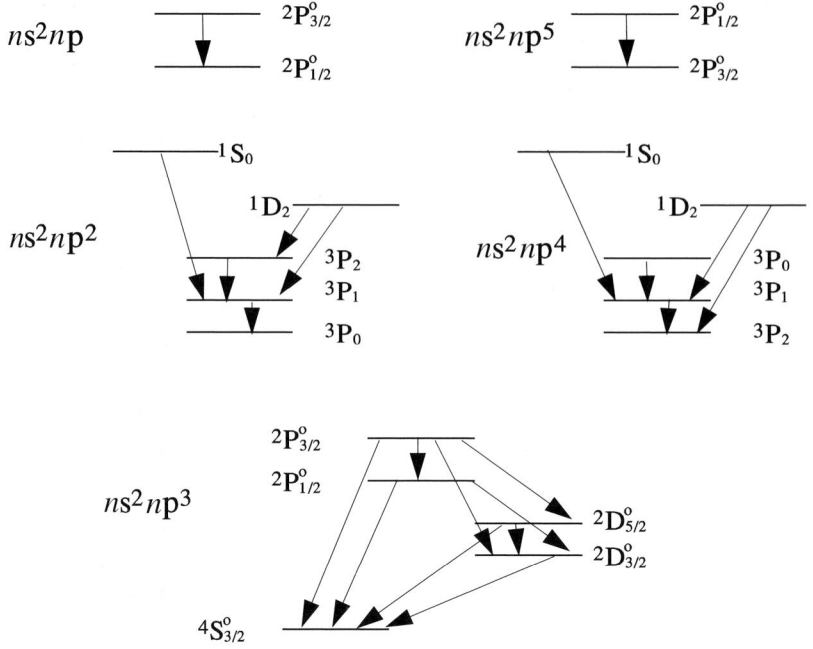

Fig. 8.1. Ground configuration Grotrian diagrams.

8.2.1 p and p^5 configurations

Transitions of this type occur in the ground configurations ns^2np and ns^2np^5. The M1 line-strength factor is given by

$$S_{M1} = 4/3 \tag{8.9}$$

and the semiempirical value for the transition probability can be computed knowing nothing more than the wavelength of the transition and the degeneracy of the upper level.

Values for the M1 transition probabilities within these ground terms can be obtained for entire isoelectronic sequences using the screening parametrization of the regular doublet law that was described in Section 3.6. This has been done for the ns^2np terms in the B ($n = 2$) [60], Al ($n = 3$) [61], Ga ($n = 4$) [61, 66], In ($n = 5$) [61], and Tl ($n = 6$) [61] sequences. It has also been done for the ns^2np^5 terms in the F ($n = 2$) [60], Cl ($n = 3$) [61], Br ($n = 4$) [61], I ($n = 5$) [61], and At ($n = 6$) [61] sequences.

The results of this analysis for the B sequence [60] are summarized in Table 8.2. Here the true ground level is the $2s^2 2p\ ^2P^o_{1/2}$, and the lifetime (neglecting E2 contributions) of the excited $2s^2 2p\ ^2P^o_{3/2}$ level varies from one year in neutral boron to 3.3 ps in B-like uranium.

Example: Fe^{13+} and Fe^{9+} (coronium)

In Al-like Fe^{13+} the configuration $3s^2 3p$ contains the ground level $^2P^o_{1/2}$, and the $^2P^o_{3/2}$ level lies 18 852.5 cm^{-1} above it. The M1 transition connecting these levels has a vacuum

Table 8.2. Calculated M1 transition rates (in ns^{-1}) for the $2s^2 2p\ ^2P^o_{1/2} - {}^2P^o_{3/2}$ transition in the B sequence. Energy-level separations (in cm^{-1}) for the entire sequence were obtained from a screening parametrization of the available database. (The full output is reproduced, although this exceeds the significant digits.)

Z	ΔE(expt)	ΔE(fit)	A(M1)	Z	ΔE(fit)	A(M1)
5	15.25(5)	15.25	3.189×10^{-17}	49	1 868 760	0.05868
6	63.41(5)	63.4	2.291×10^{-15}	50	2 040 555	0.07639
7	174.36(2)	174.4	4.766×10^{-14}	51	2 224 438	0.09896
8	386.9(7)	385.9	5.166×10^{-13}	52	2 421 034	0.1276
9	744.5(4)	744.5	3.709×10^{-12}	53	2 630 993	0.1637
10	1310(5)	1305	2.000×10^{-11}	54	2 854 988	0.2092
11	2139(6)	2133	8.722×10^{-11}	55	3 093 720	0.2662
12	3303(2)	3300	3.230×10^{-10}	56	3 347 917	0.3374
13	4903(70)	4888	1.050×10^{-9}	57	3 618 336	0.4259
14	6991(2)	6989	3.069×10^{-9}	58	3 905 762	0.5357
15	9668(45)	9703	8.212×10^{-9}	59	4 211 012	0.6714
16	13 138(7)	13 138	2.039×10^{-8}	60	4 534 937	0.8385
17	17 390(50)	17 415	4.749×10^{-8}	61	4 878 417	1.044
18	22 657(5)	22 662	1.046×10^{-7}	62	5 242 373	1.295
19		29 016	2.196×10^{-7}	63	5 627 757	1.603
20	36 643(110)	36 624	4.417×10^{-7}	64	6 035 565	1.977
21	45 637(4)	45 646	8.551×10^{-7}	65	6 466 830	2.432
22	56 240(4)	56 248	1.600×10^{-6}	66	6 922 627	2.983
23		68 608	2.904×10^{-6}	67	7 404 077	3.649
24	82 926(14)	82 915	5.125×10^{-6}	68	7 912 347	4.454
25	99 266(220)	99 367	8.821×10^{-6}	69	8 448 652	5.422
26	118 273(28)	118 176	1.484×10^{-5}	70	9 014 259	6.586
27	139 310(250)	139 560	2.444×10^{-5}	71	9 610 489	7.981
28	163 961(55)	163 754	3.948×10^{-5}	72	10 238 718	9.650
29	191 278(75)	191 000	6.265×10^{-5}	73	10 900 384	11.64
30		221 555	9.778×10^{-5}	74	11 596 988	14.02
31		255 686	1.503×10^{-4}	75	12 330 093	16.85
32		293 673	2.277×10^{-4}	76	13 102 377	20.22
33		335 810	3.405×10^{-4}	77	13 912 432	24.21
34		382 402	5.028×10^{-4}	78	14 765 164	28.94
35		433 770	7.338×10^{-4}	79	15 661 406	34.54
36		490 245	0.001059	80	16 603 115	41.15
37		552 177	0.001514	81	17 592 345	48.95
38		619 926	0.002142	82	18 631 246	58.15
39		693 870	0.003004	83	19 722 074	68.97
40		774 402	0.004175	84	20 867 195	81.70
41		861 930	0.005757	85	22 069 095	96.64
42		956 880	0.007877	86	23 330 388	114.2
43		1 059 693	0.010700	87	24 653 820	134.7
44		1 170 829	0.01443	88	26 042 283	158.8
45		1 290 767	0.01934	89	27 498 823	187.0
46		1 420 002	0.02574	90	29 026 650	219.9
47		1 559 052	0.03407	91	30 629 155	258.4
48		1 708 452	0.04483	92	32 309 914	303.3

wavelength 5304.34 Å, which leads to a transition probability

$$A_{3/2,1/2} = \frac{1}{4}\left[\frac{29.990}{5304.34}\right]^2\frac{4}{3} = 60.2 \text{ s}^{-1}. \tag{8.10}$$

Similarly, in Cl-like Fe^{9+} the configuration $3s^2 3p^5$ contains the ground level $^2P^o_{3/2}$ and the $^2P^o_{1/2}$ level lies 15 683.1 cm^{-1} above it. The M1 transition connecting these levels thus has a vacuum wavelength of 6376.29 Å, which leads to a transition probability

$$A_{1/2,3/2} = \frac{1}{2}\left[\frac{29.990}{6376.29}\right]^2\frac{4}{3} = 69.4 \text{ s}^{-1}. \tag{8.11}$$

The air wavelengths of these lines correspond to the historically important "green coronium" line at 5302.86 Å and the "red coronium" line at 6374.51 Å.

The λ5303 Å line was observed in the solar corona during the eclipse of August 7, 1869 by the American astromomers Charles Young and William Harkness, but its origin remained a puzzle for 70 years. One proposal involved the postulation of another hypothetical element "coronium," inspired by the identification of helium in the solar chromosphere by Jules Janssen and J. Norman Lockyer during the solar eclipse of 1868. The solution to the mystery was brought about by the vacuum spark spectroscopic studies [84] of Bengt Edlén at Uppsala University during the 1930s.

Through his measurements and his development of the charge-screening parametrization of the regular doublet law as described in Section 3.6, Edlén was able to identify fine structures in highly ionized systems [85]. Aided by suggestions by Walter Grotrian [111], by 1942 Edlén had identified 19 of 23 known "coronium" lines as M1 transitions between fine structure levels of ground-state configurations in highly ionized atoms of Fe, Ni and Ca. On the basis of the ionization potential needed to achieve these highly charged ions, and the high degree of thermal broadening of these lines (0.8 Å, ten times the corresponding broadening in the Fraunhofer lines), Edlén concluded that there should be a drastic upward revision of the estimate of the coronal temperature. This discovery revolutionized the understanding of the Sun, since it demonstrated that its coronal temperature is approximately 2 000 000 K, and not 6000 K as had been assumed previously.

Prior to his identification of these lines in the solar corona, Professor Edlén had abandoned his pioneering studies of highly ionized atoms in 1935 because, in his words, "Jag trodde inte att dom fanns någon annonstans på den blåa himlen eller på den gröna jorden" [89]. (I didn't believe that they existed anywhere else in the blue heavens or on the green Earth.) It is interesting to note that the spectral lines he observed in the heavens were at 5303 Å (green) and NASA photographs from space demonstrate that the Earth is blue.

Application: M1 rates in excited configurations of single-valence-electron ions

These methods can also be applied to specify the M1 rates between fine structure levels of excited terms in single-valence-electron ions. It was shown in Section 3.6 that these fine structure splittings can be predictively systematized by a screening parametrization of the regular doublet law. For example, Fig. 3.4 presents a screening parametrization that predicts the fine structure splittings for np and nd levels in the Cu isoelectronic sequence. As an illustration, Table 3.4 presents an exposition of the fine structure splitting of the 4p term in Cu-like Nd^{31+}, which is 265 110 cm^{-1}. This corresponds to an M1 transition of wavelength 377.2 Å leading to a transition probability

$$A_{3/2,1/2} = \frac{1}{4}\left[\frac{29.990}{377.2}\right]^3 \frac{4}{3} = 0.167 \ \mu s^{-1} .\tag{8.12}$$

Similar calculations can be made for the various isoelectronic and homologous counterparts of this transition.

8.2.2 Single–triplet mixing in sp or sp^5 configurations

Transitions of this type occur in the first excited configurations $nsnp$ and ns^2np^5n' of systems with ground configurations ns^2 and ns^2np^6. For these configurations it was shown in Section 7.1.3 that the singlet–triplet mixing leads to wave functions of the form

$$\begin{aligned}
|^3P_0^{o\prime}\rangle &= |^3P_0^o\rangle \\
|^3P_1^{o\prime}\rangle &= \cos\theta_1 |^3P_1\rangle - \sin\theta_1 |^1P_1^o\rangle \\
|^3P_2^{o\prime}\rangle &= |^3P_2^o\rangle \\
|^1P_1^{o\prime}\rangle &= \sin\theta_1 |^3P_1^o\rangle + \cos\theta_1 |^1P_1^o\rangle,
\end{aligned}\tag{8.13}$$

where the mixing angle θ_1 can be deduced from spectroscopic energy-level data. In intermediate coupling, the selection rules that remain are conservation of energy, no change of parity, and $\Delta J = 0, \pm 1$ (no $0 \rightarrow 0$), which allow the transitions

$$S_{M1}\left(^3P_1^{o\prime}, {}^3P_2^{o\prime}\right) = S_{M1}\left(^3P_1^o, {}^3P_2^o\right)\cos^2\theta_1 = \frac{5}{2}\cos^2\theta_1 \tag{8.14}$$

$$S_{M1}\left(^3P_0^{o\prime}, {}^3P_1^{o\prime}\right) = S_{M1}\left(^3P_0^o, {}^3P_1^o\right)\cos^2\theta_1 = 2\cos^2\theta_1 \tag{8.15}$$

$$S_{M1}\left(^3P_2^{o\prime}, {}^1P_1^{o\prime}\right) = S_{M1}\left(^3P_2^o, {}^3P_2^o\right)\sin^2\theta_1 = \frac{5}{2}\sin^2\theta_1 \tag{8.16}$$

$$S_{M1}\left(^3P_0^{o\prime}, {}^1P_1^{o\prime}\right) = S_{M1}\left(^3P_0^o, {}^3P_1^o\right)\sin^2\theta_1 = 2\sin^2\theta_1 \tag{8.17}$$

$$\begin{aligned}
S_{M1}\left(^3P_1^{o\prime}, {}^1P_1^{o\prime}\right) &= \left[\sqrt{S_{M1}\left(^3P_1^o, {}^3P_1^o\right)} - \sqrt{S_{M1}\left(^1P_0^o, {}^1P_1^o\right)}\right]^2 \sin^2\theta_1 \cos^2\theta_1 \\
&= \left[\sqrt{27/2} - \sqrt{6}\right]^2 \sin^2\theta_1 \cos^2\theta_1 = \frac{3}{2}\sin^2\theta_1 \cos^2\theta_1 .
\end{aligned}\tag{8.18}$$

Example: Be-like Xe^{50+}

It was shown in Chapters 5 and 7 that it is possible to use charge-screening parametrizations to isoelectronically extrapolate energy levels, singlet–triplet mixing angles, and E1 line strengths to very high stages of ionization with high reliability. Such predictions [63] have been made for the 2s2p configuration in the Be sequence. We shall draw here from the semiempirical predictions for the ion Xe^{50+} in that study.

Estimates of the excitation energies of the $^3P^o_0$, $^3P^o_1$, $^3P^o_2$ and $^1P^o_1$ levels are, respectively, 836 575, 1 023 080, 3 777 401 and 4 297 048 cm^{-1}. Using the methods of Section 7.1.3, the energy separations among these levels can be mapped into the Slater parameters $G_1 = 353\,076$, $\zeta_p = 1\,960\,551$ and $\zeta'_p = 1\,984\,922$ cm^{-1}. These yield a singlet–triplet mixing angle between the $^3P_1$ and $^1P_1$ levels of $\theta_1 = 29.50°$.

A screening parametrization [63] of the Be isoelectronic sequence yields, for Be-like Xe, the reduced line-strength factors $Z^2 S_r(\text{Res}) \cong 43.0$ and $Z^2 S_r(\text{Int}) \cong 35.7$. (In this sequence, unusually strong asymptotic mixing between the $2s^2$ and $2p^2$ configurations depresses the high-Z limit to 38.5, below the hydrogenic limit of 54.) For the E1 $2s^2{}^1S_0 - 2s2p\ ^1P^o_1$ resonance transition at $\lambda 23.27$ Å the rate is given by

$$A\left(^1S_0, {}^1P^o_1\right) = \frac{1}{3}\left(\frac{1265.38}{23.27}\right)^3 \frac{43.0}{54^2}\cos^2(29.50°) = 598 \text{ ns}^{-1}. \tag{8.19}$$

Similarly, for the relativistic E1 $2s^2{}^1S_0 - 2s2p\ ^3P^o_1$ intercombination transition at $\lambda 97.74$ Å the rate is given by

$$A\left(^1S_0, {}^3P^o_1\right) = \frac{1}{3}\left(\frac{1265.38}{97.74}\right)^3 \frac{35.7}{54^2}\sin^2(29.50°) = 2.15 \text{ ns}^{-1}. \tag{8.20}$$

In addition to the E1 decay mode to ground of the $^3P^o_1$ level, it has an M1 decay to the $^3P^o_0$ at $\lambda 536.2$ Å, and is repopulated by M1 cascades from the $^3P^o_2$ level at $\lambda 36.32$ Å and from the $^1P^o_1$ at $\lambda 30.54$ Å. We can use the methods of this section to determine whether or not any of these M1 rates are commensurate with the E1 rates.

Using the expressions above for the line-strength factors, the transition probability rates are given by

$$A\left(^3P^{o\prime}_1, {}^3P^{o\prime}_0\right) = \frac{1}{3}\left[\frac{29.990}{536.2}\right]^3 2\cos^2(29.50°) = 8.8 \times 10^{-5} \text{ ns}^{-1}$$

$$A\left(^1P^{o\prime}_1, {}^3P^{o\prime}_0\right) = \frac{1}{3}\left[\frac{29.990}{28.90}\right]^3 2\sin^2(29.50°) = 0.18 \text{ ns}^{-1}$$

$$A\left(^1P^{o\prime}_1, {}^3P^{o\prime}_1\right) = \frac{1}{3}\left[\frac{29.990}{30.54}\right]^3 \frac{3}{2}\sin^2(29.50°)\cos^2(29.50°) = 0.087 \text{ ns}^{-1}$$

$$A\left(^1P^{o\prime}_1, {}^3P^{o\prime}_2\right) = \frac{1}{3}\left[\frac{29.990}{192.4}\right]^3 \frac{5}{2}\sin^2(29.50°) = 0.000\,76 \text{ ns}^{-1}$$

$$A\left(^3P^{o\prime}_2, {}^3P^{o\prime}_1\right) = \frac{1}{5}\left[\frac{29.990}{36.31}\right]^3 \frac{5}{2}\cos^2(29.50°) = 0.21 \text{ ns}^{-1}. \tag{8.21}$$

These M1 branches are all negligible compared to the ground-state decay rates of the resonance and intercombination lines (598 and 2.15 ns^{-1}, respectively, from above), but the cascade from the $^3P^{o\prime}_2$ would contribute a repopulation lifetime component to the intercombination decay curve that is only ten times longer than its own.

8.2.3 Singlet–triplet mixing in p^2 or p^4 configurations

For these configurations it was shown in Section 7.1.6 that the singlet–triplet mixing leads to wave functions

$$
\begin{aligned}
|^3P_0'\rangle &= \cos\theta_0 |^3P_0\rangle - \sin\theta_0 |^1S_0\rangle \\
|^3P_1'\rangle &= |^3P_1\rangle \\
|^3P_2'\rangle &= \cos\theta_2 |^3P_2\rangle - \sin\theta_2 |^1D_2\rangle \\
|^1D_2'\rangle &= \sin\theta_2 |^3P_2\rangle + \cos\theta_2 |^1D_2\rangle \\
|^1S_0'\rangle &= \sin\theta_0 |^3P_0\rangle + \cos\theta_0 |^1S_0\rangle,
\end{aligned}
\tag{8.22}
$$

where the mixing angles θ_0 and θ_2 can be deduced from spectroscopic energy-level data. In intermediate coupling, the selection rules are reduced to conservation of energy, no change of parity, and $\Delta J = 0, \pm 1$ (no $0 \to 0$), which allows the transitions

$$
S_{M1}\left(^3P_1', {}^3P_2'\right) = S_{M1}\left(^3P_1, {}^3P_2\right)\cos^2\theta_2 = \frac{5}{2}\cos^2\theta_2
\tag{8.23}
$$

$$
S_{M1}\left(^3P_0', {}^3P_1'\right) = S_{M1}\left(^3P_0, {}^3P_1\right)\cos^2\theta_0 = 2\cos^2\theta_0
\tag{8.24}
$$

$$
S_{M1}\left(^3P_1', {}^1D_2'\right) = S_{M1}\left(^3P_1, {}^3P_2\right)\sin^2\theta_2 = \frac{5}{2}\sin^2\theta_2
\tag{8.25}
$$

$$
S_{M1}\left(^3P_1', {}^1S_0'\right) = S_{M1}\left(^3P_1, {}^3P_0\right)\sin^2\theta_0 = 2\sin^2\theta_0
\tag{8.26}
$$

$$
\begin{aligned}
S_{M1}\left(^3P_2', {}^1D_2'\right) &= \left[\sqrt{S_{M1}\left(^3P_2, {}^3P_2\right)} - \sqrt{S_{M1}\left(^1D_2, {}^1D_2\right)}\right]^2 \sin^2\theta_2 \cos^2\theta_2 \\
&= \left[\sqrt{135/2} - \sqrt{30}\right]^2 \sin^2\theta_2 \cos^2\theta_2 = \frac{15}{2}\sin^2\theta_2 \cos^2\theta_2
\end{aligned}
\tag{8.27}
$$

Ni^{12+} and Ni^{14+} (more coronium)

In siliconlike Ni^{14+} and sulfurlike Ni^{12+}, transitions within the respective ground configurations $3s^2 3p^2$ and $3s^2 3p^4$ also contribute optical lines to the spectrum of the solar corona. For Ni^{14+} the lines are from the $^3P_1$–$^3P_2$ and $^3P_0$–$^3P_1$ transitions at air wavelengths $\lambda 6702$ and $\lambda 8024$ Å, and for Ni^{12+} the lines are from the $^3P_1$–$^1D_2$ and $^3P_2$–$^3P_1$ transitions at air wavelengths $\lambda 3643$ and $\lambda 5116$ Å. The M1 transition probabilities for these transitions can easily be computed using these semiempirical methods.

For Ni^{14+}, the ground configuration consists of the levels $^3P_0$, $^3P_1$, $^3P_2$, $^1D_2$ and $^1S_0$, with energies 0, 14 917.5, 27 376.5, 62 852.1 and 115 511 cm^{-1}. This maps into the effective Slater parameters $F_1 = 5711.7$, $\zeta_{pp} = 17415$, $\zeta_{pp}'' = 18422$ and $\zeta_{pp}' = 17\,934$ cm^{-1}, corresponding to mixing angles $\theta_0 = 13.41°$ and $\theta_2 = 21.95°$. This reduction of the spectroscopic energy-level data yields the transition probabilities

$$
A\left(^1S_0', {}^3P_1'\right) = \left[\frac{29.990}{994.1}\right]^3 2\sin^2(13.41°) = 2952 \text{ s}^{-1}
$$

$$
A\left(^1D_2', {}^3P_2'\right) = \frac{1}{5}\left[\frac{29.990}{2819}\right]^3 \frac{15}{2}\sin^2(21.95°)\cos^2(21.95°) = 217 \text{ s}^{-1}
$$

$$A\left({}^{1}D'_{2}, {}^{3}P'_{1}\right) = \frac{1}{5}\left[\frac{29.990}{2086}\right]^{3}\frac{5}{2}\sin^{2}(21.95°) = 208 \text{ s}^{-1}$$

$$A\left({}^{3}P'_{2}, {}^{3}P'_{1}\right) = \frac{1}{5}\left[\frac{29.990}{8026}\right]^{3}\frac{5}{2}\cos^{2}(21.95°) = 22 \text{ s}^{-1}$$

$$A\left({}^{3}P'_{1}, {}^{3}P'_{0}\right) = \frac{1}{3}\left[\frac{29.990}{6704}\right]^{3}2\cos^{2}(13.41°) = 56 \text{ s}^{-1}. \tag{8.28}$$

For Ni^{12+}, the ground configuration consists of the levels ${}^{3}P_{2}, {}^{3}P_{1}, {}^{3}P_{0}, {}^{1}D_{2}$, and ${}^{1}S_{0}$, with energies 0, 19 541.8, 20 060, 47 032.9 and 97 836.2 cm^{-1}. This maps into the effective Slater parameters $F_{1} = 5254.2$, $\zeta_{pp} = -15\,717$, $\zeta''_{pp} = -16\,078$ and $\zeta'_{pp} = -18\,180$ cm^{-1}, corresponding to mixing angles $\theta_{0} = 17.89°$ and $\theta_{2} = 16.59°$. This reduction of the spectroscopic energy-level data yields the transition probabilities

$$A\left({}^{1}S'_{0}, {}^{3}P'_{1}\right) = \left[\frac{29.990}{1277}\right]^{3}2\sin^{2}(17.89°) = 2444 \text{ s}^{-1}$$

$$A\left({}^{1}D'_{2}, {}^{3}P'_{2}\right) = \frac{1}{5}\left[\frac{29.990}{2126}\right]^{3}\frac{15}{2}\sin^{2}(16.59°)\cos^{2}(16.59°) = 314 \text{ s}^{-1}$$

$$A\left({}^{1}D'_{2}, {}^{3}P'_{1}\right) = \frac{1}{5}\left[\frac{29.990}{3638}\right]^{3}\frac{5}{2}\sin^{2}(16.59°) = 23 \text{ s}^{-1}$$

$$A\left({}^{3}P'_{1}, {}^{3}P'_{2}\right) = \frac{1}{3}\left[\frac{29.990}{5117}\right]^{3}\frac{5}{2}\cos^{2}(16.59°) = 154 \text{ s}^{-1}$$

$$A\left({}^{3}P'_{0}, {}^{3}P'_{1}\right) = \left[\frac{29.990}{192976}\right]^{3}2\cos^{2}(17.89°) = 0.0068 \text{ s}^{-1}. \tag{8.29}$$

Clearly the observation of these lines requires that the ions be in an environment of very low density, with very low collision rates, to permit these levels to decay by the M1 process.

8.2.4 Doublet–quartet mixing in p^{3} configurations

For these configurations it was shown in Section 5.4 that the doublet-quartet mixing leads to wave functions

$$\left|{}^{2}D^{o'}_{5/2}\right\rangle = \left|{}^{2}D^{o}_{5/2}\right\rangle$$

$$\left|{}^{4}S^{o'}_{3/2}\right\rangle = T_{SS}\left|{}^{4}S^{o}_{3/2}\right\rangle + T_{SD}\left|{}^{2}D^{o}_{3/2}\right\rangle + T_{SP}\left|{}^{2}P^{o}_{3/2}\right\rangle$$

$$\left|{}^{2}D^{o'}_{3/2}\right\rangle = T_{DS}\left|{}^{4}S^{o}_{3/2}\right\rangle + T_{DD}\left|{}^{2}D^{o}_{3/2}\right\rangle + T_{DP}\left|{}^{2}P^{o}_{3/2}\right\rangle$$

$$\left|{}^{2}P^{o'}_{3/2}\right\rangle = T_{PS}\left|{}^{4}S^{o}_{3/2}\right\rangle + T_{PD}\left|{}^{2}D^{o}_{3/2}\right\rangle + T_{PP}\left|{}^{2}P^{o}_{3/2}\right\rangle$$

$$\left|{}^{2}P^{o'}_{1/2}\right\rangle = \left|{}^{2}P^{o}_{1/2}\right\rangle. \tag{8.30}$$

Example: ground configuration of Fe^{19+}

To illustrate the use of this method, consider the $2s^2 2p^3$ ground configuration of Fe^{19+} in the nitrogen isoelectronic sequence [68]. Relative to the $^2P^o_{1/2}$ level, the measured energy levels are (in cm^{-1}) $E(^4S^o_{3/2}) = -260\,090$, $E(^2D^o_{3/2}) = -121\,820$, $E(^2D^o_{5/2}) = -84\,280$, $E(^2P^o_{3/2}) = 63\,090$. For these measured values,

$$21F_2 = 318\,820$$
$$90F_2^2 - 9\zeta^2/4 = 7.5805 \times 10^9$$
$$99F_2\zeta^2/4 = -1.9990 \times 10^{15}$$
$$-6F_2 = -84280. \tag{8.31}$$

Through overdetermined adjustment of the two Slater parameters to these four equations, the values $F_2 = 9\,253\,925$ and $\zeta = 63\,568$ are obtained. This yields a transformation matrix (with rows and columns labeled in the order $^2D^o_{5/2}$, $^4S^o_{3/2}$, $^2D^o_{3/2}$, $^2P^o_{3/2}$, $^2P^o_{1/2}$)

$$T_{ij} = \begin{pmatrix} 1 & 0 & 0 & 0 & 0 \\ 0 & -0.9279 & -0.1709 & 0.3314 & 0 \\ 0 & 0.3003 & -0.8694 & 0.3924 & 0 \\ 0 & 0.2211 & 0.2211 & 0.8580 & 0 \\ 0 & 0 & 0 & 0 & 1 \end{pmatrix}. \tag{8.32}$$

The LS transition elements can be obtained as the positive square root of the S_{M1} values in Table 8.1 that were computed from Eqs. 8.6, 8.7,

$$M_{ij} = \begin{pmatrix} \sqrt{378/5} & 0 & \sqrt{12/5} & 0 & 0 \\ 0 & \sqrt{60} & 0 & 0 & 0 \\ \sqrt{12/5} & 0 & \sqrt{48/5} & 0 & 0 \\ 0 & 0 & 0 & \sqrt{80/3} & \sqrt{4/3} \\ 0 & 0 & 0 & \sqrt{4/3} & \sqrt{2/3} \end{pmatrix}. \tag{8.33}$$

If we transform these transition elements from the LS to the stabilized representation

$$M'_{mn} = \sum_{i,j} T_{mi} M_{ij} T_{jn}^{-1} = \begin{pmatrix} 8.695 & -0.265 & -1.347 & 0.718 & 0 \\ -0.265 & 7.327 & -1.026 & -0.366 & 0.383 \\ -1.347 & -1.026 & 3.835 & 1.004 & 0.453 \\ 0.718 & -0.366 & 1.004 & 4.846 & 0.991 \\ 0 & 0.718 & -0.366 & 0.991 & 0.817 \end{pmatrix}, \tag{8.34}$$

then squaring each cell yields the semiempirical values for the M1 line-strength factor

$$S_{M1}(n, m) = |M_{nm}|^2 = \begin{pmatrix} 75.60 & 0.070 & 1.814 & 0.516 & 0 \\ 0.070 & 53.68 & 1.053 & 0.134 & 0.147 \\ 1.814 & 1.053 & 14.71 & 1.008 & 0.205 \\ 0.516 & 0.134 & 1.008 & 23.49 & 0.982 \\ 0 & 0.147 & 0.205 & 0.982 & 0.667 \end{pmatrix}. \tag{8.35}$$

Table 8.3. M1 transition probabilities connecting the $2s^2 2p^3$ levels in Fe^{19+}.
(From Ref. [68].)

Transition	SE $A(s^{-1})$	Obs. $\lambda(\text{Å})$	HFR-CI $A(s^{-1})$	HFR-CI $\lambda(\text{Å})$	MCDF $A(s^{-1})$	MCDF $\lambda(\text{Å})$
$^4S_{3/2}-^2D_{3/2}$	1.88×10^4	723	1.62×10^4	710	1.75×10^4	707
$^4S_{3/2}-^2D_{5/2}$	1.71×10^3	569	1.36×10^3	548	1.33×10^3	561
$^4S_{3/2}-^2P_{1/2}$	3.48×10^4	384	3.15×10^4	385	3.26×10^4	385
$^4S_{3/2}-^2P_{3/2}$	3.05×10^4	309	2.99×10^4	308	2.91×10^4	309
$^2D_{3/2}-^2D_{5/2}$	4.31×10^2	2664	5.80×10^2	2393	3.95×10^2	2714
$^2D_{3/2}-^2P_{1/2}$	5.00×10^3	821	5.66×10^3	842	5.52×10^3	846
$^2D_{3/2}-^2P_{3/2}$	4.30×10^4	541	4.39×10^4	544	4.29×10^4	550
$^2D_{5/2}-^2P_{3/2}$	1.11×10^4	679	1.13×10^4	705	1.22×10^4	689
$^2P_{1/2}-^2P_{3/2}$	1.66×10^3	1585	1.76×10^3	1541	1.64×10^3	1577

These values are converted to A-values using the wavelength factors and Eq. 8.8 and are summarized in column 2 of Table 8.3. Also presented in Table 8.3 are two types of theoretical calculations, one by Cowan [30] using his Hartree–Fock relativistic code with configuration interaction (HFR-CI), the other by Huang [121], using a multiconfiguration Dirac–Hartree–Fock code (MCDF). Clearly this simple semiempirical model is in very good agreement with both calculations.

8.2.5 Use of screening parameter extrapolations of energy levels to obtain M1 rates

In Chapter 5 it was shown that data for the energy levels of various configurations can be interpolated and extrapolated by first reducing the measurements to effective Slater and spin–orbit energies, and then making a screening parametrization of these quantities. The isoelectronic measurements for the raw energy levels have a complicated dependence on Z, but the reduction to the various F_k, G_k, and ζ_ℓ values separates the power-law dependences on Z, which can then be studied by an empirical exposition that renders it regular and slowly varying along the sequence.

In most cases the M1 transitions can be specified knowing only the LS values for the angular momentum line-strength factors and the intermediate coupling amplitudes of the system. Since the intermediate coupling can be specified from the Slater and spin–orbit energies that are so easily and accurately specified by screening parametrization, it is possible to compute the M1 transition probabilities with no additional inputs or information.

Calculations performed using the methods described in this section are displayed for M1 transitions within the $3s^2 3p^2$ ground configuration of the Si isoelectronic sequence in Fig. 8.2(a), within the $3s^2 3p^4$ ground configuration of the S isoelectronic sequence in Fig. 8.2(b), and within the $3s^2 3p^3$ ground configuration of the P isoelectronic sequence in

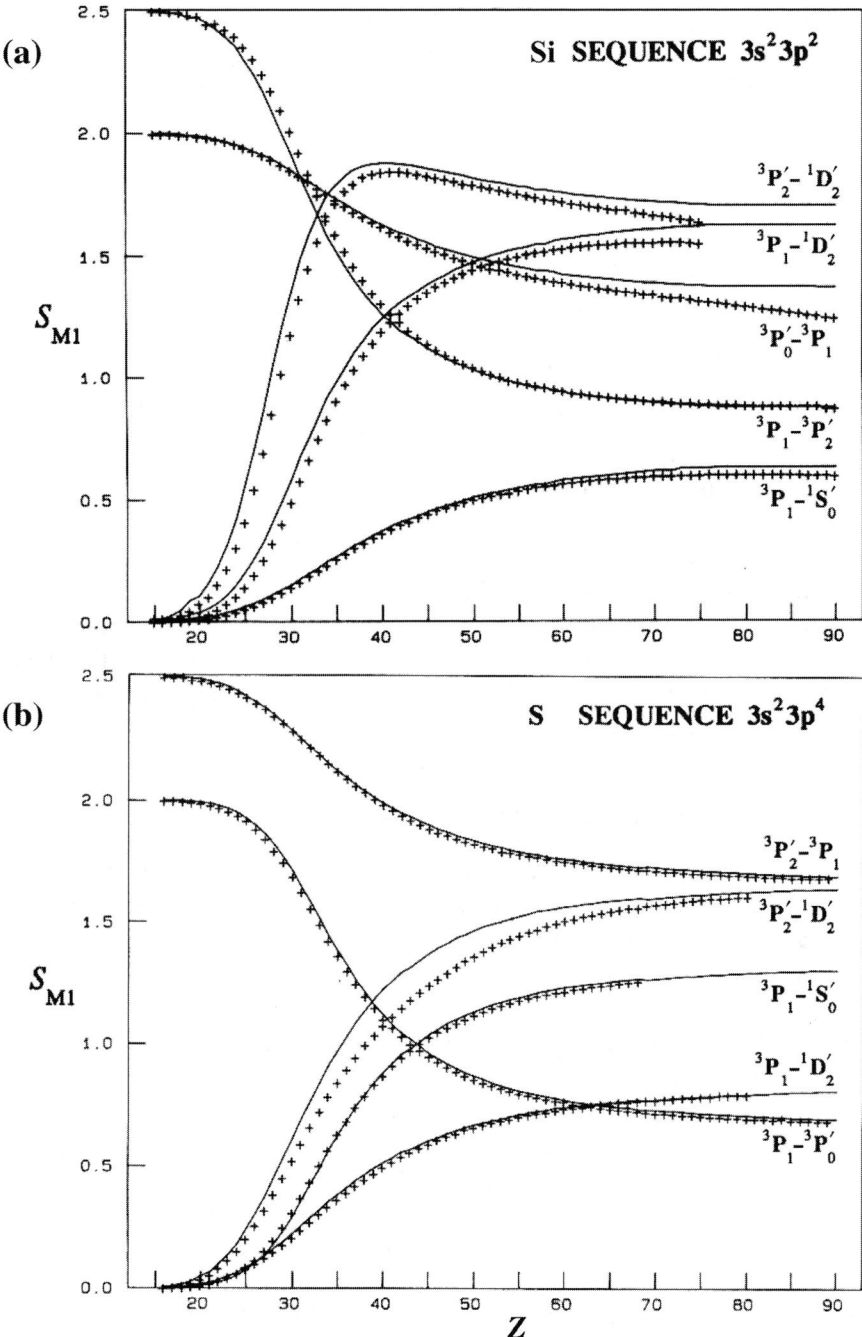

Fig. 8.2. Line strengths for the M1 transitions within (a) the $3s^2 3p^2$ ground configuration for the Si isoelectronic sequence and (b) the $3s^2 3p^4$ ground configuration for the S isoelectronic sequence. The solid lines denote calculations made using Eqs. 8.23–8.27, with mixing angles obtained from energy-level data. The symbols (+) represent *ab initio* MCDF calculations [122, 175]. (From Ref. [48].)

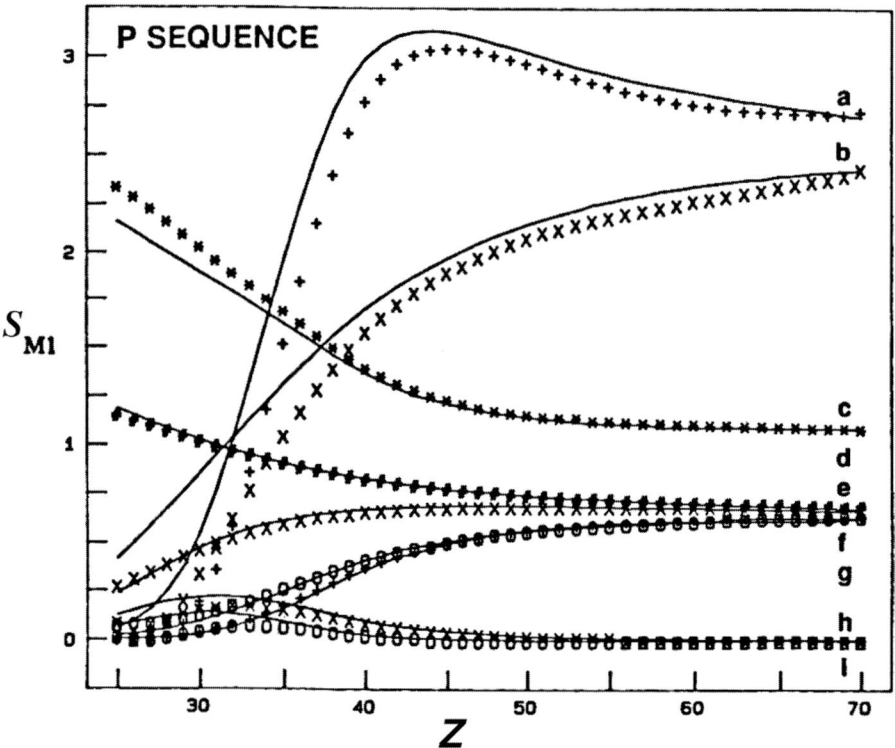

Fig. 8.3. Line strengths for the M1 transitions within the $3s^2 3p^3$ ground configuration the P isoelectronic sequence. The solid lines denote calculations made using the methods of Section 8.2.4, with mixing amplitudes obtained from energy-level data. The symbols represent *ab initio* MCDF calculations [121]. Transitions are labeled as follows: (a) $^4S_{3/2}-^2D_{3/2}$, (b) $^2D_{3/2}-^2P_{3/2}$, (c) $^2D_{3/2}-^2D_{5/2}$, (d) $^2P_{1/2}-^2P_{3/2}$, (e) $^2D_{5/2}-^2P_{3/2}$, (f) $^2S_{3/2}-^2P_{1/2}$, (g) $^4S_{3/2}-^2D_{5/2}$, (h) $^2D_{3/2}-^2P_{1/2}$, and (i) $^4S_{3/2}-^2P_{3/2}$. (From Ref. [68].)

Fig. 8.3. Plotted on these same figures are multiconfiguration Dirac–Hartee–Fock calculations (Si sequence [122], S sequence [175], P sequence [121]), for these same quantities. The agreement is excellent, and for the small discrepancies that do exist, it is possible that the semiempirical results may be preferable. The semiempirical results are based on the measured energy levels, hence may yield intermediate coupling amplitudes that are more accurate than the approximations of the *ab initio* calculations.

9

Absorption of radiation

Backward, turn backward, O time in your flight. Make me a child again, just for tonight.
                                                    – Elizabeth Akers Allen

Since the time of Kirchhoff it has been known that, when light is passed through an atomic gas, those wavelengths are observed that would be emitted if the gas were incandescent. If the gas is sufficiently cold, then the wavelengths observed are limited to ground-state transitions. As the temperature of the sample is elevated, wavelengths corresponding to transitions between excited states become absorbing, and balances between emission and absorption occur.

The study of the central wavelengths of emission lines or absorption notches is known as first-order spectroscopy, and it provides information on the energy-level structure of the atom. The study of the shape of these lines in frequency space is known as second-order spectroscopy, and provides information on the lifetime of the level and the collision rates and temperature of the gas. Thus, whereas first-order spectroscopy shows that emission and absorption measurements yield the same central wavelengths, second-order spectroscopy shows that the natural linewidth for emission and absorption are both specified by the level lifetime, and that the intensity of emission and absorption features both involve the line-strength factor (through the emission transition probability rate and the absorption oscillator strength).

The connection between the lifetime and the linewidth can be made plausible by a simple semiclassical model. When an electron is excited to a specific orbit in an atom, its binding energy is established through the exchange of virtual photons with the effective central core. If the perturbations that eventually cause the electron to make a transition from that orbit are weak, the meanlife will be long. In this case an equilibrium is established with the virtual-photon field, leading to a precisely determined binding energy. If the perturbations are strong and lead quickly to a transition from that orbit, the meanlife will be short. In this case the electron has little time to establish an equilibrium with the virtual-photon field, and the binding energy is less precisely determined. The same picture can apply to the lower orbit to which the transition is made, and both contribute to the energy spread of the emitted photons. While the uncertainty principle is a much more general concept than this simple picture, the model nonetheless illustrates the connection between the lifetime and the linewidth. When freed of initial and final conditions and the λ and λ^3 antenna factors, absorption can be viewed as a time-reversed version of this picture.

9.1 Driven damped harmonic oscillator

The absorption of radiation by a material medium can be modeled classically in terms of a driven damped harmonic oscillator. If we consider an electron in an atom as a one-dimensional oscillator with a Hooke's law restoring force, velocity-dependent viscous drag force, and a jerks-dependent radiation reaction force, then $F = ma$ becomes

$$-eE(t) - m\omega_0^2 x - 2m\gamma \frac{dx}{dt} - \frac{m}{\omega_0^2 \tau_0} \frac{d^3 x}{dt^3} = m \frac{d^2 x}{dt^2}. \tag{9.1}$$

For simplicity in taking derivatives, we adopt complex notation. Provided the equations remain linear, the sinusoidal solution can be recovered by taking the real part. Here $E(t)$ is the oscillatory electric field strength, ω_0 is the natural frequency of the oscillator, γ is the collisional damping constant, and $1/\tau_0$ is the reciprocal radiative lifetime of a simple harmonic oscillator $2Ke^2\omega_0^2/3mc^3$. Making the periodic ansatz

$$E(t) = E_0 e^{-i\omega t}; \qquad x(t) = x_0 e^{-i\omega t} \tag{9.2}$$

forces the condition

$$-eE(t)/m = \left[(\omega_0^2 - \omega^2) - i(2\gamma\omega + \omega^3/\omega_0^2 \tau_0) \right] x(t). \tag{9.3}$$

If we restrict the consideration to optical frequencies, then $|\omega_0 - \omega| \ll \omega$, so we can make the approximation

$$(\omega_0^2 - \omega^2) = (\omega_0 - \omega)(\omega_0 + \omega) \cong (\omega_0 - \omega)2\omega, \tag{9.4}$$

and the displacement can be written as

$$x(t) = -\frac{eE(t)}{2m\omega} \frac{1}{(\omega_0 - \omega) - i(\gamma + \omega^2/2\omega_0^2 \tau_0)}. \tag{9.5}$$

This is for a single frequency. Let us extend the consideration to a set of N_i atoms/volume, each with a set of frequencies ω_{ij} and an oscillator strength f_{ij}. This oscillatory motion will generate an oscillating dipole moment $-ex_{ij}(t)$ in the atom. If there is an ensemble of these oscillators, then the electric field P produced by this polarization is

$$P = N_i \sum_j f_{ij}[-ex_{ij}(t)] = \epsilon_0 \chi_e E(t) \tag{9.6}$$

where χ_e is the electric susceptibility, obtained by combining Eqs. 9.5 and 9.6,

$$\chi_e = \frac{N_i e^2}{2\epsilon_0 m\omega} \sum_j \frac{f_{ij}}{(\omega_{ij} - \omega) - i(\gamma + \omega^2/\omega_{ij}^2 \tau_i)}. \tag{9.7}$$

The permittivity ϵ is defined

$$\epsilon \equiv \epsilon_0(1 + \chi_e). \tag{9.8}$$

The index of refraction is the square root of the dielectric constant (the ratio of the permittivity to that of free space). Since $\chi_e \ll 1$ this can be binomial expanded

$$n = \sqrt{\epsilon/\epsilon_0} \cong 1 + \frac{1}{2}\chi_e = 1 + \pi \frac{N_i K e^2}{m\omega} \sum_j \frac{f_{ij}}{(\omega_{ij} - \omega) - i(\gamma + \omega^2/2\omega_{ij}^2 \tau_i)}, \tag{9.9}$$

where $K \equiv 1/4\pi\epsilon_0$. Rationalizing the denominator, the index of refraction has a real part

$$n_R = 1 + \pi \frac{N_i K e^2}{m\omega} \sum_j f_{ij} \frac{(\omega_{ij} - \omega)}{(\omega_{ij} - \omega)^2 + (\gamma + \omega^2/2\omega_{ij}^2 \tau_i)^2} \tag{9.10}$$

which is frequency dependent and therefore exhibits dispersion. Near an absorption line where $\omega_{ij} \cong \omega$ this becomes "anomalous dispersion" which forms the basis for the measurement of oscillator strengths by the "hook" method (the interference fringes acquire an oblique structure with characteristic hook-shaped patterns about their centers). The index of refraction also has an imaginary part

$$n_I = \pi \frac{N_i K e^2}{m\omega} \sum_j f_{ij} \frac{(\gamma + \omega^2/2\omega_{ij}^2 \tau_i)}{(\omega_{ij} - \omega)^2 + (\gamma + \omega^2/2\omega_{ij}^2 \tau_i)^2} \tag{9.11}$$

which causes the material to absorb light. This can be seen by considering a propagating wave

$$E = E_0 e^{i(kx - \omega t)} = E_0 e^{i\omega(n_R x/c - t)} e^{-n_I \omega x/c}, \tag{9.12}$$

for which the intensity is given by

$$I(x) = E^* E = I(0) e^{-2\omega n_I x/c}. \tag{9.13}$$

This can be compared to the cross section in the equation

$$I(x) = I(0) e^{-N_i \sigma x} \tag{9.14}$$

to obtain

$$\sigma = \frac{2\omega}{N_i c} n_I = 2\pi \frac{K e^2}{mc} \sum_j f_{ij} \frac{(\gamma + \omega^2/2\omega_{ij}^2 \tau_i)}{(\omega_{ij} - \omega)^2 + (\gamma + \omega^2/2\omega_{ij}^2 \tau_i)^2}. \tag{9.15}$$

If we rewrite this in terms of energies instead of frequencies

$$E_{ij} = \hbar\omega_{ij}; \qquad E = \hbar\omega, \tag{9.16}$$

and express the damping term as a full width at half maximum (FWHM),

$$\Gamma_{ij}/2 = \hbar(\gamma + \omega^2/2\omega_{ij}^2 \tau_i), \tag{9.17}$$

then this has a form

$$\sigma = 2\pi \frac{K e^2 \hbar}{mc} \sum_j f_{ij} \left[\frac{\Gamma_{ij}/2}{(E_{ij} - E)^2 + (\Gamma_{ij}/2)^2} \right]. \tag{9.18}$$

The bracketed portion is the Lorentzian lineshape, which is characteristic of both collisional broadening (γ) and natural linewidth ($\omega^2/2\omega_{ij}^2 \tau_i$).

The partial cross section σ_{ij} for absorption by a specific spectral transition can be written, introducing atomic constants and normalizing factors, as

$$\sigma_{ij} = (2\pi)^2 \alpha a_0^2 Ry f_{ij} \left[\frac{1}{\pi} \frac{\Gamma_{ij}/2}{(E_{ij} - E)^2 + (\Gamma_{ij}/2)^2} \right]. \tag{9.19}$$

In this form the bracketed Lorentzian profile integrates to unity, the units of the energy are set by those of Ry, and the length-squared nature of the cross section is made explicit by the a_0^2 units.

The constants appearing in various forms of this expression can differ by factors of 2π, $\hbar$, and c/λ^2, depending on whether the dispersion is represented as a function of the energy, the radian/s frequency, the Hertz frequency, or the wavelength, since

$$\int dE\, \sigma_1(E) = \hbar \int d\omega\, \sigma_2(\omega) = 2\pi\hbar \int d\nu\, \sigma_3(\nu) \approx \frac{2\pi\hbar c}{\lambda_0^2} \int d\lambda\, \sigma_4(\lambda). \tag{9.20}$$

Here the choice depends on taste and the nature of the dispersive element (e.g., dispersion by frequency for a prism, dispersion by wavelength for a grating). Provided the line is sufficiently narrow that the λ^2 correction can be made using the central wavelength, these multiplicative factors specify the relative absorption of different features.

The Lorentzian function that occurs here can be generalized to an arbitrary dispersion function $g_{ij}(E)$ that has unit normalization

$$1 = \int_0^\infty dE\, g_{ij}(E), \tag{9.21}$$

with the corresponding partial cross section

$$\sigma_{ij} = (2\pi)^2 \alpha a_0^2 Ry f_{ij} g_{ij}(E). \tag{9.22}$$

9.2 Doppler broadening

Because of factors such as the thermal motion of the atoms in the sample, there are other sources of broadening. If the source has a temperature T, then the velocity distribution of the absorbing and emitting atom will follow a Boltzmann function

$$dN/dE \propto e^{-E/kT} = e^{-mv^2/2kT}. \tag{9.23}$$

Along the direction of motion, the light emitted or absorbed will have a shift in frequency (nonrelativistically)

$$\hbar\omega = \hbar\omega_0(1 + v/c); \tag{9.24}$$

hence, the Boltzmann distribution can be rewritten in terms of

$$v = c\frac{\hbar\omega - \hbar\omega_0}{\hbar\omega_0}, \tag{9.25}$$

to obtain

$$dN/dt \propto \exp\left[-\frac{(\hbar\omega - \hbar\omega_0)^2}{2kT(\hbar\omega_0)^2/mc^2} \right]. \tag{9.26}$$

Rewriting this using the corresponding energies and the equivalent Gaussian standard deviation,

$$E = \hbar\omega; \qquad E_0 = \hbar\omega_0; \qquad \sigma_G^2 \equiv kT(\hbar\omega_0)^2/mc^2, \tag{9.27}$$

yields a Gaussian lineshape

$$dN/dt \propto \exp\left[-\frac{(E - E_0)^2}{2\sigma_G^2}\right]. \tag{9.28}$$

9.3 Comparison of the Lorentzian and Gaussian functions

The Lorentzian function

$$L_{ab}(x) = \frac{b/\pi}{(x - a)^2 + b^2} \tag{9.29}$$

has a number of useful properties. It has a central height

$$L_{ab}(a) = \frac{b/\pi}{(a - a)^2 + b^2} = \frac{1}{\pi b}, \tag{9.30}$$

and attains half that value at a distance b from its center

$$L_{ab}(a \pm b) = \frac{b/\pi}{b^2 + b^2} = \frac{1}{2\pi b}. \tag{9.31}$$

The integral of the area under the curve

$$\int_0^\infty dx\, L_{ab}(x) = \int_0^\infty dx \frac{b/\pi}{(x - a)^2 + b^2} = \frac{1}{\pi}\left[\tan^{-1}\left(\frac{x - a}{b}\right)\right]_0^\infty \tag{9.32}$$

is not symmetric because it is bounded from below, but approaches unity for $a \ll b$. The fraction of the curve contained within n units of the FWHM is

$$\int_{a-nb}^{a+nb} dx\, L_{ab}(x) = \frac{2}{\pi} \tan^{-1}(n). \tag{9.33}$$

The Gaussian standard deviation can be written in terms of a corresponding full width at half maximum

$$e^{-(x-a)^2/2\sigma_G^2} \to 1/2; \qquad (x - a) = \sqrt{2\ln 2}\, \sigma_G \tag{9.34}$$

so, in terms of the normalization, the fraction of the curve contained within one unit of the FWHM is

$$\frac{1}{2\pi\sigma_G} \int_{-\sqrt{2\ln 2}\sigma_G}^{\sqrt{2\ln 2}\sigma_G} dx\, e^{-x^2/2\sigma_G^2}. \tag{9.35}$$

Table 9.1 indicates the percentages of the Lorentzian and Gaussian distributions that are contained within various numbers of FWHMs. It is clear from this comparison that the Lorentzian profile has much wider wings than that of the Gaussian.

Table 9.1. Comparison of the wings of the
Lorentzian and Gaussian functions.

Number of FWHMs	Included in Lorentzian (%)	Included in Gaussian (%)
1	50.0	76.1
2	70.5	98.1
3	79.5	99.95
4	84.4	100
5	87.4	100
10	93.7	100
100	99.4	100

9.4 Convolutions of lineshape functions

In general, a measured lineshape is a mixture of many different broadening processes. In the development of the Lorentzian lineshape we have already seen that it can include the effects of both radiation and collisional broadening. Since a transition connects two levels, the effective linewidth includes the radiation broadening of both the upper and lower levels, each possessing its own Lorentzian profile. Similarly, the thermal broadening affects both the source and the absorber, which may be at different temperatures, and thus possess differing Gaussian profiles. Thus it is necessary to perform convolution integrals of Lorentzian profiles with other Lorentzian profiles, Gaussian profiles with other Gaussian profiles, and Lorentzian profiles with Gaussian profiles. These integrals can be performed very effectively through the use of the convolution theorem of Fourier transform theory.

The Fourier transform and its inverse can be defined

$$f(k) = \frac{1}{\sqrt{2\pi}} \int_{-\infty}^{\infty} \mathrm{d}x \, \mathrm{e}^{-ikx} F(x)$$

$$F(x) = \frac{1}{\sqrt{2\pi}} \int_{-\infty}^{\infty} \mathrm{d}k \, \mathrm{e}^{ikx} f(k). \tag{9.36}$$

The convolution theorem can be deduced by considering the quantity

$$\frac{1}{\sqrt{2\pi}} \int_{-\infty}^{\infty} \mathrm{d}k \, \mathrm{e}^{ikx} f(k) g(k) = \int_{-\infty}^{\infty} \mathrm{d}k \, \mathrm{e}^{ikx} f(k) \left[\frac{1}{\sqrt{2\pi}} \int_{-\infty}^{\infty} \mathrm{d}x' \, \mathrm{e}^{-ikx'} G(x') \right]$$

$$= \int_{-\infty}^{\infty} \mathrm{d}x' G(x') \left[\frac{1}{\sqrt{2\pi}} \int_{-\infty}^{\infty} \mathrm{d}k \, \mathrm{e}^{ik(x-x')} f(k) \right] \tag{9.37}$$

where, in the second step, we have interchanged the orders of integration. This can be identified as

$$\frac{1}{\sqrt{2\pi}} \int_{-\infty}^{\infty} \mathrm{d}k \, \mathrm{e}^{ikx} f(k) g(k) = \int_{-\infty}^{\infty} \mathrm{d}x' G(x') F(x - x'). \tag{9.38}$$

Thus the convolution of two displaced functions F and G (on the right hand side) can be written as the inverse transform of the product of these transforms (on the left hand side). This will be applied below.

9.4.1 Convolution of two Lorentzians

Consider the two normalized Lorentzian functions

$$F(x) = \frac{b_1/\pi}{x^2 + b_1^2}; \qquad G(x) = \frac{b_2/\pi}{x^2 + b_2^2}. \tag{9.39}$$

Their Fourier transforms can be shown to be

$$f(k) = \frac{1}{\sqrt{\pi}} e^{-kb_1}; \qquad g(x) = \frac{1}{\sqrt{\pi}} e^{-kb_2}. \tag{9.40}$$

The desired integral is given by

$$\int_{-\infty}^{\infty} dx' G(x') F(x - x') = \frac{1}{\pi} \int_{-\infty}^{\infty} dk e^{ikx} e^{-kb_1} e^{-kb_2} = (ix - b_1 - b_2)^{-1}. \tag{9.41}$$

Rationalizing the denominator and taking the real part,

$$\int_{-\infty}^{\infty} dx' \left[\frac{b_1/\pi}{(x')^2 + b_1^2}\right]\left[\frac{b_2/\pi}{(x - x')^2 + b_2^2}\right] = \frac{(b_1 + b_2)/\pi}{x^2 + (b_1 + b_2)^2}, \tag{9.42}$$

and the convolution of two Lorentzians is a Lorentzian of width given by the linear sum of the two widths comprising it. The process can be repeated for any number of convolved Lorentzians. Thus, for a transition that has finite lifetimes for both its upper and lower levels and collisional broadening, the total width is given by the sum of the width of the upper level, the width of the lower level, and the collisional width.

9.4.2 Convolution of two Gaussians

Consider the two normalized Gaussian functions

$$F(x) = \frac{1}{\sqrt{2\pi}\,\sigma_1} e^{-x^2/2\sigma_1^2}; \qquad G(x) = \frac{1}{\sqrt{2\pi}\,\sigma_2} e^{-x^2/2\sigma_2^2}. \tag{9.43}$$

Their Fourier transforms can be shown to be

$$f(k) = \frac{1}{\sqrt{2\pi}} e^{-k^2\sigma_1^2/2}; \qquad g(x) = \frac{1}{\sqrt{2\pi}} e^{-k^2\sigma_2^2/2}. \tag{9.44}$$

The desired integral is given by

$$\int_{-\infty}^{\infty} dx' \, G(x') F(x - x') = \frac{1}{2\pi} \int_{-\infty}^{\infty} dk e^{ikx} e^{-k^2\sigma_1^2/2} e^{-k^2\sigma_1^2/2}. \tag{9.45}$$

We can complete the square of the exponential terms on the right hand side

$$-\frac{(\sigma_1^2 + \sigma_1^2)k^2 - (2i)kx}{2} = -\frac{1}{2}\left[\sqrt{\sigma_1^2 + \sigma_1^2}\,k - \frac{ix}{\sqrt{\sigma_1^2 + \sigma_1^2}}\right]^2 - \frac{x^2}{2(\sigma_1^2 + \sigma_1^2)}. \tag{9.46}$$

The bracketed portion integrates to the norm, leaving

$$\int_{-\infty}^{\infty} dx' \left[\frac{e^{-(x')^2/2\sigma_1^2}}{\sqrt{2\pi}\sigma_1} \right] \left[\frac{e^{-(x-x')^2/2\sigma_2^2}}{\sqrt{2\pi}\sigma_1} \right] = \frac{1}{\sqrt{2\pi\left(\sigma_1^2 + \sigma_1^2\right)}} \exp\left[-\frac{x^2}{2\left(\sigma_1^2 + \sigma_1^2\right)} \right] \quad (9.47)$$

and the convolution of two Gaussians is a Gaussian of width given by the quadrature sum of the two widths comprising it.

9.4.3 Convolution of a Lorentzian with a Gaussian

The convolution of a Lorentzian and a Gaussian profile has been extensively studied using approximation techniques and numerical methods. The intensity distribution is called the Voigt profile, and its mathematical representation is called the Voigt function. It was first investigated [188] in 1912 by Göttingen Professor Woldemar Voigt. Voigt made many notable contributions that are sometimes overlooked, and are worthy of recall. It was Voigt who, in an 1887 study [187] of the Doppler shift, first deduced the relativistic transformation equations that now bear the name of Hendrik Antoon Lorentz (from a 1905 application of Voigt's equations to conduction in metals). It was also Voigt who, in his studies of elasticity, developed the methods and gave the name to tensor mathematics. Interestingly, Voigt was also one of the foremost musicians of the time.

Consistent with the fact that a convolution of Lorentzians is a Lorentzian and a convolution of Gaussians is a Gaussian, it follows that a convolution of Voigt profiles yields another Voigt profile. The spectroscopist usually wishes to deduce the Lorentzian and Gaussian widths from observed lineshape data, and at the same time to test whether the observed profile can be satisfactorily represented by a convolution of these two types of mathematical functions. Many tabulations and graphical expositions of these functions exist in the literature (e.g., Ref. [79]) with various choices for how the widths and relative intensities of the two functions are presented. These tabulations have largely been superceded by standardized computer software that is now available.

9.4.4 The inverse Lorentzian function

The inverse Lorentzian function is a very useful quantity that has applications in spectrum simulation. In any real spectrum, the data will cluster about the line center with a Lorentzian distribution. The probability $P(x, b)$ that a point will lie a distance x or greater from the line center is given by

$$P(x, b) = \int_x^{\infty} dx' \frac{b/\pi}{x'^2 - b^2} = \frac{1}{\pi} \left[\tan^{-1}\left(\frac{x'}{b}\right) \right]_x^{\infty} = \frac{1}{2} - \frac{1}{\pi} \tan^{-1}\left(\frac{x}{b}\right), \quad (9.48)$$

so the value of x that corresponds to a given probability is

$$x = b \tan\left[\pi \left(\frac{1}{2} - P \right) \right]. \quad (9.49)$$

The probability is uniformly distributed from 0 to 1, so the argument of the tangent varies between $-\pi/2$ and $+\pi/2$, corresponding to a value of x between $-\infty$ and $+\infty$. A large positive or negative excursion of x corresponds to a small value for the tangent, and vice versa.

If a spectral line has a central energy E_0 and a full width at half maximum Γ, a set of Lorentzian distributed points $E(i)$ can be obtained by the procedure

$$E(i) = E_0 + \Gamma \tan\left[\pi\left(\frac{1}{2} - \text{RANDOM}(i)\right)\right] \tag{9.50}$$

where RANDOM(i) is a set of numbers from 0 to 1 produced by a random-number generator. This is the procedure by which Fig. 1.1 was produced.

A similar procedure can be used to simulate a Gaussian distribution, separately, or in conjunction with a Lorentzian distribution to obtain a Voigt profile. Standard numerically-based computer programs exist that return values for the inverse Gaussian (the value of x corresponding to a given value for the probability).

9.5 Equivalent width

There are many reasons why a measured spectral distribution can deviate significantly from a Voigt profile. For emission measurements, the detection apparatus may not be linear, and saturation may occur which truncates the peak. In absorption measurement, a very strong absorption feature may cause extinction of the light in the center of the line, leading to a truncated distribution with only the wings remaining. Various degrees of absorption-line saturation are illustrated in the upper panel of Fig. 9.1.

In emission there can be effects known as self-absorption and self-inversion, if the optical path between the source and detector passes through regions of differing temperature. The hot regions will be dominated by emission due to the copious population of the excited states, whereas absorption will occur in the cooler regions because of the large population of the ground state. Moreover, the Doppler width of the light from a hot emission region will be broader than that of the absorption feature in a cooler region. This absorption can produce a notch in the emission line, either at or displaced from its center, depending on the presence of convection between the two regions. (This observed crater in the peak can be mistaken for a doublet!).

Another possibility is "hole burning" in absorption. Here, the intensity of the light at the center of the line is sufficient to completely depopulate the lower level, creating a transparency at that wavelength.

Since there are many possibilities for the shapes of absorption lines, it is important to develop a means for characterizing spectral absorption that is independent of the mathematical form of the distribution. If we consider a beam of white light of intensity I_0 passing through a gas that contains N_i absorbers/volume, the integrated intensity lost after passing through a thickness L of the sample is

$$\int_0^\infty dE\,\Delta I(E) = I_0 \int_0^\infty dE\left[1 - e^{-N_i \sigma_{ij}(E)L}\right] \cong I_0 N_i L \int_0^\infty dE\,\sigma_{ij}(E) \tag{9.51}$$

where the approximation assumes that the sample optically thin, hence the exponential can be approximated as $e^{-x} \cong 1 - x$. Inserting the general expression for the partial absorption

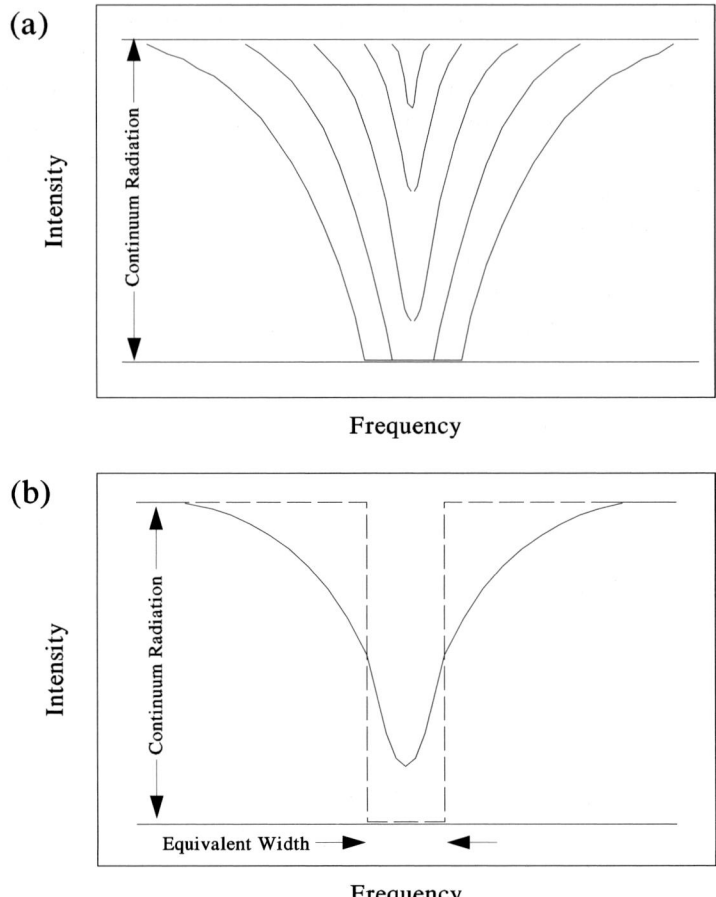

Fig. 9.1. Saturation effects and equivalent width. (After Ref. [2].)

cross section for an isolated spectral line given in Eq. 9.22, the equivalent width W_{ij} can be computed (noting that the integral of the spectral distribution $g_{ij}(E)$ is normalized to unity in Eq. 9.21)

$$W_{ij} \equiv \frac{\int_0^\infty dE \, \Delta I(E)}{I_0} = (2\pi)^2 \alpha a_0^2 Ry N_i L f_{ij}. \tag{9.52}$$

This concept of the equivalent width is illustrated in the lower panel of Fig. 9.1. This is the width of a hypothetical absorption line with rectangular profile that has zero intensity along its entire width, and represents the same subtraction from the spectrum as the actual line. To within overall constants, the equivalent width is proportional to the product of the column density $N_i L$ (the number of atoms/m^2 in that state along the path) and the absorption oscillator strength, and thus is given by

$$W_{ij} \propto N_i L f_{ij}. \tag{9.53}$$

If either of these quantities is known, the other can be determined from the measured equivalent width. For this proportionality to hold, the assumption of an optically thin source is essential. If the number of absorbing atoms is large, or the magnitude of the oscillator strength is large, or both, the absorption feature is driven into saturation, and no light remains to be absorbed except along the wings. The generalized study of the relationship between absorption and equivalent width is called the "curve-of-growth" method, and the optically thin condition is called "the linear portion of the curve-of-growth." There are many applications of this technique in astrophysics, and this illustrates the virtue of a knowledge of the oscillator strengths of very weak lines, since they are likely to be unsaturated in absorption. A few illustrative examples will be given below.

9.5.1 Example: column densities of interstellar atoms along the line-of-sight from a star

Absorption spectra in continuum light from a distant star can be used to specify the relative abundances of elements in the interstellar medium. Suppose that astronomical line-of-sight measurements indicate that the ratio of the equivalent width of the unsaturated absorption line for the $2s^2 2p^2\ ^3P$–$2s2p^3\ ^3D^o$ transition in neutral carbon at 1561 Å to that of the hydrogen Lyman-α transition at 1216 Å is given by

$$\frac{W_\omega(1561)}{W_\omega(1216)} = \frac{W_\lambda(1561)}{W_\lambda(1216)} \left[\frac{1216\ \text{Å}}{1561\ \text{Å}} \right]^2 = 7.2 \times 10^{-5} \tag{9.54}$$

where a wavelength correction is needed if the dispersion is according to wavelength, but not if it is according to frequency or energy. Independent laboratory measurements of the oscillator strengths for these transitions have yielded the values $f(1561) = 0.0810$ and $f(1216) = 0.4154$. From these data, the combination of astrophysical and laboratory measurements specify the ratio of the column densities to be

$$\frac{N(\text{C})L}{N(\text{H})L} = \frac{W_\omega(1561)/f(1561)}{W_\omega(1216)/f(1216)} = (7.2 \times 10^{-5})\frac{(0.4154)}{(0.0810)} = 3.7 \times 10^{-4}. \tag{9.55}$$

Astronomical abundances are usually quoted in parts/trillion parts hydrogen on a $\log_{10}$(dex, or decimal exponent) scale, so

$$[\text{C}] = \log_{10}[10^{12} N(\text{C})/N(\text{H})] = 12 - 3.34 = 8.57\ \text{dex}. \tag{9.56}$$

9.5.2 Example: stellar temperature determination from relative absorption among levels in a given atom

If it can be assumed that local thermodynamic equilibrium exists in a star, then the populations of the various levels in a hydrogen atom should follow the Boltzmann distribution

$$N_{n\ell} = N_0(2\ell + 1)\exp(-E_n/k_\text{B}T). \tag{9.57}$$

Suppose that absorption spectra from a certain star indicates that the equivalent width for Lyman-α (Lyα) is 10^6 times larger than the equivalent width for Balmer (Hα). The oscillator strengths for hydrogen are as follows: $f_{1s,2p} = 0.4162$; $f_{2s,3p} = 0.4349$; $f_{2p,3s} = 0.01359$; $f_{2p,3d} = 0.6958$. Taking into account the degeneracies of the s- and p- levels, the ratio of the equivalent widths is given by

$$\frac{W(\mathrm{H}\alpha)}{W(\mathrm{Ly}\alpha)} = \frac{1(f_{2s,3p}) + 3(f_{2p,3s} + f_{2p,3d})}{1(f_{1s,2p})} \exp[-(E_2 - E_1)/k_\mathrm{B}T] \tag{9.58}$$

or,

$$10^{-6} = \frac{(0.4349) + 3(0.01359 + 0.6958)}{(0.04162)} \exp(-10.2\,\mathrm{eV}/k_\mathrm{B}T), \tag{9.59}$$

which yields $k_\mathrm{B}T = 0.652$, or $T = 7600$ K.

If stars of increasing temperature are studied, it is to be expected that the absorption of Hα would increase (as the population of the $n = 2$ levels grow) and the absorption of Lyα should decrease (since the increased population of the excited states occurs at the expense of that of the ground state).

Cecilia Payne-Gaposchkin used [164] this technique to classify stars according to their temperature. Rather than using hydrogen to probe both ground and excited states (the Lyα UV radiation was inaccessible prior to satellite-borne spectrographs), Payne-Gaposchkin used the Ca$^+$ 4s–4p H and K resonance lines, because the low excitation of these $\Delta n = 0$ transitions causes the population ratio to be very sensitive to the temperature. In low-temperature stars the populations are dominantly in the ground state, so the H and K lines are strongly absorbing, whereas the stellar envelope is transparent to Hα. In high-temperature stars the atomic ground states are depleted, so the H and K absorption is diminished whereas the Hα absorption is significant.

While the work of Kirchhoff and Bunsen had shown that the elements in the Sun and stars could be identified by spectroscopic observations, it remained for Payne-Gaposchkin to make the first detailed determination of the abundances of the elements in the stars. She recognized that the spectroscopic classifications of star groups developed [24] by Annie Jump Cannon were in fact correlated to the degree of ionization and hence to the stellar temperature. The strength of a given absorption line depends on the oscillator strength, the amount of that element present, and the population of the lower level. The Boltzmann equation predicts the relative populations within a given ion, but it was the extension of the equilibrium formula to prescribe the distribution of ionization by Meg Nad Saha in 1920 that provided [174] the key. Payne-Gaposchkin's use of the Saha equation to specify the population from the temperature enabled her to specify the stellar abundances, and her 1925 thesis [164] provided the first detailed determination of the elemental abundances of the stars.

Payne-Gaposchkin's work showed that the gross differences among stellar spectral types are due primarily to temperature, and that differences in chemical composition are minor effects. It is interesting to note that her discovery that hydrogen is the overwhelmingly dominant element in stars was treated with great initial skepticism by Sir Arthur Eddington, Henry Norris Russell, and other leading experts of the time [110].

9.6 Atomic derivation of the Planck radiation law

Einstein's 1917 formulation of "The quantum theory of radiation" [92] was the first formulation of a number of different physical processes. It was the first definition of the Einstein A and B coefficients that govern spontaneous emission, stimulated emission, and absorption. It was the first consideration of stimulated emission (Einstein referred to it as "induced emission of radiation"), which provided the basis for the maser and laser long before their development. It was the first demonstration that the Planck continuum radiation law can be deduced from the Bohr theory of the atom and the Boltzmann statistical law for an ensemble in thermal equilibrium. A major portion of the paper was devoted to the recognition that, while the emission of radiation can be considered as consisting of spherical waves, the absorption of radiation is a "fully directed event" involving a plane wave. Thus, while emission could be treated through conservation of energy, Einstein showed that absorption required the consideration of the directed momentum of the photon and the recoil of the absorbing atom. Thus the paper also contained the first formulation of what was later to be known as the Compton effect.

Einstein's introduction of the A and B coefficients was conceptual, assuming (in analogy to nuclear decay processes) that the various rates were proportional to the instantaneous number of atoms multiplied by a constant coefficient. Since this was done before the relationships of these coefficients to the electric dipole matrix element was formulated, Einstein had to deduce the ratio of A to B empirically, by forcing it to conform to the known quantities that occur in the Planck law. Since these relationships are now known, it is possible to generalize Einstein's formulation, obtaining the Planck law from first principles without recourse to the empirical specification of any factors.

Einstein's A coefficient is still in use, and is the standard spontaneous transition probability rate A_{ij}. The use of the B coefficient has largely been replaced by the oscillator strength f_{ji}, which is a dimensionless quantity that relates the absorption of radiation by a classical simple harmonic oscillator to that of the corresponding quantum mechanical system. The Einstein B coefficient differs from the A coefficient in that it involves the energy/time rather than photons/time, which also reflects itself in the definition of the radiation density $\rho_E(\omega)$ as an energy density rather than a photon density. (Thus the units of ρ_E are J-s/m^3 and those of B are photons/s per unit ρ_E.) The relationship between the Einstein B coefficient and the oscillator strength is given by

$$\pm \hbar \omega B_{ji} = \frac{\pi K e^2}{m} f_{ji}, \qquad (9.60)$$

where the sign is plus for stimulated emission and minus for absorption. (The definition of the oscillator strength contains the sign implicitly, whereas for the B coefficient the sign is put into the population equation explicitly). The quantity $\pi K e^2 / m$ is the corresponding value for a classical simple harmonic oscillator [159]. Correspondingly, the radiation density will also be formulated on the basis of photons rather than joules, using

$$\rho_{\text{ph}}(\omega) \equiv \frac{\rho_E(\omega)}{\hbar\omega}. \tag{9.61}$$

Here $\rho_E(\omega)$ is the standard radiation density in J-s/m^3, and $\rho_{\text{ph}}(\omega)$ is the photon density in photon-s/m^3. Thus, we can make the substitution in the Einstein formulation

$$B_{ji}\rho_E(\omega) = \frac{\pi K e^2}{m} f_{ji}\rho_{\text{ph}}(\omega). \tag{9.62}$$

In terms of these quantities, the population equation for an arbitrary pair of levels a and b with populations N_a and N_b (where $E_a > E_b$) is

$$-\frac{dN_a}{dt} = \frac{dN_b}{dt} = N_a A_{ab} + \frac{\pi K e^2}{m}[N_a f_{ab} + N_b f_{ba}]\rho_{\text{ph}}(\omega), \tag{9.63}$$

where the three contributions on the right hand side are the spontaneous emission, the stimulated emission, and the absorption.

At equilibrium $-dN_a/dt = dN_b/dt = 0$, and this equation can be solved for $\rho_{\text{ph}}(\omega)$

$$\rho_{\text{ph}}(\omega) = \frac{m N_a A_{ab}}{\pi K e^2 (N_b f_{ba} + N_a f_{ab})}. \tag{9.64}$$

The populations can be eliminated by use of the Boltzmann distribution

$$\frac{N_b}{N_a} = \frac{g_b}{g_a}\exp[-(E_b - E_a)/kT] = \frac{g_b}{g_a}e^{-\hbar\omega/k_B T}. \tag{9.65}$$

The relationship between absorption and stimulated emission oscillator strengths involves a sign and the degeneracies

$$g_b f_{ba} = -g_a f_{ab}. \tag{9.66}$$

This yields

$$\rho_{\text{ph}}(\omega) = \frac{m}{\pi K e^2}\frac{g_a A_{ab}}{g_b f_{ba}}\frac{1}{(e^{\hbar\omega/k_B T} - 1)}. \tag{9.67}$$

Using the relationships from Eqs. 6.4, 6.5, and 6.18

$$g_a A_{ab} = \frac{3}{\tau_{\text{SHO}}}g_b f_{ba} = \frac{2 K e^2 \omega^2}{mc^3}g_b f_{ba}, \tag{9.68}$$

from which we deduce the Planck distribution in photon space

$$\rho_{\text{ph}}(\omega) = \frac{2\omega^2}{\pi c^3}\frac{1}{(e^{\hbar\omega/k_B T} - 1)}. \tag{9.69}$$

It should be emphasized that, while this description treats two levels a and b, there is no requirement that these be bound states. The development is equally valid for continuum states, and one of the most important applications of the Planck law is to continuum radiation emitted by free electrons in a plasma or in a metal.

In order to connect this to the standard formulations of the Stefan–Boltzmann T^4 law and the Wien displacement law, the expression can be converted from a photon density to an energy density by multiplying by $\hbar\omega$

$$\rho_E(\omega) = \frac{2\hbar\omega^3}{\pi c^3} \frac{1}{(e^{\hbar\omega/k_{\mathrm{B}}T} - 1)}. \tag{9.70}$$

10

Time-resolved measurements

Time is that great gift of nature which prevents everything from happening at once.
– Clarence J. Overbeck [158]

The key position played by the field of atomic physics in the development of modern quantum theory is owed in large part to the high precision with which the energy-level structure of the atom can be measured by the methods of high wavelength-resolution optical spectroscopy. Wavelength and frequency measurement accuracies that exceed parts in 10^8 are not only obtainable, but are required if the database is to be useful for diagnostic applications. By contrast, the measurement accuracies that can be obtained for other types of atomic structure properties is much lower. For lifetimes, transition probabilities, and oscillator strengths, extraordinary effort is required to achieve accuracies better than one percent. For cross section measurements, one must often be content with order-of-magnitude determinations, but the range of possible values makes reliable measurements to this accuracy valuable. While great strides have been made in *ab initio* theoretical methods, the attainable measurement accuracies for these quantities still exceeds the general reliability of calculations for cases involving complex many-electron atoms. Moreover, the accurate specification of wavelength and energy-level data does not ensure correct predictions of transition probabilities and lifetimes.

Measurements of lifetimes are particularly important, since they provide absolute rate values necessary to normalize relative transition probabilities obtained by time-integrated techniques. The availability of a comprehensive database for atomic transition probability rates has a significant impact on progress in other fields of science and technology, e.g., in fundamental physics and precise measurements; in the generation of coherent light; in atomic analysis in complex environments; in solar and astrophysics; and in plasma diagnostics.

The most direct method for the experimental determination of level-population lifetimes is through the time-resolved measurement of the free decay of the fluorescence radiation following a cutoff of the source of excitation. An important factor limiting the accuracy is the repopulation of the level of interest by cascade transitions from higher-lying levels, which can produce multiexponential decay curves that are difficult to analyze with precision by standard curve-fitting methods. For this reason, two types of alternative methods are often employed. One involves selective excitation of the level of interest, thus eliminating cascading altogether; another uses correlations between cascade connected decays to account for the effects of cascades.

While selective excitation methods totally eliminate the effects of cascade repopulation, they are generally limited to levels in neutral and singly ionized atoms that can be accessed from the ground state by strongly absorptive E1 transitions, and the selectivity itself is a limitation.

More general access can be obtained by nonselective excitation methods, such as pulsed electron beam bombardment of a gas cell or a gas jet, or in-flight excitation of a fast ion beam as it traverses a thin foil. Pulsed electron excitation methods are well suited to measurements in neutral and near neutral ions (although for very long lifetimes in ionized species, the decay curves can be distorted if particles escape from the viewing volume through the Coulomb explosion effect [57]). Nonselective excitation techniques can also be applied to measurements such as the phase-shift method and the Hanle effect, in which case cascade repopulation also can become a serious problem [33]. For highly ionized atoms, the most generally applicable method of lifetime determination is via excitation of a fast ion beam by its passage through a thin foil (beam–foil excitation).

10.1 Time dependence of measured decay curves

The meanlife determinations discussed in this chapter involve a measurement of the time dependence of the radiation emitted by a relaxing atomic level. The meanlife of the level is directly inferred by this decay curve through analytical expressions that will be developed in this section. These expressions are applicable not only to beam–foil measurements, but also to decay curves generated under a wide class of excitation conditions.

10.1.1 Solution of the driven coupled linear rate equations

Most techniques for the direct measurement of atomic meanlives involve a study of the time dependence of the radiation emitted by a sample during or subsequent to an external source of excitation $Q(t)$. Usually low densities are used so that collisional effects and radiation trapping within the sample are negligible. In such cases the instantaneous population $N_n(t)$ of a state or level n is governed by the equation

$$\mathrm{d}N_n/\mathrm{d}t = \sigma_n Q(t) + \sum_j N_j(t) A_{jn} - N_n(t)\alpha_n, \qquad (10.1)$$

where σ_n is the excitation cross section, A_{jn} is the transition probability rate for a cascade from the state or level j, and $\alpha_n \equiv 1/\tau_n$ is the reciprocal meanlife of the level n

$$\alpha_n = \sum_k A_{nk}. \qquad (10.2)$$

For convenience the levels are labeled in increasing order of excitation energy.

The population equation has an integrating factor $\exp(\alpha_n t)$ that converts it to the form

$$\frac{\mathrm{d}}{\mathrm{d}t}[N_n(t)\mathrm{e}^{\alpha_n t}] = \mathrm{e}^{\alpha_n t}\left[\sigma_n Q(t) + \sum_j N_j(t) A_{jn}\right]. \qquad (10.3)$$

To avoid infinite sums, assume that the number of cascades is finite, and equal to $m - n$.

If we further assume that $N_n(-\infty) = 0$, this expression can be integrated

$$N_n(t) = \int_{-\infty}^{t} dt' e^{-\alpha_n(t-t')} \left[\sigma_n Q(t') + \sum_{j=n+1}^{m} N_j(t') A_{jn} \right]. \tag{10.4}$$

A similar equation holds for each of the cascade levels, so this can be successively iterated

$$N_n(t) = \sigma_n \int_{-\infty}^{t} dt' e^{-\alpha_n(t-t')} Q(t')$$

$$+ \sum_{j=n+1}^{m} \sigma_j A_{jn} \int_{-\infty}^{t} dt' e^{-\alpha_n(t-t')} \int_{-\infty}^{t'} dt'' e^{-\alpha_j(t'-t'')} Q(t'') + \cdots. \tag{10.5}$$

Notice that the integral is a Laplace transform of shifted variable $T \equiv t - t'$

$$\int_{-\infty}^{t} dt' e^{-\alpha_n(t-t')} Q(t') = \int_{0}^{\infty} dT\, e^{-\alpha_n T} Q(t - T). \tag{10.6}$$

This operation has useful properties on successive application. If we define the operator $L_j(t - T)$: symbolically as

$$L_j(t - T) : Q(T) \equiv \int_{-\infty}^{t} dT e^{-\alpha_j(t-T)} Q(T), \tag{10.7}$$

then the nesting of two of these operators yields

$$L_k : L_j : Q = \int_{-\infty}^{t} dt' e^{-\alpha_k(t-t')} \int_{-\infty}^{t'} dT\, e^{-\alpha_j(t'-T)} Q(T)$$

$$= e^{-\alpha_k t} \int_{-\infty}^{t} dt' e^{(\alpha_k - \alpha_j)t'} \int_{-\infty}^{t'} dT e^{\alpha_j T} Q(T). \tag{10.8}$$

This can be rewritten using a parts integration with dv being the first integrand and u being the second integral, to yield $vdu = d(uv) - udv$ given by

$$L_k : L_j : Q = e^{-\alpha_k t} \left\{ \left[\frac{e^{(\alpha_k - \alpha_j)t'}}{(\alpha_k - \alpha_j)} \int_{-\infty}^{t'} dT e^{\alpha_j T} Q(T) \right]_{t'=-\infty}^{t'=t} \right.$$

$$\left. - \int_{-\infty}^{t} dt' \frac{e^{(\alpha_k - \alpha_j)t'}}{(\alpha_k - \alpha_j)} e^{\alpha_j t'} Q(t') \right\}. \tag{10.9}$$

This reduces to

$$L_k : L_j : Q = \frac{L_j : Q}{(\alpha_k - \alpha_j)} + \frac{L_k : Q}{(\alpha_j - \alpha_k)}. \tag{10.10}$$

This can be repeated any number of times, always yielding the result of a single operation with a constant coefficient. Therefore all nested integrations in the iterative expansion above can be reduced to a linear expression in this operator. Note that for $i = j$, L'Hôpital's rule yields $-\partial L_i / \partial \alpha_j$ for the right-hand side of this equation.

Examples of excitation functions $Q(t)$ and their corresponding transform functions $L(t - T) : Q(T)$ is given in Table 10.1.

Table 10.1. Transform functions.

Excitation		$Q(t)$	$L(t - T) : Q(T)$
Impulsive		$Q_0 \delta(t)$	$Q_0\, e^{-\alpha_j t}$
Stepwise	$(t < 0)$	0	0
	$(t > 0)$	Q_0	$Q_0(1 - e^{-\alpha_j t})/\alpha_j$
Rectangular	$(t < 0)$	0	0
	$(0 \le t \le a)$	Q_0	$Q_0(1 - e^{-\alpha_j t})/\alpha_j$
	$(t > a)$	0	$Q_0(e^{\alpha_j a} - 1)e^{-\alpha_j t}/\alpha_j$
Gaussian		$\dfrac{Q_0}{\sqrt{2\pi}\,w} e^{-t^2/2w^2}$	$Q_0 e^{-\alpha_j t + w^2 \alpha_j^2/2}$
			$\times \frac{1}{2}\left[1 + \mathrm{erf}\left(\frac{t - w^2\alpha_j}{\sqrt{2}w}\right)\right]$
Modulated		$Q_0 + M\cos(\omega t)$	$\dfrac{Q_0}{\alpha_j} + M\,\mathrm{Real}\left[\dfrac{e^{i\omega t}}{\alpha_j + i\omega}\right]$

An equation formed from nested summations of these linear transforms can be iteratively generated for each level of the system. The simultaneous solution of the coupled equations connecting a specific level with its cascades can be used to completely specify the instantaneous populations. The solution can conveniently be written in a closed-form series decomposition, in which the individual cascade terms are classified according to the number of intermediate levels they pass through in their chain of transitions to the primary level. This can be written symbolically in the form (shorthanded as $L_i = L_i{:}Q$)

$$N_n(t) = \sigma_n L_n(t) + \sum_j \{j \to n\} + \sum_k \sum_j \{k \to j \to n\} + \sum_l \sum_k \sum_j \{l \to k \to j \to n\}$$

$$+ \cdots + \sum \cdots \sum \{m \to (r{-}\text{steps}) \to n\} \tag{10.11}$$

where the quantities in braces are a diagrammatic mnemonic for a generic function for a chain of a given order. Successive operation with the transform operator shows these function to be given by

$$\{j \to n\} \equiv \sigma_j A_{jn} \left[\frac{L_n(t)}{(\alpha_j - \alpha_n)} + \frac{L_j(t)}{(\alpha_n - \alpha_j)} \right] \tag{10.12}$$

$$\{k \to j \to n\} \equiv \sigma_k A_{kj} A_{jn} \left[\frac{L_n(t)}{(\alpha_j - \alpha_n)(\alpha_k - \alpha_n)} + \frac{L_j(t)}{(\alpha_k - \alpha_j)(\alpha_n - \alpha_j)} \right.$$

$$\left. + \frac{L_k(t)}{(\alpha_n - \alpha_k)(\alpha_j - \alpha_k)} \right] \tag{10.13}$$

$$\{l \to k \to j \to n\} \equiv \sigma_l A_{lk} A_{kj} A_{jn} \left[\frac{L_n(t)}{(\alpha_j - \alpha_n)(\alpha_k - \alpha_n)(\alpha_l - \alpha_n)} \right.$$

$$+ \frac{L_j(t)}{(\alpha_k - \alpha_j)(\alpha_l - \alpha_j)(\alpha_n - \alpha_j)} + \frac{L_k(t)}{(\alpha_l - \alpha_k)(\alpha_n - \alpha_k)(\alpha_j - \alpha_k)}$$

$$\left. + \frac{L_l(t)}{(\alpha_n - \alpha_l)(\alpha_j - \alpha_l)(\alpha_k - \alpha_l)} \right], \tag{10.14}$$

which can be generalized to the form

$$\{m \to (r\text{-steps}) \to n\} \equiv \sigma_m A_{mb} \cdots A_{an} \sum_{i=n}^{m} \left[\frac{L_n(t)}{\prod_{j \neq i} (\alpha_j - \alpha_i)} \right]. \tag{10.15}$$

Here i and j range over the $r + 1$ cascade and primary states or levels, and the product of transition probabilities is over the r-step cascade chain. This can be refactored into the form

$$N_n(t) = \left[\sigma_n - \sum_i \beta_{ni} \right] L_n(t) + \sum_i \beta_{ni} L_i(t), \tag{10.16}$$

where the sum includes one term for the primary level n, and one term for every level that cascades, either directly or indirectly, into it.

10.1.2 Applications

Through these relationships it is possible to obtain directly a specific expression for the population of any arbitrarily cascaded and driven level. This can be applied to each level separately to obtain a set of coupled equations.

Example: three-level system with two cascades

As an illustration, consider an impulsively driven three-level system, in which a primary level (labeled 1) is repopulated by two cascades (labeled 2 and 3).

$$N_1(t) = [N_1(0) - \beta_{12} - \beta_{13}] e^{-\alpha_1 t} + \beta_{12} e^{-\alpha_2 t} + \beta_{13} e^{-\alpha_3 t}. \tag{10.17}$$

There are two possible schemes by which this can occur: direct (simultaneous); and indirect (sequential) cascading.

Case 1: Direct cascades – both 2 and 3 have transitions to 1, but are not themselves repopulated. The coefficients are

$$\beta_{12} = \frac{N_2(0) A_{21}}{(\alpha_1 - \alpha_2)}$$

$$\beta_{13} = \frac{N_3(0) A_{31}}{(\alpha_1 - \alpha_3)}. \tag{10.18}$$

Case 2: Indirect cascades – 3 cascades into 2, which cascades into 1, but 3 is not itself repopulated. The coefficients are

$$\beta_{12} = \frac{N_2(0) A_{21}}{(\alpha_1 - \alpha_2)} + \frac{N_3(0) A_{32} A_{21}}{(\alpha_3 - \alpha_2)(\alpha_1 - \alpha_2)}$$

$$\beta_{13} = \frac{N_3(0) A_{32} A_{21}}{(\alpha_1 - \alpha_3)(\alpha_2 - \alpha_3)}. \tag{10.19}$$

These equations can be used to illustrate the "growing-in" ambiguity in the assignment of meanlives to levels. In this situation, one or more of the coefficients of an exponential term is negative, and a maximum in the intensity of the decay curve may occur at a time other than $t = 0$. An examination in the absence of indirect cascades ($N_3(0) = 0$) yields

$$N_1(t) = \left[N_1(0) - \frac{N_2(0)A_{21}}{\alpha_1 - \alpha_2} \right] e^{-\alpha_1 t} + \left[\frac{N_2(0)A_{21}}{\alpha_1 - \alpha_2} \right] e^{-\alpha_2 t}, \tag{10.20}$$

which admits two such possibilities: a "fast-follower" primary; and an "expiring" cascade. The first is a short-lived primary $\alpha_1 > \alpha_2$ of low initial population such that

$$\frac{N_1(0)}{N_2(0)} < \frac{A_{21}}{\alpha_1 - \alpha_2}, \tag{10.21}$$

so that it first builds up a population and then tracks the cascade in time dependence. The second is a cascade level that is shorter lived than the primary, $\alpha_1 < \alpha_2$ and thus expends its population and ultimately leaves the primary unrepopulated. The situation is much more complicated when indirect cascading is included, and indicates the dangers in specifying lifetimes from a single decay curve in a cascade coupled decay scheme.

Example: the phase-shift method

As another illustrative example, we can consider the phase-shift method. Here, a modulated source of excitation

$$Q(t) = Q + M \cos \omega t \tag{10.22}$$

is provided, and the emitted radiation

$$N_n(t) = B_0 + B_1 \cos(\omega t + \phi) \tag{10.23}$$

will exhibit a phase shift ϕ relative to the driver that is frequency dependent, and provides a probe of the lifetimes of the levels in the sample. For compactness of notation, we denote here $\alpha_i \equiv 1/\tau_i$. For a singly cascaded system this becomes.

$$N_1(t) = Q_0 \tau_1 (\sigma_1 + \sigma_2 A_{21} \tau_1) + M \, \mathrm{Real} \left[\frac{\sigma_1 \tau_1 e^{i\omega t}}{(1 + i\omega\tau_1)} + \frac{\sigma_2 A_{21} \tau_2 \tau_1 e^{i\omega t}}{(1 + i\omega\tau_2)(1 + i\omega\tau_1)} \right]. \tag{10.24}$$

Taking the real part of the time-dependent portion, the phase shift is given by

$$\phi = \arg \left[\frac{(1 + i\omega\tau_2 + \beta)}{(1 + i\omega\tau_2)(1 + i\omega\tau_1)} \right], \tag{10.25}$$

where β is the "cascade fraction"

$$\beta \equiv \sigma_2 \tau_2 A_{21} / \sigma_1, \tag{10.26}$$

and this yields

$$\phi = \tan^{-1}(\omega\tau_1) + \tan^{-1}(\omega\tau_2) - \tan^{-1}\left(\frac{\omega\tau_2}{1 + \beta} \right). \tag{10.27}$$

By similar considerations [56], the contributions of any number of cascades can be formally described.

10.2 <u>A</u>djusted <u>n</u>ormalization of <u>d</u>ecay <u>c</u>urve (ANDC) method

In measurements that use nonselective excitation, the level populations (and hence the decay curves) are affected by cascade repopulation. Thus the decay curve involves a sum of many exponentials, one corresponding to the primary level, and one to each level that cascades (either directly or indirectly) into it. Decay exponentials do not comprise a complete set of functions to span this space, and the representation of an infinite sum by a finite sum through curve-fitting methods can, in unfavorable cases, lead to significant errors. Thus, while cascades that differ significantly in lifetime from that of the primary do not pose a serious problem, cascades that have lifetimes similar to that of the primary can seriously distort a decay curve fit.

Fortunately, alternative methods to exponential curve fitting exist, which permit the accurate extraction of lifetimes to be made from correlated sets of nonselectively populated decay curves. Situations in which cascading is dominated by a few strong decay channels are ideally suited to the adjusted normalization of decay curve (ANDC) method [65]. This method exploits dynamical correlations among the cascade-related decay curves that arise from the rate equation that connects the population of a given level to those of the levels that cascade directly into it. The instantaneous population of each level is, to within constant factors involving the transition probabilities and detection efficiencies, proportional to the intensity of the radiation emitted in any convenient decay branch. Through a joint correlated analysis of the decay curves of the primary level and those of the levels that directly repopulate it, this analysis yields both the primary lifetime and the intensity normalizations of the cascades relative to that of the primary.

To develop the basis for the ANDC method, let us examine the population equation for an impulsively excited system.

$$\frac{dN_n}{dt}(t) = \sum_i N_i(t)A_{in} - N_n(t)\alpha_n. \tag{10.28}$$

The intensity of the observed radiation is given by

$$I_{ij}(t) = N_i(t)A_{ij}\eta_{ij}, \tag{10.29}$$

where η_{ij} is the detection efficiency. If this quantity is constant in time, the same time dependence will reside in both the measured radiation and the instantaneous population of the levels. Any convenient decay branch can be chosen for each of the various decays, since all branches exhibit the same time variation, which is that of the upper-level population, so

$$I_{in} = I_{ik}\frac{A_{in}\eta_{in}}{A_{ik}\eta_{ik}}. \tag{10.30}$$

In terms of measured decay curves, the population equation becomes

$$\frac{dI_{nj}}{dt}(t) = \sum_i \xi_i I_{ik}(t) - I_{nj}(t)\alpha_n, \tag{10.31}$$

where

$$\xi_i \equiv \frac{\eta_{nj}A_{in}A_{nj}}{\eta_{ik}A_{ik}} \tag{10.32}$$

is a normalizing constant relating the cascade and primary decay curves. Subject to the formation of a numerical differentiation (or, if one chooses to integrate both sides of the equation, a numerical integration), the constants ξ_i and α_n can be deduced from a linear fit to the decay curves.

These quantities can be written in terms of a linear equation

$$y(t_p) = \alpha_n - \sum_i \xi_i x_i(t_p), \tag{10.33}$$

where

$$y(t_p) \equiv -\frac{1}{I_{nj}(t_p)}\frac{dI_{nj}(t_p)}{dt}$$

$$x_i(t_p) \equiv \frac{I_{ik}(t_p)}{I_{nj}(t_p)}. \tag{10.34}$$

Here t_p denotes a set of time coordinates ($p = 1, 2, 3, \ldots$), each of which corresponds to a "panel" of points on the decay curve, of sufficient number to determine a local numerical derivative. The panels may either be overlapping or nonoverlapping, dependent upon whether statistical independence is desired.

For a single repopulating cascade, the parameters can be determined from the slope and intercept on a plot of $y(t_p)$ vs $x(t_p)$. An example is shown in Fig. 10.1, in which the lifetime of the 4p $^2P_{3/2}$ level in Cu-like Kr VIII is determined by joint ANDC analysis of the measured 4s–4p and 4p–4d decay curves. Since the repopulation of the 4d is primarily via the yrast chain, the inclusion of this single direct cascade automatically includes the very strong and unbranched repopulation along the chain of circular orbits.

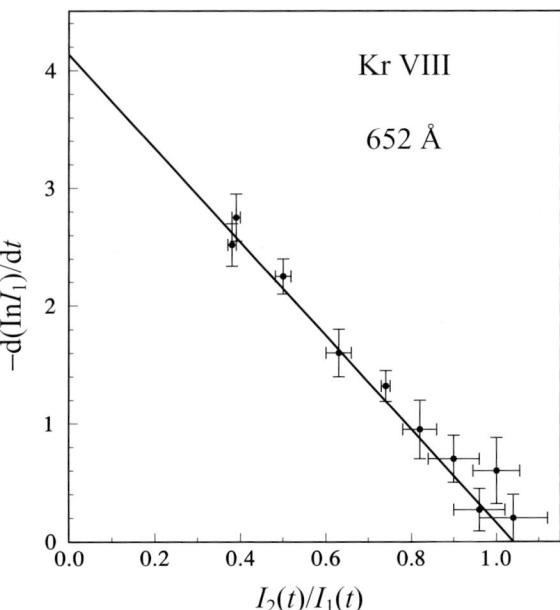

Fig. 10.1. ANDC plot of measured 4s–4p and 4p–4d decay curves in Kr VIII.

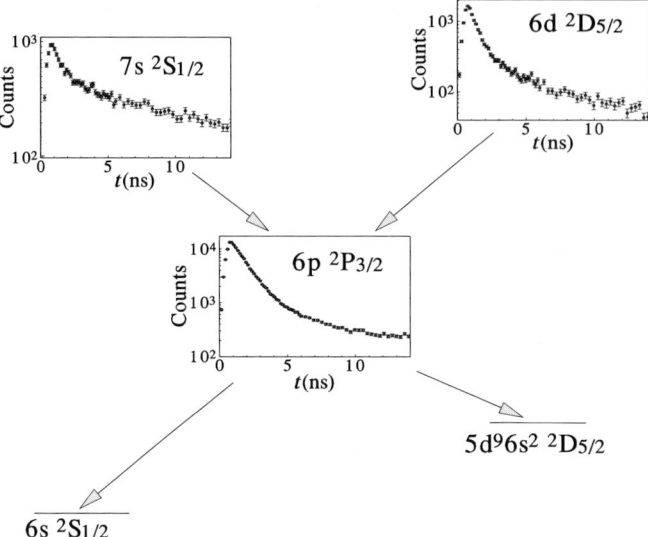

Fig. 10.2. Schematic representation of the application of the ANDC method to the determination of the Tl III 6p ^{2}P$_{3/2}$ lifetime. (After Ref. [119].)

The same procedure can be applied to any number of cascades. A schematic representation of the ANDC method, juxtaposing decay curves on a Grotrian diagram, is shown in Fig. 10.2. Here the method is applied to the resonance transition of Tl III in the Au isoelectronic sequence. The inset decay curves indicate that the arbitrarily normalized decay curves $I_{6p}(t)$ of the 6p ^{2}P$_{3/2}$ level and its cascades from the 6d ^{2}D$_{5/2}$ and 7s ^{2}S$_{1/2}$ are jointly analyzed using the population equation expressed in the form

$$\tau_{6p}\frac{\mathrm{d}I_{6p}}{\mathrm{d}t}(t_i) = \xi_{6d}I_{6d}(t_i) + \xi_{7s}I_{7s}(t_i) - I_{6p}(t_i)\,. \tag{10.35}$$

Here the lifetime τ_{6p} of the 6p-level and the relative normalizations ξ_{6d} and ξ_{7s} for the cascades and primary decay curves are determined by simultaneous solution of the large set of relationships provided by evaluating Eq. 10.35 at each of the common points t_i on the decay curves. If all significant direct cascades have been included, the goodness-of-fit will be uniform for all time subregions, indicating reliability. If important cascades have been omitted or blends are present, the fit will vary over time subregions, indicating a failure of the analysis. Very rugged algorithms have been developed that permit accurate lifetimes to be extracted even in cases where statistical fluctuations are substantial.

10.3 Differential lifetime measurements

Because of the effects of intermediate coupling (IC) and configuration interaction (CI), additional decay channels are sometimes opened to a specific fine structure level in a term. If the transition probabilities for the decay channels that are common to all members of the fine structure manifold are of similar magnitude, then it is sometimes possible to deduce the transition probability of the additional channel through differential decay curve measurements. There are two intriguing cases where this procedure has been successfully applied. One involves intermediate coupling, in which one of the fine structure levels has a channel to the ground state opened by singlet–triplet mixing. A second involves configuration interaction, in which one of the fine structure levels has a channel to the continuum (via autoionization) opened by doublet–quartet mixing.

Measurements of this type are particularly interesting because of the development of position-sensitive detectors. These now make possible the simultaneous measurement of the decay curves of lines that are closely spaced in wavelength. Since these techniques can register the intensities at a given decay time on all of the decay curves at the same time, systematic experimental errors that are common to all will be cleanly removed by the differential measurement. Thus high accuracies in small differences should be possible. An example of each of these cases will be described below.

10.3.1 Radiative intercombination rates in the Be-like triplets

Transition probabilities for the $2s^2\ ^1S_0$–$2s3p\ ^3P_1$ intercombination lines in the Be isoelectronic sequence have been a challenging subject for theoretical study because they are E1-forbidden in pure LS coupling, but become E1-allowed when the spin–orbit interaction and other relativistic effects cause mixing between the $2snp\ ^3P_1$ and $^1P_1$ series. Moreover, the lower stages of ionization of this sequence are strongly affected by CI, whereas the higher charge states are strongly affected by IC. Consequently, no energetically allowed transition from the $2s3p\ ^3P$ level can automatically be ruled out from making a significant contribution. Thus, this intercombination lifetime provides a useful test of theory, since the system is simple enough to be tractable, but the calculation must include configuration interaction, intermediate coupling, and relativistic effects.

The intercombination transition is not the only decay channel open to the $2snp\ ^3P_1$ level. A partial energy-level diagram that is typical of ions in this sequence is shown in Fig. 10.3, and indicates additional channels that are available, and dominate at low Z. To deduce the intercombination transition probability from a direct measurement of the lifetime, the branching fraction would be required, and an intensity calibration at these wavelengths is not feasible.

However, the intercombination transition probability for the $2s3p\ ^3P_1$ level can be accurately determined for much lower states of ionization by this method of differential lifetime measurements. As can be seen from Fig. 10.3, the $2s3p\ ^3P_J$ levels have spin-allowed decay channels to the $2s3s\ ^3S_1$ level, but the lifetimes have a strong J-dependence caused by the $J = 0$ ground-state intercombination decay channel that is accessible to the $J = l$ level, but is E1-forbidden to the $J = 0$ level (no $J = 0 \rightarrow 0$) and $J = 2$ level ($\Delta J = 0, \pm 1$).

Measured decay curves for the ^3P_J levels for $J = 0, 1, 2$ are shown in Fig. 10.4. The effect is quite large in this ion, but because of the advantages offered by differential measurements,

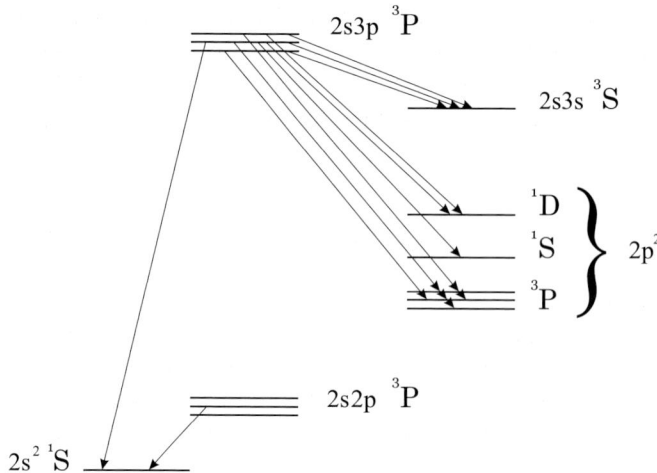

Fig. 10.3. Grotrian diagram for the decay scheme for the 2s3p levels in the Be isoelectronic sequence.

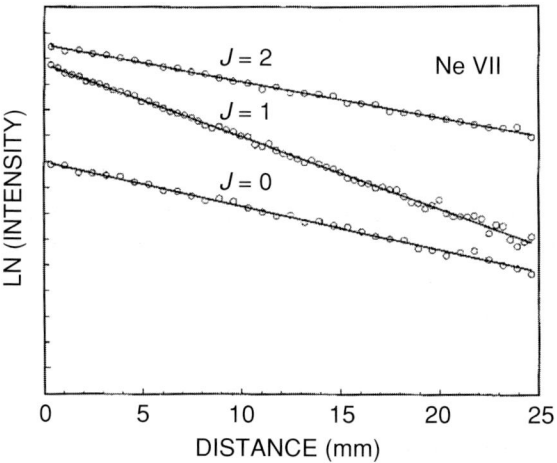

Fig. 10.4. J dependence of the 2s3p ^3P_J decay curves in Ne VII. (From Ref. [116].)

it is possible to measure much smaller contributions from the intercombination channel for lower ions in this sequence.

10.3.2 Autoionization rates in doubly excited levels in Li-like quartet levels

The doubly excited 1s2p^2 ^{4}P term in the Li isoelectronic sequence provides an autoionization analogue to the radiative intercombination channels described above, since it undergoes transitions to the lower-lying 1s2s2p ^{4}P^o doubly excited state. Lifetimes of fine structure components thus exhibit a "differential metastability" in which a weak mixing of the 1s2p^2 ^{4}P$_{5/2}$ level with the rapidly autoionizing ^{2}D$_{5/2}$ level of the same configuration causes a substantial shortening in its lifetime relative to those of the ^{4}P$_{3/2}$ and ^{4}P$_{1/2}$.

The difference between the reciprocal lifetime of the $J = \frac{5}{2}$ level and that of either the $J = \frac{3}{2}$ or the $J = \frac{1}{2}$ levels thus yields the Auger decay rate of the $J = \frac{5}{2}$ level.

10.4 Hanle effect

The Hanle effect provides a method for determining lifetimes from an aligned source without the need for resolving the time coordinate. This involves the impression of a magnetic field on the sample, causing it to precess. Although the decay curve is not directly observed, the logarithmic decrement of the radiation on the time scale specified by the rotational motion provides a measure of the lifetime. Quantum mechanically this corresponds to a mixing of the magnetic substates of the excitation Hamiltonian, but there is a simple classical model for this process that we shall present below.

As was shown in Section 6.1.1, the emitted radiation of an aligned source is given by

$$\frac{dI}{d\Omega} = \frac{I}{4\pi}\left[\frac{1 - P\cos^2\vartheta}{1 - P/3}\right], \tag{10.36}$$

where

$$I = \frac{3}{2}(I_\pi + I_\sigma); \qquad P \equiv \frac{I_\pi - I_\sigma/2}{I_\pi + I_\sigma/2}. \tag{10.37}$$

If the axis of symmetry of the (either continuous or impulsive) excitation is along the z-axis, the σ dipoles will be in the x–y plane, and the π dipole will be along the z-axis. If the system is subjected to an external magnetic field B in the x-direction, the system will precess about the field with the Larmor frequency

$$\omega = g\frac{eB}{2m}. \tag{10.38}$$

As shown in Fig. 10.5, the system rotates with an angle $\vartheta = \omega t$, and a detector sighted along the z-axis alternately views I_π and $I_\sigma/2$ radiation as the system precesses. If the transition observed decays with an exponential meanlife τ, both component intensities will decay at the same rate according to

$$I_\pi(t) = I_\pi^0 e^{-t/\tau}; \qquad I_\pi(t) = I_\pi^0 e^{-t/\tau}. \tag{10.39}$$

Since the linear polarization factor P contains the same exponentials in the numerator and denominator, that quantity will be time independent

$$P \equiv \frac{I_\pi^0 - I_\sigma^0/2}{I_\pi^0 + I_\sigma^0/2}, \tag{10.40}$$

but the amplitude of the total radiation will involve the exponential. Thus the time-integrated

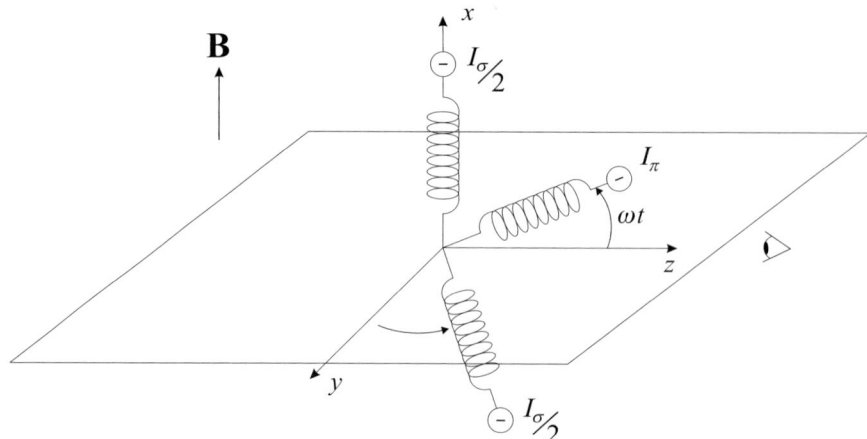

Fig. 10.5. Classical model for the Hanle effect.

radiation emitted will be

$$\left(\frac{dI}{d\Omega}\right)_{\text{int}} = \frac{I}{4\pi(1 - P/3)}\int_0^\infty dt\,e^{-t/\tau}[1 - P\cos^2\omega\tau]$$

$$= \frac{I\tau}{1 - P/3}\left[\left(1 - \frac{P}{2}\right) - \frac{P/2}{1 + (2\omega\tau)^2}\right]. \tag{10.41}$$

This simplifies to

$$\left(\frac{dI}{d\Omega}\right)_{\text{int}} = \frac{I\tau}{8\pi}\frac{(1 - P/2)}{(1 - P/3)}\left[1 + \left(\frac{P}{2 - P}\right)\frac{1}{1 + (2\omega\tau)^2}\right], \tag{10.42}$$

which corresponds to a Lorentzian notch centered about zero frequency. The full width at half maximum is

$$\omega_{\text{FWHM}} = 1/\tau. \tag{10.43}$$

By varying the frequency (achieved by sweeping the strength of the magnetic field), the lifetime can be specified from the FWHM. Since the Lorentzian profile is centered on zero field, the Hanle effect is sometimes called a "zero-field level crossing."

11

Hyperfine structure

To find meaning, I looked inside the atom, and found it almost empty.

Atomic interactions are usually described in terms of three different types of interactions. The gross structure refers to the predictions of the Kepler–Coulomb–Schrödinger nonrelativistic electrostatic model in which the electron moves in a simple $1/r$ central potential. The fine structure refers to the relativistic correction to that picture due to: the relativistic momentum; various interactions between the magnetic moment of the electron with that of other electrons; and the relative motion of the static charge of the nucleus. The quantum electrodynamic corrections due to the interaction of the electron with the radiation field are often included with the fine structure. The hyperfine structure refers to a general class of interactions that arise as a result of the finite mass, size, charge distribution, and charge circulation of the nucleus.

11.1 The origins of hyperfine structure observations

Hyperfine structure was discovered by Albert A. Michelson in what might be called the second disappointment of the Michelson interferometer. Precision optical measurements were Michelson's lifelong passion, as evidenced by his pursuit of additional significant digits in the value for the speed of light. He began this quest in 1878, and by 1882 had a value good to within 0.02 percent. In 1926 he improved that measurement to just over one part in 10^5.

In 1881 Michelson began the construction of his "interferential refractometer" (the Michelson interferometer) in the hope of making a precision measurement of the motion of the Earth through the imagined *luminiferous aether*. The results of this attempt (jointly with Edward W. Morley) were declared a "failure" in 1887. (It should be noted that the correct explanation for the null result was given by Woldemar Voigt that same year in a little-noticed paper [187] that first presented the equations that were later called the "Lorentz transformations.") Although legend now places this experiment as a keystone of the restructuring of classical physics into modern physics, the null result was a failure to the metrologist Michelson. No matter how accurately the experiment is performed, if one is fated to measure zero then the percentage uncertainty is not small.

Although discouraged by this "failure," Michelson was determined to apply his instrument to other uses, since the interferometer possessed a sensitivity of a few parts in 10^{10}. He therefore undertook a metrology project to determine the length of the standard meter scribed on the platinum–iridium bar at the International Bureau of Weights and Measures in Sévres France. Michelson attempted to use the bright yellow sodium resonance doublet of Melvill (see Section 3.1) as a standard, but under higher resolution he found [156] that each member of the doublet broke up into a manifold of lines (due to the energy-level structure shown in Fig. 11.1, see p. 215). This is not restricted to sodium. With rare exceptions, each spectral line is not a single entity but is instead composed of a number of lines differing very slightly in wavelength from each other. Michelson's frustration with this "defect" in the spectra is clear from the following extract from his Nobel Prize acceptance speech.

One of the most serious difficulties encountered in the attempt to carry into practice the method of counting the alternations of light and darkness in the interference method, is the defect in homogeneity of the light employed. This causes indistinctness of the interference rings when the distance is greater than a few centimeters. The light emitted by various kinds of gases and metallic vapors, when made luminous by the electric discharge, differ enormously in this respect. A systematic search showed that among some forty or more radiations nearly all were defective, some being represented by a spectrum of broad hazy "lines", others being double, triple, or even more highly complex. But the red light emitted by luminous vapor of metallic cadmium was found to be almost ideally adapted for the purpose. [156]

Thus, once again, Michelson's historic new discovery was an impediment to the primary goal of his measurement. Fortunately, the structure usually consists of a limited set of principal components with a small number of satellites. As indicated above, Michelson found a promising candidate in the cadmium 5s5p $^1P_1$–5s5d $^1D_2$ red line, and succeeded in measuring the length of the standard meter bar to be 1 553 163.5 wavelengths of this line. Cadmium has eight stable isotopes, two of which have nonzero nuclear spin, but ^{114}Cd provides a suitable standard.

11.2 Magnetic dipole moment of the nucleus

The property of the nucleus that has the most significant effect on atomic spectra is the magnetic dipole moment. Whereas other properties can shift the energies of the atomic energy levels relative to what their positions would be in a point nucleus, the magnetic dipole moment changes the quantum numbers that characterize the system. In a complex atom, the only quantities that dependably characterize a level are the energy, the parity (excluding weak interactions), and the total angular momentum, since these are based on firm conservation laws. The magnetic dipole moment of the nucleus changes the total angular moment that describes the atom.

For a nucleus of intrinsic spin $\mathbf{I}$, the total angular momentum $\mathbf{F}$ of the nucleus plus orbital electron is given by

$$\mathbf{F} = \mathbf{I} + \mathbf{J}. \tag{11.1}$$

For a nucleus with magnetic g-factor g_I, the magnetic moment $\boldsymbol{\mu}_I$ is

$$\boldsymbol{\mu}_I = g_I \frac{e}{2M_{\mathrm{p}}} \mathbf{I}, \tag{11.2}$$

where M_{p} is the proton mass, which here yields the nuclear magneton unit.

11.2.1 Vector model formulation

There is a magnetic interaction between the nucleus and the various electrons. The magnetic field $\mathbf{B}$ produced by the electrons at the site of the nucleus consists of two parts: one part $\mathbf{B}_{\mathrm{o}}$ is produced by the combined orbital motions $\mathbf{L}$ of the various electrons; the other part $\mathbf{B}_{\mathrm{s}}$ is produced by the combined intrinsic spins $\mathbf{S}$ of the various electrons.

The field due to the orbital motion of the electrons is very similar to that computed in Section 2.4.9 for the interaction of the electron's orbital field with its own spin, except there the field was evaluated in the frame of the electron (due to the apparent motion of the nucleus). That transformation introduced factors of Z and the Thomas precession into the expression which are not present here. Noting that the coordinate of the nucleus relative to the electron is $-\mathbf{r}$ and the electron charge $-e$, the field due to the orbital motion of the electrons is given by the Biot–Savart law as

$$\mathbf{B}_{\mathrm{o}} = \frac{\mu_0}{4\pi} \frac{e(\mathbf{v} \times \mathbf{r})}{r^3} = -\frac{\mu_0}{4\pi} \frac{e\mathbf{L}}{mr^3}, \tag{11.3}$$

where we have identified the orbital angular momentum $\mathbf{L} = m(\mathbf{r} \times \mathbf{v})$.

The intrinsic magnetic moment of the electron is given by (see Eq. 2.93)

$$\boldsymbol{\mu}_{\mathrm{s}} = -g_{\mathrm{e}} \frac{e}{2m} \mathbf{S}, \tag{11.4}$$

and the magnetic dipole field arising from this electron spin moment is, in the asymptotic limit of large r, given by

$$\mathbf{B}_{\mathrm{s}} = \frac{\mu_0}{4\pi} \frac{1}{r^3} \left[\frac{3(\mathbf{r} \cdot \boldsymbol{\mu}_{\mathrm{s}})\mathbf{r}}{r^2} - \boldsymbol{\mu}_{\mathrm{s}} \right] = -\frac{\mu_0}{4\pi} \frac{eg_{\mathrm{e}}}{2mr^3} \left[\frac{3(\mathbf{r} \cdot \mathbf{S})\mathbf{r}}{r^2} - \mathbf{S} \right]. \tag{11.5}$$

If we combine these two contributions, the total magnetic field at the nucleus due to the orbital electron is

$$\mathbf{B} = \mathbf{B}_{\mathrm{o}} + \mathbf{B}_{\mathrm{s}} = -\frac{\mu_0}{4\pi} \frac{e}{mr^3} \left[\mathbf{L} - \frac{g_{\mathrm{e}}}{2} \left(\mathbf{S} - \frac{3(\mathbf{r} \cdot \mathbf{S})\mathbf{r}}{r^2} \right) \right]. \tag{11.6}$$

In the Schrödinger formulation, this expression is not valid for s-states for two reasons. First, the $\langle r^{-3} \rangle$ term in the orbital contribution has a factor of ℓ in the denominator, and the Schrödinger s wave function does not vanish at $r = 0$, so the contribution diverges. Second, the dipole field of the spin contribution is an asymptotic expression for large r, which breaks down if the electron has a significant position probability density near $r = 0$. Therefore, we shall temporarily restrict the consideration to states with $\ell \neq 0$, and return later to the consideration of the "Fermi contact interaction" (the nuclear counterpart of the Darwin term [78] (see Section 2.4.9) in atomic fine structure).

If we denote the vector portion of the field as **N**, where

$$\mathbf{N} \equiv \mathbf{L} - \frac{g_e}{2}\left[\mathbf{S} - \frac{3(\mathbf{r}\cdot\mathbf{S})\mathbf{r}}{r^2}\right], \tag{11.7}$$

then the energy of interaction between the electron and the nucleus is given by

$$\Delta E = -\boldsymbol{\mu}_I \cdot \mathbf{B} = -\left[g_I\frac{e}{2M_\mathrm{p}}\mathbf{I}\right]\cdot\left[-\frac{\mu_0}{4\pi}\frac{e}{mr^3}\mathbf{N}\right]. \tag{11.8}$$

This can be put into atomic units by noting that $\mu_0 e^2/4\pi = Ke^2/c^2 = 2Ry\,a_0/c^2$ and $\hbar^2/m = mc^2\alpha^2 a_0^2$, and can be written

$$\Delta E = g_I Ry\alpha^2\frac{m}{M_\mathrm{p}}\frac{a_0^3}{r^3}\frac{(\mathbf{I}\cdot\mathbf{N})}{\hbar^2}. \tag{11.9}$$

Since this corresponds to a perturbation calculation based on an unperturbed representation that is diagonal in **J**, we use the Wigner–Eckart theorem [83, 193] and the vector model to first project **N** onto **J**, and then **J** onto **I**

$$\langle\mathbf{N}\cdot\mathbf{I}\rangle = \frac{\langle\mathbf{N}\cdot\mathbf{J}\rangle\langle\mathbf{J}\cdot\mathbf{I}\rangle}{\langle\mathbf{J}\cdot\mathbf{J}\rangle}. \tag{11.10}$$

To evaluate this, we must consider the various terms in the quantity

$$\begin{aligned}
\mathbf{N}\cdot\mathbf{J} &= \left[\mathbf{L} - \frac{g_e}{2}\left(\mathbf{S} - \frac{3(\mathbf{r}\cdot\mathbf{S})\mathbf{r}}{r^2}\right)\right]\cdot(\mathbf{L}+\mathbf{S}) \\
&= \mathbf{L}^2 - \left(1 - \frac{g_e}{2}\right)(\mathbf{L}\cdot\mathbf{S}) - \frac{g_e}{2}\left[\mathbf{S}^2 - \frac{3(\mathbf{r}\cdot\mathbf{S})(\mathbf{r}\cdot\mathbf{L})}{r^2} - \frac{3(\mathbf{r}\cdot\mathbf{S})^2}{r^2}\right].
\end{aligned} \tag{11.11}$$

Since $\mathbf{r}\perp\mathbf{L}$, the term containing their dot product vanishes. The spin is given in terms of the Pauli spin matrices

$$\mathbf{S} = \frac{\hbar}{2}\left[\begin{pmatrix}0 & 1\\ 1 & 0\end{pmatrix}\mathbf{e}_x + \begin{pmatrix}0 & -\mathrm{i}\\ \mathrm{i} & 0\end{pmatrix}\mathbf{e}_y + \begin{pmatrix}1 & 0\\ 0 & -1\end{pmatrix}\mathbf{e}_x\right]. \tag{11.12}$$

Thus

$$\mathbf{S}\cdot\mathbf{S} = \frac{3\hbar^2}{4}\begin{pmatrix}1 & 0\\ 0 & 1\end{pmatrix} \tag{11.13}$$

and

$$\frac{\mathbf{r}\cdot\mathbf{S}}{r} = \frac{\hbar}{2}\left[\sin\vartheta\cos\varphi\begin{pmatrix}0 & 1\\ 1 & 0\end{pmatrix} + \sin\vartheta\sin\varphi\begin{pmatrix}0 & -\mathrm{i}\\ \mathrm{i} & 0\end{pmatrix} + \cos\vartheta\begin{pmatrix}1 & 0\\ 0 & -1\end{pmatrix}\right], \tag{11.14}$$

which yields

$$\frac{\mathbf{r}\cdot\mathbf{S}}{r} = \frac{\hbar}{2}\begin{pmatrix}\cos\vartheta & \sin\vartheta\,\mathrm{e}^{-\mathrm{i}\varphi}\\ \sin\vartheta\,\mathrm{e}^{\mathrm{i}\varphi} & -\cos\vartheta\end{pmatrix}. \tag{11.15}$$

So,

$$\frac{3(\mathbf{r}\cdot\mathbf{S})(\mathbf{r}\cdot\mathbf{S})}{r^2} = \frac{3\,\hbar^2}{4}\begin{pmatrix}1 & 0 \\ 0 & 1\end{pmatrix}. \tag{11.16}$$

Substituting Eqs. 11.13 and 11.16 into Eq. 11.11, the two terms involving them cancel. Since $\mathbf{r}\cdot\mathbf{L} = 0$, the entire bracketed expression in Eq. 11.11 vanishes, leaving

$$(\mathbf{N}\cdot\mathbf{J}) = \mathbf{L}^2 + (1 - g_e/2)(\mathbf{L}\cdot\mathbf{S}). \tag{11.17}$$

Thus, in the approximation of the Dirac g-factor ($g_e = 2$), the dependence of the operator on $\mathbf{S}$ vanishes, and the interaction energy becomes

$$\Delta E = g_I R y \alpha^2 \frac{m}{M_p}\left\langle\frac{a_0^3}{r^3}\right\rangle\frac{\langle\mathbf{L}\cdot\mathbf{L}\rangle}{\langle\mathbf{J}\cdot\mathbf{J}\rangle}\frac{\langle\mathbf{I}\cdot\mathbf{J}\rangle}{\hbar^2}. \tag{11.18}$$

In the experimental application to complex atoms, the interaction energy is written as

$$\Delta E = A(\mathbf{I}\cdot\mathbf{J})/\hbar^2. \tag{11.19}$$

where A is the magnetic dipole hyperfine splitting constant that has the theoretical form

$$A \equiv g_I R y \alpha^2 \frac{m}{M_p}\left\langle\frac{a_0^3}{r^3}\right\rangle\frac{\ell(\ell+1)}{j(j+1)}. \tag{11.20}$$

In Section 4.5.1 a convention was adopted whereby single-particle quantum numbers are denoted by lower-case letters, and multiparticle quantum numbers are denoted by upper-case letters. Here there is a single nucleus, but there can be one or more electrons. In this section lower-case letters will be used to denote all quantum numbers, with the understanding that these can describe either the single-electron or the multielectron case.

The quantity $(\mathbf{I}\cdot\mathbf{J})/\hbar^2$ can be evaluated using

$$\mathbf{F}^2 = (\mathbf{I}+\mathbf{J})^2 = \mathbf{I}^2 + \mathbf{J}^2 + 2(\mathbf{I}\cdot\mathbf{J}), \tag{11.21}$$

so

$$(\mathbf{I}\cdot\mathbf{J}) = (\mathbf{F}^2 - \mathbf{I}^2 - \mathbf{J}^2)/2 = [f(f+1) - j(j+1) - i(i+1)]\hbar^2/2. \tag{11.22}$$

11.2.2 Application to hydrogen

A hydrogenic example can elucidate the extension of these considerations to s-states. Combining the radial and angular expectation values from the Schrödinger model yields

$$\left\langle\frac{a_0^3}{r^3}\right\rangle\frac{\langle\mathbf{L}\cdot\mathbf{L}\rangle}{\hbar^2} = \frac{Z^3\ell(\ell+1)}{n^3\ell\left(\ell+\frac{1}{2}\right)(\ell+1)}. \tag{11.23}$$

For $\ell = 0$, this expression has zeros in both numerator and denominator, and is thus indeterminate. However, if the semiclassical EBK value is computed, the expression becomes

$$\left\langle\frac{a_0^3}{r^3}\right\rangle\frac{\langle\mathbf{L}\cdot\mathbf{L}\rangle}{\hbar^2} = \frac{Z^3\left(\ell+\frac{1}{2}\right)^2}{n^3\left(\ell+\frac{1}{2}\right)^3} = \frac{Z^3}{n^3\left(\ell+\frac{1}{2}\right)}. \tag{11.24}$$

Since the Maslov index produces a nonvanishing orbital periapsis, the divergence at the origin is avoided. This is the classical counterpart of the Zitterbewegung of the Foldy–Wouthuysen transformation [98], which removes the divergence by delocalizing the mean-position variable of the electron.

The angular factors can be simplified for the case of $i = \frac{1}{2}$, and yield

$$2(\mathbf{I} \cdot \mathbf{J})/\mathbf{J}^2 = 1/(j+1) \qquad (f = j + 1/2)$$
$$= -1/j \qquad (f = j - 1/2). \qquad (11.25)$$

In analogy with the Maslov index, this yields a general result $\pm 1/(f + \frac{1}{2})$. Thus for $\ell \neq 0$ the energy for a hydrogenic system is given by

$$\Delta E = g_I Ry\alpha^2 \frac{m}{M_{\mathrm{p}}} \frac{Z^3}{2n^3} \frac{2(f - j)}{\left(f + \frac{1}{2}\right)\left(\ell + \frac{1}{2}\right)}. \qquad (11.26)$$

The computation of the Fermi contact interaction for s-states requires a knowledge of the wave function of the electron at $r = 0$. For a hydrogenic atom, a perturbation calculation [14] yields

$$\Delta E = \frac{4}{3} g_e g_I \, Ry\alpha^2 \frac{m}{M_{\mathrm{p}}} \frac{Z^3}{n^3} [f(f + 1) - 3/2]. \qquad (11.27)$$

Except for the factor of g_e that occurs in this spin–spin interaction, setting $\ell = 0$ causes the two expressions to agree for the values $f = 1$ and $f = 0$. This agreement is specific to hydrogen and not generally applicable.

If this is applied to the ground state of hydrogen ($Z = 1, n = 1$), the energy separation between the $f = 1$ and $f = 0$ states is

$$\Delta E(f = 1) - \Delta E(f = 0) = \frac{1}{21.106} \; (\mathrm{cm}^{-1}). \qquad (11.28)$$

This separation is both measurable and calculable to within parts in 10^{11}, so it provides both a frequency standard and a probe of fundamental theory. It has historical significance in that discrepancies between experiment and Dirac theory led to the 1947 suggestion [18] by Breit that the value of g_e differs slightly from 2.

This is the famous "21-cm line" that has been used in radio astronomy to map the density distribution of hydrogen in the interstellar medium, and also to determine temperatures in gas clouds. The decay rate for this transition can easily be determined. Spontaneous decay between the excited $F = 1$ level and the $F = 0$ ground state will occur by M1 radiation, which was shown in Section 8.2 to be given by

$$(2f + 1)A_{10}(\mathrm{ns}^{-1}) = \left[\frac{29.990}{\lambda_{10}(\text{Å})}\right]^3 S_{\mathrm{M1}}(1, 0) \qquad (11.29)$$

where the degeneracy is now that of f rather than j. For these hyperfine levels the line strength is given [178] by Sobelman

$$S_{\mathrm{M1}} = (2f + 1)(2f' + 1) \left\{ \begin{array}{ccc} j & f & i \\ f' & j & 1 \end{array} \right\}^2 j(j + 1)(2j + 1). \qquad (11.30)$$

For $f = 1$, $f' = 0$, $j = i = \frac{1}{2}$, this expression yields $S_{M1} = 3$. The transition probability rate for the 21-cm transition is therefore

$$A_{21} = 10^9 \left[\frac{29.990}{21.106 \times 10^8} \right]^3 = 2.87 \times 10^{-15} \text{ s}^{-1}.$$ (11.31)

This corresponds to a meanlife of 11 million years. Because of this very long meanlife, collisions are the dominant mechanism that populate the states, so the line strength indicates both the abundance and the temperature. Since the long lifetime also results in a negligible natural linewidth, the linewidth also yields the temperature.

11.3 Electric quadrupole moment of the nucleus

The electrostatic potential $\phi(\mathbf{r})$ for an arbitrary distribution of charge $\rho(\mathbf{r})$ is given by

$$\phi(\mathbf{r}) = K \int \int \int d\mathbf{r}' \frac{\rho(\mathbf{r}')}{|\mathbf{r} - \mathbf{r}'|}$$ (11.32)

which can be expanded for $r > r'$

$$\phi(\mathbf{r}) = K \int \int \int d\mathbf{r}' \rho(\mathbf{r}') \sum_{n=0}^{\infty} \frac{(r')^{n+1}}{r^n} P_n(\cos \vartheta)$$

$$= K \left[\frac{q}{r} + \frac{\mathbf{p} \cdot \mathbf{r}}{r^3} + \frac{\mathbf{r} \cdot \mathsf{Q} \cdot \mathbf{r}}{2r^5} + \cdots \right].$$ (11.33)

Here q is the monopole moment, $\mathbf{p}$ is the dipole moment (zero for an eigenstate of parity) and Q is the quadrupole tensor

$$Q_{ij} = \int \int \int d\mathbf{r}' \rho(\mathbf{r}')[3x_i' x_j' - \delta_{ij}(r')^2].$$ (11.34)

A simple model for a structured nucleus is given by a uniformly charged ellipsoid of revolution of semiaxes a and b

$$\rho = \frac{3q}{4\pi ab^2} \qquad \left(\frac{x^2 + y^2}{b^2} + \frac{z^2}{a^2} \leq 1 \right)$$

$$= 0 \qquad \left(\frac{x^2 + y^2}{b^2} + \frac{z^2}{a^2} > 1 \right).$$ (11.35)

In cylindrical coordinates

$$\eta^2 \equiv x^2 + y^2$$
$$\varphi \equiv \tan^{-1}(y/x).$$ (11.36)

The $i = j = 3$ element of the quadrupole moment is

$$Q_{33} = \frac{3q}{4\pi ab^2} \int_{-a}^{a} dz' \int_{0}^{2\pi} d\varphi \int_{0}^{b\sqrt{1-z^2/a^2}} d\eta \, \eta(2z^2 - \eta^2)$$

$$= 2q(a^2 - b^2)/3.$$ (11.37)

This single element is sufficient to specify the entire tensor, as can be seen from the following symmetry considerations. Because the charge distribution has inversion symmetry, Q is a diagonal tensor with only three nonvanishing components. Since the charge distribution has symmetry between x and y, $Q_{11} = Q_{22}$. By its definition, Q has a vanishing trace, so $Q_{11} + Q_{22} = -Q_{33}$. Thus

$$Q_{ij} = \delta_{ij} Q_{ij}; \qquad Q_{11} = Q_{22} = -Q_{33}/2. \tag{11.38}$$

The energy of interaction of the multipole moments with an external electric field $\phi_0(\mathbf{r})$ is given by

$$\Delta E = q\phi_0 + \sum_i p_i \frac{\partial \phi_0}{\partial x_i} + \frac{1}{6} \sum_i \sum_j Q_{ij} \frac{\partial^2 \phi_0}{\partial x_i \partial x_i} + \cdots. \tag{11.39}$$

Since the nucleus is an eigenstate of parity, the electric dipole moment p_i vanishes. For a quadrupole charge distribution with these symmetry properties this becomes

$$\begin{aligned}
\Delta E_Q &= \frac{1}{6} \left[Q_{11} \frac{\partial^2 \phi_0}{\partial x^2} + Q_{22} \frac{\partial^2 \phi_0}{\partial y^2} + Q_{33} \frac{\partial^2 \phi_0}{\partial z^2} \right] \\
&= \frac{Q_{33}}{6} \left[\frac{\partial^2 \phi_0}{\partial z^2} - \frac{1}{2} \left(\frac{\partial^2 \phi_0}{\partial x^2} + \frac{\partial^2 \phi_0}{\partial y^2} \right) \right].
\end{aligned} \tag{11.40}$$

Assuming $\nabla^2 \phi_0 = 0$, this reduces to

$$\Delta E_Q = \frac{Q_{33}}{4} \frac{\partial^2 \phi_0}{\partial z^2}. \tag{11.41}$$

If the external field were the result a single orbital electron, the potential and the field would be given by

$$\phi_0 = Ke/r; \qquad E_z = -\partial \phi_0 / \partial x = Kez/r^3, \tag{11.42}$$

so

$$\frac{\partial^2 \phi_0}{\partial z^2} = Ke \left[\frac{3z^2}{r^5} - \frac{1}{r^3} \right] = \frac{Ke}{r^3} (3 \cos^2 \vartheta - 1). \tag{11.43}$$

The energy then becomes (denoting Q_{33} dimensionlessly as Qea_0^2, and substituting $Ke^2 = 2Rya_0$)

$$\Delta E_Q = Ry \frac{a_0^3}{2r^3} Q(3 \cos^2 \vartheta - 1). \tag{11.44}$$

This has a quantum mechanical analogue of the form

$$\Delta E_Q = \frac{B}{4} \frac{\left[3(\mathbf{I} \cdot \mathbf{J})^2 + \frac{3}{2}(\mathbf{I} \cdot \mathbf{J}) - \mathbf{I}^2 \mathbf{J}^2 \right]}{2I\left(I + \frac{1}{2}\right)\left(J + \frac{1}{2}\right)}, \tag{11.45}$$

where the angular portion involves averaging by Racah algebra and B contains a radial expectation value calculation and the quadrupole moment

$$B = 2 Ry \left\langle \frac{a_0^3}{r^3} \right\rangle Q. \tag{11.46}$$

11.4 Example: hyperfine splitting of the 4p term in ^{23}Na

The relationships for the magnetic dipole and electric quadrupole splitting can be evaluated for specific values of the quantum numbers. Let us consider the 4p excited configuration of the ^{23}Na atom. In the absence of hyperfine interaction, this would consist of the 4p $^2P_{1/2}$ and 4p $^2P_{3/2}$ levels, hence $L = 1$, $S = \frac{1}{2}$, and $J = 1, 2$. Since the nuclear spin is $I = \frac{3}{2}$, the $J = \frac{3}{2}$ fine structure level splits into the $F = 0, 1, 2, 3$ hyperfine structure levels, and the $J = \frac{1}{2}$ fine structure level splits into the $F = 1, 2$ hyperfine structure levels. A diagram of these splittings is given in Fig. 11.1. The coefficients of A and B for each of the levels can be computed from the formulae in the previous two sections, and are also labeled on Fig. 11.1.

Measured hyperfine data for these splittings are given in Table 11.1. The sources of the data are Ref. [199] (for $^2P_{3/2}$) and Ref. [3] (for $^2P_{1/2}$). The values for A and B can be deduced from these relationships, and are also given in Table 11.1.

11.5 Isotope shifts

The effects of nuclear properties on atomic spectra can be separated into two classes: isotope shifts; and hyperfine structure. Isotope shifts can be considered in systems in which the basic energy-level structure (quantum numbers, number of levels, etc.) of two or more isotopes is the same, but the lines are shifted slightly due to effects arising from the mass, size, or shape of the nucleus. In the case of hyperfine structure the quantum numbers characterizing the state are altered by the existence of a nuclear magnetic moment.

The energy levels of two different isotopes of the same element (that have the same nuclear spin and hence the same energy-level structure) can differ because of several different effects, with their origins in the finite nuclear mass and the finite nuclear size. Techniques now exist that permit highly precise measurements of these shifts to be made, and very sophisticated but highly specialized methods have been developed for their analysis. Since excellent

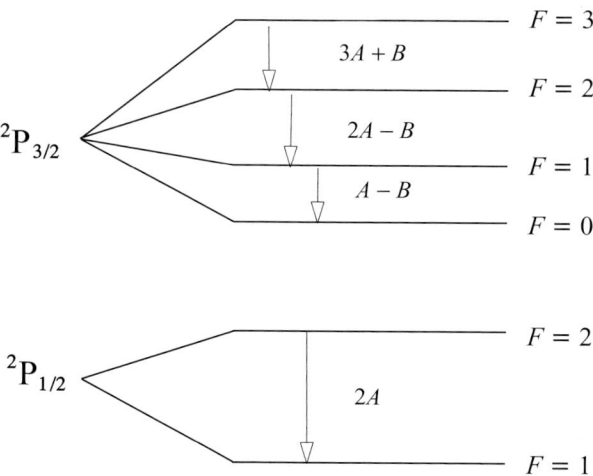

Fig. 11.1. Hyperfine splitting of an $L = 1$, $S = \frac{1}{2}$, $I = \frac{3}{2}$ term.

Table 11.1. Measurements of the 4p hyperfine structure splittings in ^{23}Na.

Term	F	F'	ΔE(MHz)	Formula	A(MHz)	B(MHz)
4p ^{2}P$_{3/2}$					18.534(15)	2.724(30)
	3	2	58.33	$3A + B$		
	3	1	92.67	$5A$		
	3	0	108.48	$6A - B$		
	2	1	34.34	$2A - B$		
	2	0	50.15	$3A - 2B$		
	1	0	15.81	$A - B$		
4p ^{2}P$_{1/2}$					94.3(1)	
	2	1	188.6	$2A$		

reviews of this subject exist (e.g., Ref. [131]), only a few brief comments will be made here. Isotope shifts are often discussed in terms of the "mass effect," or the "field effect" depending on whether the corrections arise from the nuclear mass, or the size and shape of its charge distribution.

11.5.1 Finite nuclear mass

A commonly cited example of this type of isotope shift is provided by hydrogen, deuterium, and tritium. Since these systems are all two-body binary systems, they can be treated by the reduced-mass transformation, and presented in terms of the reduced Rydberg constant $R_Z = Ry/(1 + m/AM)$ where their atomic numbers are $A = 1, 2, 3$. Unfortunately, this case is better suited to a course in elementary quantum mechanics than to the study of atomic structure. Since the basic reduced-mass concept is valid only for a two-body system, hydrogen is certainly not the simplest case of a general atomic system, but in a distinct class by itself. As was discussed in Section 3.2.3, for a multielectron atom the reduced-mass approach breaks down. For relatively simple systems it is sometimes possible to compute the "coupling effect" among the center-of-mass motions of the various electron–nucleus pairings, but this is very difficult to do accurately in heavy, complex atoms.

11.5.2 Finite nuclear size

If one assumes that the nuclear charge is distributed in some manner throughout the nuclear volume, and the wave function for a given electron orbital has an overlap with the nuclear volume, then the electron will be less tightly bound for a nucleus of finite size than it would be for a point nucleus. Moreover, the reduction will be greater the larger the size of the nucleus. This is especially important for s-states, where the overlap can be large. Some care must be exercised in making these calculations, and relativistic and higher-order perturbation calculations are often necessary. It was discussed both in Chapters 2 and 4 that the Darwin term (and its nuclear analogue, the Fermi contact interaction) is necessary to include the effect of Zitterbewegung, which smears out the wave function of the s-electron in requiring modifications in the zeroth-order electron density at the nucleus. With this caveat, a very simple illustration of the penetration of a model nucleus by a 1s hydrogenic electron is presented below.

As a rudimentary model of the nucleus, consider a hollow sphere of charge of radius r_0. The electrostatic potential is then given by

$$V(r) = -ZKe^2/r \qquad (r > r_0)$$
$$= -ZKe^2/r_0 \qquad (r \leq r_0). \tag{11.47}$$

This can be treated by perturbation theory using $H = H_0 + \Delta V$, where the unperturbed Hamiltonian is the standard hydrogenic case

$$H_0 = p^2/2m - ZKe^2/r, \tag{11.48}$$

and the perturbation is

$$\Delta V = 0 \qquad\qquad (r > r_0)$$
$$= -ZKe^2(1/r_0 - 1/r) \qquad (r \leq r_0). \tag{11.49}$$

Applying this to the 1s-state in a hydrogenic atom, the radial wave function is

$$R_{1s}(u) = 2(Z/a_0)^{3/2}e^{-u/2}, \tag{11.50}$$

where $u \equiv 2Zr/a_0$. Denoting $u_0 \equiv 2Zr_0/a_0$ the correction is

$$\Delta E = \langle \Delta V \rangle = -\frac{4Z^2Ke^2}{a_0} \int_0^{u_0} du\, u^2 \left(\frac{1}{u_0} - \frac{1}{u}\right) e^{-u}. \tag{11.51}$$

These are incomplete gamma functions, which can be evaluated by a formula given in Eq. 5.47. The constants can be converted using $Ry \equiv Ke^2/2a_0$ to obtain

$$\Delta E = 2RyZ^2 \left[\left(1 - \frac{2}{u_0}\right) + \left(1 + \frac{2}{u_0}\right)e^{-u_0}\right], \tag{11.52}$$

which can be evaluated by expanding the exponential for small u_0

$$\Delta E \cong 2RyZ^2 \left[\left(1 - \frac{2}{u_0}\right) + \left(1 - u_0 + \frac{u_0^2}{2} - \frac{u_0^3}{3!} + \dots\right)\right.$$
$$\left. + \left(\frac{2}{u_0} - 2 + u_0 - \frac{u_0^2}{3} + \dots\right)\right]. \tag{11.53}$$

This simplifies to

$$\Delta E \cong RyZ^2u_0^2/3 = 4RyZ^4r_0^2/3a_0^2. \tag{11.54}$$

Assuming an incompressible nucleus of atomic mass number A, the nuclear radius is given approximately by $r_0 = (1.5\ \text{fm})A^{1/3}$, so the correction is

$$\Delta E \cong (1.1 \times 10^{-9})RyZ^4A^{2/3}. \tag{11.55}$$

While this quantity is very small for neutral hydrogen, the value for a hydrogenlike (one-electron) uranium atom is $\Delta E(Z = 92, A = 238) \cong 3Ry$.

More-sophisticated models for the size and shape of the nucleus can be used, and wave functions for the electrons can be computed for multielectron systems by a variety of theoretical techniques.

11.6 Hyperfine quenching

One of the most interesting phenomena in the theory of highly forbidden transitions is the effect of hyperfine quenching, whereby mixing by the hyperfine interaction can significantly alter the lifetimes of the levels [103]. The phenomenon was first discussed [17] in 1930 by Bowen, who pointed out that the substantial strength that was observed in the $6s^2\,{}^1S_0$–$6s6p$ ${}^3P_2$ line at 2270 Å in the spectrum of Hg I was primarily due to E1 radiation caused by coupling with the nuclear spin (and not to possible higher-order multipole radiation as had been suggested). Thus both S and J labels are only nominal here. Bowen's conclusion was confirmed [157] by Mrozowski, who experimentally observed the $6s^2\,{}^1S_0$–$6s6p$ ${}^3P_0$ line at 2656 Å in Hg I. This transition would be rigorously forbidden to all multipole orders of single-photon decay in an atom with a spinless nucleus by the no $J = 0 \rightarrow J' = 0$ selection rule of angular momentum conservation. Mrozowski attributed the appearance of this line to coupling between the magnetic moments (spin and orbital) of the electron and the spin of the nucleus.

Unlike two-valence-electron systems in more complex atoms, the lowest ${}^3P_0$ state in the helium sequence is not metastable, having allowed E1 transitions to the $1s2s\,{}^3S_1$ level. With increasing Z, variously forbidden transitions to the $1s^2\,{}^1S_0$ ground state have an increasingly important effect on the lifetimes of the individual $1s2p\,{}^3P_J$ fine structure levels, as can be seen from an isoelectronic plot of these lifetimes in Fig. 11.2.

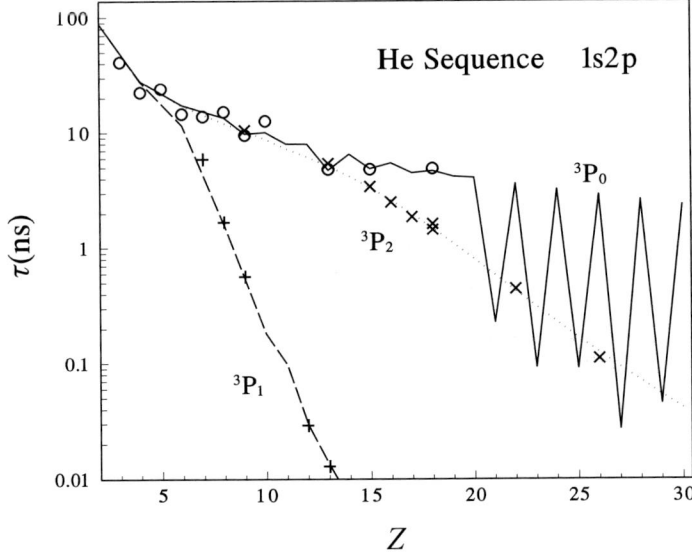

Fig. 11.2. Lifetimes of the $1s2p\,{}^3P_J$ levels in the He isoelectronic sequence, indicating the hyperfine quenching of the $J = 0$ levels that occurs in odd-Z nuclei for $Z > 21$. (After Ref. [42].)

At low Z all three $1s2p\,^3P_J$ levels have similar lifetimes, since the E1 transition to $1s2s\,^3S_1$ dominates. With increasing Z the lifetime of the $^3P_1$ level is drastically shortened, since the $\Delta S = 0$ selection rule is valid only for pure relativistic LS coupling, and spin mixing causes the intercombination transition to the ground state to become E1-allowed. A similar but less drastic shortening of the $^3P_2$ lifetime takes place because of its M2 transition to ground. For systems with nonzero nuclear spin I, both the $^3P_2$ and $^3P_0$ lifetimes are affected by hyperfine induced E1 transitions to the ground state. This is particularly striking for the $^3P_0$ level, because it has no other decay channel to ground, and because the effect is not smeared over a multiplicity of values of $F = I + J$.

As can be seen from Fig. 11.2, there is a sharp "turning-on" of the hyperfine quenching effect at $Z = 21$. There is also an alternation between zero and nonzero values for I in the most abundant isotopes of nuclei of even and odd Z, which causes the hyperfine quenching to vanish for even Z. The ratio of the $^3P_0$ to $^3P_2$ lifetimes for $Z > 20$ thus provides a sensitive test of this theory.

12

Electrostatic polarizabilities and long-range interactions

The whole is more than the sum of the parts.

The core polarization model provides a means for treating an atom with a highly excited electron as a single particle outside an effective distortable core of charge. Semiempirical methods for characterizing such systems in terms of adiabatic and nonadiabatic polarizabilities were described in Section 3.4. It is also possible to compute these quantities using *ab initio* theoretical models involving long-range interactions.

12.1 Rayleigh–Schrödinger perturbation theory

All realistic quantitative formulations of atomic properties have recourse to approximation methods. One of the most commonly used approximative approaches involves the use of standard Rayleigh–Schrödinger perturbation theory [155], in which the Hamiltonian is decomposed into an "unperturbed" and a "perturbed" portion

$$H = H_0 + \lambda \Delta V. \tag{12.1}$$

H_0 is an unperturbed Hamiltonian

$$H_0 = -\frac{\hbar^2}{2m} \nabla^2 + V_0 \tag{12.2}$$

corresponding to a soluble problem

$$H |n\rangle = \varepsilon_n |n\rangle \tag{12.3}$$

which has known eigenfunctions and eigenenergies $|n\rangle$ and ε_n. These solutions will be assumed to be nondegenerate (that is, states with different quantum numbers do not have the same energy). If this is not the case, the discussion that follows is restricted to those levels that are not degenerate.

The quantity $\lambda \Delta V$ is assumed to be a small correction, where $0 \leq \lambda \leq 1$ is a physically motivated dimensionless parameter characterizing the magnitude of the correction. The Schrödinger equation to be solved is

$$H \Psi_n = E_n \Psi_n. \tag{12.4}$$

The eigenvectors and eigenenergies Ψ_n and E_n are then sought in terms of the representation of the unperturbed basis states. This is done using the two ansatz expansions

$$E_n = E_n^{(0)} + \lambda E_n^{(1)} + \lambda^2 E_n^{(2)} + \cdots \tag{12.5}$$

and

$$\Psi_n = \Psi_n^{(0)} + \lambda \Psi_n^{(1)} + \lambda^2 \Psi_n^{(2)} + \cdots . \tag{12.6}$$

To accomplish this, Eqs. 12.1, 12.5 and 12.6 are inserted into Eq. 12.4, and the resulting expression is refactored as coefficients of powers of λ. Terms of like power of λ on opposite sides of the equation are separately equated to obtain

$$H_0 \Psi_n^{(0)} = E_n^{(0)} \Psi_n^{(0)} \tag{12.7}$$

$$H_0 \Psi_n^{(1)} + \Delta V \Psi_n^{(0)} = E_n^{(0)} \Psi_n^{(1)} + E_n^{(1)} \Psi_n^{(0)} \tag{12.8}$$

$$H_0 \Psi_n^{(2)} + \Delta V \Psi_n^{(1)} = E_n^{(0)} \Psi_n^{(2)} + E_n^{(1)} \Psi_n^{(1)} + E_n^{(1)} \Psi_n^{(1)} \tag{12.9}$$

$$\vdots$$

$$H_0 \Psi_n^{(k)} + \Delta V \Psi_n^{(k-1)} = \sum_{i=0}^{k} E_n^{(i)} \Psi_n^{(k-i)} . \tag{12.10}$$

The perturbed wave functions are expanded on the unperturbed basis set

$$\Psi_n^{(k)} = \sum_m C_{nm}^{(k)} |m\rangle . \tag{12.11}$$

The zeroth-order contribution of Eq. 12.7 is identical with the unperturbed problem of Eq. 12.3, and indicates only that $E_n^{(0)} = \varepsilon_n$ and $C_{nm}^{(0)} = \delta_{nm}$.

Inserting the basis expansion of Eq. 12.11 into Eq. 12.10, operating from the left with $\langle p|$, and using the orthogonality of the basis set, a recursion relation is obtained

$$(\varepsilon_p - \varepsilon_n) C_{np}^{(k)} + \sum_j C_{nj}^{(k-1)} \langle p|\Delta V|j\rangle = \sum_{i=1}^{k} E_n^{(i)} C_{np}^{(k-i)} . \tag{12.12}$$

This expression yields relationships both for $j \neq n$ and for the diagonal ($j = n$) form

$$\sum_j C_{nj}^{(k-1)} \langle n|\Delta V|j\rangle = \sum_{i=1}^{k} E_n^{(i)} C_{nn}^{(k-i)} . \tag{12.13}$$

After some manipulation, these equations yield the following expressions for the energies, expressed to zeroth, first, second, and third orders of the quantity λ

$$E_n^{(0)} = \varepsilon_n \tag{12.14}$$

$$E_n^{(1)} = \langle n|\Delta V|n\rangle \tag{12.15}$$

$$E_n^{(2)} = \sum_{i \neq n} \frac{\langle n|\Delta V|i\rangle \langle i|\Delta V|n\rangle}{(\varepsilon_n - \varepsilon_i)} \tag{12.16}$$

$$E_n^{(3)} = \sum_{i \neq n} \sum_{j \neq n} \frac{\langle n|\Delta V|i\rangle \langle i|\Delta V|j\rangle \langle j|\Delta V|n\rangle}{(\varepsilon_n - \varepsilon_i)(\varepsilon_n - \varepsilon_j)} - \langle n|\Delta V|n\rangle \sum_{k \neq n} \frac{\langle n|\Delta V|k\rangle \langle k|\Delta V|n\rangle}{(\varepsilon_n - \varepsilon_k)^2}.$$

(12.17)

Here the summation notation is interpreted to also include an integration over the continuum.

12.2 The Dalgarno–Lewis operator

A weakness in the Rayleigh–Schrödinger approach lies in the fact that infinite sums arise in all orders beyond the first. This leads to a situation in which one must consider the orders of two different expansions. One is the number of terms included in the perturbation expansion, and the other is the number of states of the complete set that have been included in the sum. In addition, the fact that the diagonal matrix elements are excluded from the sums prevents the direct use of completeness conditions,

An ingenious and elegant solution to this problem was developed [76] by Dalgarno and Lewis. This involves the definition of a function F that has the commutation property

$$[F, H_0] \equiv \Delta V.$$

(12.18)

If it is possible to solve this relationship for F, then sums can be eliminated through the closure property. The virtue of this quantity can be seen by considering the matrix element

$$\langle n | F H_0 - H_0 F | i \rangle = \langle n | \Delta V | i \rangle.$$

(12.19)

The diagonal elements yield the first-order correction $E_n^{(1)}$. If these first-order corrections are nonvanishing, they can be eliminated by subtraction. The off-diagonal elements yield

$$\langle n | F | i \rangle = \frac{\langle n | \Delta V | i \rangle}{(\varepsilon_n - \varepsilon_i)}.$$

(12.20)

Substitution of this into the second- and third-order energy terms above yields

$$E_n^{(2)} = \sum_{i \neq n} \langle n | F | i \rangle \langle i | \Delta V | n \rangle$$

(12.21)

$$E_n^{(3)} = - \sum_{i \neq n} \sum_{j \neq n} \langle n | F | i \rangle \langle i | \Delta V | j \rangle \langle j | F | n \rangle.$$

(12.22)

These sums could be performed through the closure property were it not for the exclusion of the $i = n$ and $j = n$ terms, which can be cured by extending the sum and subtracting them off. The second-order term

$$E_n^{(2)} = \sum_{i} \langle n | F | i \rangle \langle i | \Delta V | n \rangle - \langle n | F | n \rangle \langle n | \Delta V | n \rangle$$

(12.23)

reduces to

$$E_n^{(2)} = \langle n | F \Delta V | n \rangle - \langle n | F | n \rangle E_n^{(1)}, \tag{12.24}$$

which is known as the Lennard-Jones sum rule [127]. The third-order term

$$E_n^{(3)} = \sum_i \sum_j \langle n | F | i \rangle \langle i | \Delta V | j \rangle \langle j | F | n \rangle - \sum_i \langle n | F | i \rangle \langle i | \Delta V | n \rangle \langle n | F | n \rangle$$
$$- \sum_j \langle n | F | n \rangle \langle n | \Delta V | j \rangle \langle j | F | n \rangle + \langle n | F | n \rangle \langle n | \Delta V | n \rangle \langle n | F | n \rangle \tag{12.25}$$

reduces to

$$E_n^{(3)} = \langle n | F^2 \Delta V | n \rangle - 2 \langle n | F \Delta V | n \rangle \langle n | F | n \rangle + \langle n | F | n \rangle^2 \langle n | \Delta V | n \rangle. \tag{12.26}$$

This can be simplified by use of the expression for $E_n^{(2)}$ to obtain

$$E_n^{(3)} = \langle n | F^2 \Delta V | n \rangle - 2 \langle n | F | n \rangle E_n^{(2)} - \langle n | F | n \rangle^2 E_n^{(1)}. \tag{12.27}$$

These expressions provide a method by which exact results can be obtained for the various orders of the perturbation expansion, provided it is possible to find, by whatever means possible, the operator F. One means for doing this is to operationally expand the commutation relation

$$(F H_0 - H_0 F) \phi_n = -\frac{\hbar^2}{2m} [F \nabla^2 \phi_n - \nabla^2 (F \phi_n)]$$
$$= \frac{\hbar^2}{2m} [\phi_n \nabla^2 F + 2 \nabla F \cdot \nabla \phi_n], \tag{12.28}$$

and insert this into Eq. 12.18 to obtain the differential equation

$$\phi_n \nabla^2 F + 2 \nabla F \cdot \nabla \phi_n = \frac{2m}{\hbar^2} \Delta V \phi_n. \tag{12.29}$$

Solutions can be sought for specific cases of ϕ_n and ΔV.

12.3 Application: ground-state polarizabilities

These methods are well suited to the description of an atom or ion in the ground state that is in the field of a distant charge. This can describe, for example, either a heliumlike atom or ion in which one electron is in the 1s-state and the other is in a high Rydberg state, or the molecular bonding between a hydrogen atom and a proton. Both of these cases can be described, with small modifications, by the same development. We choose to begin with the case of the heliumlike atom, which consists of a nucleus of charge Ze, a 1s-electron, and a high Rydberg electron.

12.3.1 Formulation

If the ground-state electron is at a vector distance $\mathbf{r}$ from the nucleus and the high Rydberg electron is at a vector distance $\mathbf{R}$ from the nucleus, the Hamiltonian is given by

$$H = -\frac{\hbar^2}{2m}\nabla_r^2 - \frac{KZe^2}{r} - \frac{\hbar^2}{2m}\nabla_R^2 - \frac{KZe^2}{R} + \frac{Ke^2}{|\mathbf{r}-\mathbf{R}|}. \qquad (12.30)$$

If $R \gg r$, then the interelectron repulsion can be written

$$\frac{Ke^2}{|\mathbf{r}-\mathbf{R}|} \approx Ke^2 \sum_{m=0}^{\infty} \frac{r^m}{R^{m+1}} P_m(\cos\vartheta)$$

$$= \frac{Ke^2}{R} + Ke^2 \sum_{m=1}^{\infty} \frac{r^m}{R^{m+1}} P_m(\cos\vartheta). \qquad (12.31)$$

where we have split off the monopole contribution, so that the Hamiltonian becomes

$$H = -\frac{\hbar^2}{2m}\nabla_r^2 - \frac{KZe^2}{r} - \frac{\hbar^2}{2m}\nabla_R^2 - \frac{K(Z-1)e^2}{R} + Ke^2 \sum_{m=1}^{\infty} \frac{r^m}{R^{m+1}} P_m(\cos\vartheta). \qquad (12.32)$$

This can be separated into a standard two-independent-particle unperturbed Hamiltonian

$$H_0 = -\frac{\hbar^2}{2m}\nabla_r^2 - \frac{KZe^2}{r} - \frac{\hbar^2}{2m}\nabla_R^2 - \frac{K(Z-1)e^2}{R}, \qquad (12.33)$$

and a perturbation

$$\Delta V = Ke^2 \sum_{m=1}^{\infty} \frac{r^m}{R^{m+1}} P_m(\cos\vartheta). \qquad (12.34)$$

If we express this in atomic units $r' \equiv Zr/a_0$, $R' \equiv ZR/a_0$, $E' \equiv E/RyZ^2$ these become

$$H_0' = -\nabla_{r'}^2 - \frac{2}{r'} - \nabla_{R'}^2 - \frac{2(Z-1)}{ZR'}, \qquad (12.35)$$

and

$$\Delta V' = \frac{2}{Z} \sum_{m=1}^{\infty} \frac{(r')^m}{(R')^{m+1}} P_m(\cos\vartheta). \qquad (12.36)$$

In these units the wave function of the ground state is

$$\psi_{1s}(\mathbf{r}) - \frac{1}{\sqrt{\pi}} e^{-r'}. \qquad (12.37)$$

Since the perturbation excludes the monopole term, the first-order correction $E_n^{(1)}$ vanishes. Let us assume a spherically symmetric form for $F(\mathbf{r})$,

$$F(\mathbf{r}) = \sum_{m=0}^{\infty} f_m(r) P_m(\cos\vartheta). \qquad (12.38)$$

Thus

$$\nabla F \cdot \nabla \phi_n = -\phi_n \sum_{m=0}^{\infty} \frac{\mathrm{d} f_m}{\mathrm{d} r} P_m(\cos \vartheta), \tag{12.39}$$

so ϕ_n is an overall multiplicative factor in Eq. 12.29. Thus the differential equation for f_n becomes

$$\frac{\mathrm{d}^2 f_m}{\mathrm{d} r'^2} - 2 \left(1 - \frac{1}{r'}\right) \frac{\mathrm{d} f_m}{\mathrm{d} r'} - \frac{m(m+1)}{(r')^2} f_m = \frac{2(r')^m}{Z(R')^{m+1}}. \tag{12.40}$$

This yields quickly to the ansatz

$$f_m(r') \equiv \frac{2}{Z} \frac{(r')^m}{(R')^{m+1}} \sum_p b_p(r')^p; \tag{12.41}$$

leading to the condition

$$\sum_p [(p+1)(p+2m+2)b_{p+1} - 2(p+m)b_p](r')^{p-1} = 1. \tag{12.42}$$

The bracketed quantity must vanish unless $p = 1$, leading to the conditions $b_1/b_0 = m/(m+1)$, $b_1 = -1/(2m+2)$, and $b_p = 0$ for $p > 1$. The solution is

$$f_m(r') = -\frac{1/Z}{(R')^{m+1}} \left(\frac{(r')^{m+1}}{m+1} + \frac{(r')^m}{m}\right), \tag{12.43}$$

so the expression for the Dalgarno–Lewis function is

$$F(\mathbf{r}') = -\sum_{m=1}^{\infty} \frac{1/Z}{(R')^{m+1}} \left(\frac{(r')^{m+1}}{m+1} + \frac{(r')^m}{m}\right) P_m(\cos \vartheta). \tag{12.44}$$

The second-order correction is given by

$$\langle ns|F\Delta V|ns\rangle$$
$$= -\frac{2}{Z^2} \left\langle 1s \left| \left\{\sum_{m=1}^{\infty} \frac{1}{(R')^{m+1}} \left(\frac{(r')^{m+1}}{m+1} + \frac{(r')^m}{m}\right) P_m\right\} \left\{\sum_{p=0}^{\infty} \frac{(r')^p}{(R')^{p+1}} P_p\right\} \right| 1s \right\rangle. \tag{12.45}$$

The angular integral yields $\delta_{mp}/(2m+1)$, collapsing the p sum

$$\langle 1s|F\Delta V|1s\rangle = -\frac{2}{Z^2} \sum_{m=1}^{\infty} \frac{1}{(2m+1)(R')^{2m+2}} \left\langle 1s \left| \frac{(r')^{2m+1}}{m+1} + \frac{(r')^{2m}}{m} \right| 1s \right\rangle. \tag{12.46}$$

The expectation values of powers of r for the hydrogenic 1s-state are known and given by

$$\langle 1s|(r')^k|1s\rangle = (k+2)!/2^{k+1}, \tag{12.47}$$

and are inserted to yield the desired correction

$$E_{1s}^{(2)} = \langle 1s|F\Delta V|1s\rangle = \frac{1}{Z^2} \sum_{m=1}^{\infty} \frac{1}{(R')^{2m+2}} \frac{(2m+2)!(m+2)}{m(m+1)2^{2m+1}}. \tag{12.48}$$

Table 12.1. Hydrogenlike values for the ξ_k coefficients in the expansion of long-range interactions $\Delta E_k = Ry\xi_k\langle(a_0/R)^k\rangle$. Here, K is the Coulombic electrostatic constant.   (After Ref. [59].)

	$K\xi_k$				
k	2nd order	3rd order	4th order	Nonadiabatic	Retardation
4	$\frac{9}{2}\frac{a_0^3}{Z^4}$				
5					$-\frac{99\alpha}{4\pi}\frac{a_0^4}{Z^3}$
6	$15\frac{a_0^5}{Z^6}$			$-\frac{129}{4}\frac{a_0^5}{Z^6}$	$\frac{129}{4}\frac{a_0^5}{Z^6}$
7		$-\frac{213}{2}\frac{a_0^6}{Z^8}$			
8	$\frac{525}{4}\frac{a_0^7}{Z^8}$		$\frac{3555}{32}\frac{a_0^7}{Z^{10}}$	$-\frac{1605}{8}\frac{a_0^7}{Z^8}$	
9		$-1773\frac{a_0^8}{Z^{10}}$			
10	$\frac{2835}{4}\frac{a_0^9}{Z^{10}}$		$\frac{80\,379}{8}\frac{a_0^9}{Z^{12}}$	$-\frac{22\,855}{8}\frac{a_0^9}{Z^{10}}$	

Paying back the substitutions into atomic units that were made in Eq. 12.35, and comparing this to the form of the multipole expansion

$$E_{1s}^{(2)} = -Ry\sum_{m=1}^{\infty}\frac{\alpha_{2m}}{R^{2m+2}}, \tag{12.49}$$

we can identify α_{2m}, the electrostatic multipole polarizability (in units consistent with those defined in Eq. 3.38), as

$$\alpha_{2m} \equiv \frac{(2m+2)!(m+2)}{m(m+1)2^{2m+1}}\frac{a_0^{2m+1}}{KZ^{2m+2}}. \tag{12.50}$$

Similar expressions obtained for higher-order terms in the perturbation expansion can be computed [75] by the same methods. The various corrections for long-range interactions have the form

$$\Delta E = Ry\sum_{k=4}\xi_k\langle(a_0/R)^k\rangle. \tag{12.51}$$

Although this power series expansion would diverge mathematically if summed to infinity, physical limitations quench high reciprocal powers of R, truncating the series at a finite upper limit. Numerical values for the coefficients ξ_k up to tenth order for the long-range interactions of the hydrogenlike core of a heliumlike system are given in Table 12.1. These expressions can be used to characterize high Rydberg states in helium and heliumlike systems. The lowest-order term was computed [190] in 1926 by Ivar Waller and many of these specific formulae were published [146] in 1934 by Guido Ludwig.

These expressions can also be used to describe the hydrogen molecular ion H_2^+, if Z is set equal to unity and the sign of the perturbation is changed to positive (since the distant charge is then a proton rather than an electron). This change in the sign of the potential does not affect the values for even orders of perturbation, but does enter for the odd orders. Thus the signs for the heliumlike case alternate with order of the perturbation correction, whereas for the H_2^+ case all perturbation orders have the same sign.

For purposes of mathematical rigor, it should be noted that this procedure for representing long-range forces by a series expansion in reciprocal powers of R yields an infinite series that is divergent for all R. It has been demonstrated [76] that this mathematical divergence cannot be removed, but that the formulation can be used to obtain accurate results if the expansion is truncated at a suitable order of the perturbation theory. Without addressing specific questions of mathematical convergence, it is not difficult to point to physical and experimental constraints that could quench these higher orders of the perturbation expansion.

12.3.2 Convergence of the sum

The Dalgarno–Lewis method has utilized the completeness of the eigenfunctions to compute the dipole polarizability of the ground state of the hydrogen atom to infinite order, yielding the result $\alpha_d = \frac{9}{2}$ (in units $a_0^3/K Z^4$). To illustrate the power of this method, let us consider the number of terms in the sum/integral that would be necessary to compute this by standard summation methods. The oscillator strength of the 1s–np transitions in hydrogen is given by

$$f_{1s,np} = \frac{2^8}{3} \frac{n^5}{(n^2-1)^4} \left(\frac{n-1}{n+1}\right)^{2n}. \tag{12.52}$$

This factorization is favorable for calculational purposes, since raising a ratio less than unity to a power avoids the computational overflows that would occur with very large numbers in both the numerator and denominator. The wavelength of the transition is

$$\frac{1}{\lambda} = Ry\left[1 - \frac{1}{n^2}\right] = Ry\frac{(n^2-1)}{n^2} \tag{12.53}$$

(where Ry is in cm^{-1}). The sum of the oscillator strengths is given by the Thomas–Reiche–Kuhn sum rule

$$1 = \sum_n^N f_{1s,np} = \frac{2^8}{3} \sum_n^N \frac{n^5}{(n^2-1)^4} \left(\frac{n-1}{n+1}\right)^{2n}, \tag{12.54}$$

where the summation symbol also implicitly includes an integral over continuum states. We wish to compare the contributions, both to the total oscillator strength and to the sum determining the dipole polarizability, for inclusions of various numbers of bound and continuum states. The dipole polarizability is given by the relationship

$$\alpha_d = 4Ry \sum_n^N f_{1s,np}\lambda_{1s,np}^2 = \frac{2^{10}}{3} \sum_n^N \frac{n^9}{(n^2-1)^6} \left(\frac{n-1}{n+1}\right)^{2n}. \tag{12.55}$$

Table 12.2. Convergence of the oscillator strength and dipole polarizability summations for hydrogen.

N	$\sum_n f_{1,n}$	$4Ry^2 \sum_n f_{1,n}\lambda^2_{1,n}$
2	0.416 197	2.959 621
10	0.557 846	3.634 364
100	0.564 927	3.662 948
1000	0.565 003	3.663 255
10 000	0.565 004	3.663 258
Continuum	1.000 000	4.500 000

Sums of these quantities for values of $N = 2, 10, 100, 1000, 10\,000$ and the continuum are listed in Table 12.2. The sum converges slowly in the bound spectrum, and much of the contribution to both total quantities is contained in the continuum contribution. This clearly illustrates the power of the Dalgarno–Lewis method, which automatically includes all bound state and continuum contributions.

It can be seen that for this application, a heliumlike ion with hydrogenlike core is a very difficult case, since all transitions in the core Rydberg series have $\Delta n \geq 1$. This can be compared with the example of the Na-like core of the Mg atom that was treated by semiempirical methods in Section 6.4, and summarized in Table 6.1. In contrast to a hydrogenlike core, the lowest resonance transition for an alkali-metallike core is an intrashell ns–np transition. Since this has an f-value near unity, in this case the lowest transition dominates the sum. This application to hydrogen demonstrates the value of the Dalgarno–Lewis method, which is applicable to complex atoms as well.

12.4 Nonadiabatic correlations

The various atomic polarizabilities computed above are adiabatic in nature, and assume that the charge polarization occurs instantaneously. There are also nonadiabatic correlations [22, 134, 146] which are a measure of the inability of the electron core to instantaneously follow the motion of the electron. These effects can be taken into account by the inclusion of an interaction known as the *distortion potential*, and the combination of the ordinary polarization potential with the distortion potential is known as the *extended potential*.

The lowest-order correction of this type is of the form

$$\Delta E_{\text{nonadiabatic}} = Ry\xi_6 \langle (a_0/R)^6 \rangle, \tag{12.56}$$

where

$$\xi_6 = -3 \sum_n f_{0n} \left(\frac{2Ry}{\varepsilon_n - \varepsilon_0} \right)^3 \tag{12.57}$$

and for which the hydrogenic value is $\frac{3}{4}(43)$. For historical reasons this is often written in terms of a quantity β that is smaller by a factor of the power of R. Thus the notation $\xi_6 = 6\beta$ often occurs in the literature.

Calculations have been made [22] using the extended potential that yield a general expression for the various orders of these nonadiabatic corrections. These can be written in the form

$$\xi_{2m+4} = -\frac{(2m+1)!}{2^{2m+1}m^2}(2m^4 + 11m^3 + 18m^2 + 10m + 2). \qquad (12.58)$$

Values of the product $K\xi_k$ are also listed in Table 12.1.

12.5 Casimir–Polder retardation corrections

It has long been suspected that there are subtle additional dynamical processes embedded within the long-range interactions of atoms and molecules [130]. It was asserted in 1873 by Johannes van der Waals [186] that experimentally observed deviations from the ideal gas law indicated that there is an attractive force at large distances between neutral atoms. This was given a quantum mechanical basis in 1930 by London [144], who showed that the interaction between two hydrogen atoms behaved as $1/R^6$ at distances large compared to the Bohr radius.

In 1948 Casimir and Polder demonstrated [23] that the finiteness of the speed of light introduces retardation effects. This classic paper showed that the interaction potential between two hydrogen atoms in the ground state changes from the van der Waals–London form $1/R^6$, previously thought to be valid for all large distances, to a dependence $1/R^7$. This effect has been confirmed in bulk matter, and there have been fascinating attempts to observe its presence in studies of atomic spectra.

High Rydberg states in the helium atom and heliumlike ions provide an excellent test bench for investigating these effects [59]. This can be done in terms of a simple classical model [44]. Consider a heliumlike system with a nucleus of charge Z with one electron in a low-lying state and another in a state of high n and high ℓ. The time scale for retardation effects can be estimated classically by considering the period of rotation of the 1s-electron, and the time it takes for a virtual photon to make a round trip from the outer electron to the inner electron and back.

Using Eq. 2.65, the EBK value for the semimajor axis of either electron is given by $a = a_0 n^2 / \zeta$. Here $\zeta = Z$ for the inner electron and $\zeta = Z - 1$ for the outer electron. From Eq. 2.27, Kepler's third law states that the period of the orbit is given by

$$T = 2\pi \sqrt{\frac{m}{-KqQ}a^3}. \qquad (12.59)$$

Since there is no screening of the nuclear charge for the inner electron, $-qQ = Ze^2$, the product of the charges of the core electron and the unscreened nucleus. For a core electron of principal quantum number n_c, the semimajor axis is given by $a = a_0 n_c^2 / Z$. Converting to atomic units using $Ke^2/m = a_0(\alpha c)^2$, (obtained from $2Ry = Ke^2/a_0 = mc^2\alpha^2$), the period of the core electron is given by

$$T_c = 2\pi \frac{a_0}{\alpha c} \frac{n_c^3}{Z^2}. \tag{12.60}$$

For a Rydberg electron in a nearly circular orbit, the time Δt_R required for a virtual photon to make the round trip between the Rydberg electron and the core electron (which is assumed close to the nucleus) is

$$\Delta t_R \approx \frac{2a}{c}. \tag{12.61}$$

As indicated above, this outer electron (of principal quantum number n_R) sees the nucleus screened by one electron, so $\zeta = Z - 1$, yielding

$$\Delta t_R \approx \frac{2a_0}{c} \frac{n_R^2}{(Z - 1)}. \tag{12.62}$$

Retardation effects can be expected to be important if $\Delta t_R \geq T_c$, so we consider the ratio

$$\frac{\Delta t_R}{T_c} = \frac{\alpha}{\pi} \frac{Z^2}{(Z - 1)} \frac{n_R^2}{n_c^3}. \tag{12.63}$$

For neutral helium ($Z = 2$) with the lower electron in the 1s-state ($n_c = 1$) this yields $\Delta t_R / T_c = 4\alpha n_R^2 / \pi$, and the condition for matching the two time intervals is $n_R = \frac{1}{2}\sqrt{\pi/\alpha} = 10.37$.

It is interesting to note that a very precise measurement [28] of the $n_R = 10$ level structure has been performed that did not find evidence of the Casimir–Polder retardation. This simple classical model yields a frequency resonance at $n_R = 10$, in which the Kepler period of the 1s-electron almost exactly matches the round trip interval of a virtual photon that connects the two electrons. Although this classical analogue is only an intriguing curiosity, such models often have quantum mechanical counterparts. It has been suggested [44] that the charge factors in Eq. 12.63 might provide advantages in the detection of this effect if heliumlike ions were used instead of neutral helium. For example, if $Li^+(Z = 3)$ were studied, the two periods would match at $n_R = \frac{2}{9}\sqrt{\pi/\alpha} = 4.61$, with harmonic overtones at 6.52 and 7.99. This would provide accessible levels that are both on-resonance and off-resonance classically that could be tested and compared.

12.6 Quadratic Stark effect

Certain aspects of this formalism can be applied directly to a similar type of system, that of an atom in a uniform external electric field E. Here the Hamiltonian can be written

$$H = \frac{p^2}{2m} - \frac{KZe^2}{r} + e\mathsf{E}z. \tag{12.64}$$

The unperturbed portion is

$$H_0 = \frac{p^2}{2m} - \frac{KZe^2}{r}, \tag{12.65}$$

and the perturbation is

$$\lambda \Delta V = e\mathsf{E}z = e\mathsf{E}r\, P_1(\cos \vartheta). \tag{12.66}$$

The linear Stark effect is described by the first-order correction

$$E_n^{(1)} = e^2\mathsf{E}^2 |\langle n|z|n\rangle|^2, \tag{12.67}$$

which is odd under reflection, and vanishes for nondegenerate states. The quadratic Stark effect can be computed as the second-order correction $E_n^{(2)}$. This can be obtained from the Dalgarno–Lewis function $F(\mathbf{r})$, through the solution of the commutator relation of Eq. 12.29. If the ansatz of Eq. 12.38 is made, the determining equation is given by

$$\sum_m \left[\frac{\mathrm{d}^2 f_m}{\mathrm{d}r'^2} - 2\left(1 - \frac{1}{r'}\right)\frac{\mathrm{d}f_m}{\mathrm{d}r'} - \frac{m(m+1)}{(r')^2} f_m \right] P_m(\cos \vartheta) = \frac{e\mathsf{E}a_0}{Z} r' P_1(\cos \vartheta). \tag{12.68}$$

The solution to this equation is similar to that obtained in Eq. 12.43 except for the fact that the angular integration yields a kronecker delta δ_{m1} that collapses the m-summation. The solution is therefore

$$F(\mathbf{r}') = -\frac{e\mathsf{E}a_0}{Z}\left[\frac{(r')^2}{2} + r'\right] P_1(\cos \vartheta). \tag{12.69}$$

The energy correction is

$$E_{1s}^{(2)} = -\frac{1}{2}\alpha_2\mathsf{E}^2, \tag{12.70}$$

where α_2 is the dipole polarizability, in units a_0^3/KZ^2. For a hydrogenic atom in the 1s-state, $\alpha_2 = \frac{9}{2}$.

12.7 Indices of refraction for inert gases

Many of the sums that have been encountered are special cases of a generalized multipole sum quantity defined as

$$S(-k) \equiv \sum_i f_{0i}\left(\frac{Ry}{E_i - E_0}\right)^k = \sum_i f_{0i}(Ry\lambda_{0i})^k \tag{12.71}$$

(using the standard spectroscopic practice of expressing energies as reciprocal lengths). In terms of this definition, we can write the Thomas–Reiche–Kuhn sum rule as $N_e = S(0)$, the dipole polarizability as $\alpha_d = 4S(-2)$, and the nonadiabatic correlation as $\beta = 4S(-3)$. Another application of these quantities is in the specification [77] of the indices of refraction of gases.

To make this formulation, we can reconsider the equations developed in Section 9.1, applying them here to a different special case. If, as in Eq. 9.2, we consider an electromagnetic wave of electric field $\mathsf{E}(t) = \mathsf{E}_0 e^{i\omega t}$ incident upon a classical simple harmonic oscillator of natural frequency $\omega_0 = \sqrt{k/m}$, and neglect both the viscous drag ($\gamma = 0$) and the radiation damping ($1/\tau_0 = 0$), Eq. 9.1 ($F = ma$) becomes

$$-e\mathsf{E}(t) - m\omega_0^2 x = m\frac{d^2 x}{dt^2}. \tag{12.72}$$

Making the periodic ansatz $x(t) = x_0 e^{i\omega t}$, Eq. 12.72 becomes

$$-e\mathsf{E}(t) = m(\omega_0^2 - \omega^2)x(t). \tag{12.73}$$

Solving this for $x(t)$ gives

$$x(t) = \frac{-e\mathsf{E}(t)}{m(\omega_0^2 - \omega^2)}. \tag{12.74}$$

To extend this from the classical to the quantum mechanical case, consider an ensemble of N_i oscillators/volume with a set of natural frequencies $\omega_0 \rightarrow \omega_{ij}$ and the corresponding oscillator strength f_{ij}. As in Eq. 9.6, the electric polarization vector P is

$$\mathsf{P} = N_i \sum_j f_{ij}[-ex(t)] = \epsilon_0 \chi_e \mathsf{E}(t). \tag{12.75}$$

In the example of Section 9.1, the assumption was made that $\omega \approx \omega_{ij}$, so that the frequency of the light was commensurate with the natural frequencies of the atom. In the present case the alternative assumption will be made that $\omega \ll \omega_{ij}$. With this assumption the energy of the photons will be insufficient to elevate the ground-state electron to the first excited state, so all excitations will be virtual (perturbing).

In inert-gas atoms, the magnetic coupling of the closed subshells in the ns^2np^6 ground state depresses its energy, and excitation to the first excited state requires an energetic vacuum ultraviolet photon. Thus, the present approximation is particularly applicable to the inert gases.

Under these assumptions, the electric susceptibility is given by

$$\chi_e = \frac{N_i e^2}{\epsilon_0 m} \sum_j \frac{f_{ij}}{\omega_{ij}^2 - \omega^2}. \tag{12.76}$$

If we convert to energy units $E_{ij} = \hbar\omega_{ij}$ and note that $e^2/4\pi\epsilon_0 = 2Ry\,a_0$ and $\hbar^2/m = 2Ry\,a_0^2$, this can be rewritten

$$\chi_e = 16\pi N_i a_0^3 \sum_j \frac{f_{ij}(Ry)^2}{E_{ij}^2 - E^2}. \tag{12.77}$$

The dimensionless constant in front of this sum can be evaluated as a function of temperature T and pressure P (relative to the standard conditions T_0 and P_0), using the ideal gas law, as

$$16\pi N_i a_0^3 = 16\pi \left(\frac{6.023 \times 10^{23}}{22\,400\ \text{cm}^3}\right)(0.529 \times 10^{-8}\text{cm})^3 \frac{P}{P_0}\frac{T_0}{T}$$

$$\approx 2.00 \times 10^{-4}\ \frac{P(\text{torr})}{760}\frac{273}{T(\text{K})}. \tag{12.78}$$

Table 12.3. Computed sums $S(-k) = \sum_j f_{0j}(Ry\lambda_{0j})^k$ for hydrogen and the inert-gas atoms. (From Ref. [9].)

k	H	He	Ne	Ar	Kr	Xe
2	1.125 000	0.344 975 56	0.666 679 44	2.771 567	4.194 123	6.767 940
3	1.343 750	0.176 425 69	0.318 040 83	2.083 067	3.623 313	7.288 098
4	1.661 458	0.096 208 94	0.181 519 37	1.746 240	3.448 920	8.454 768
5	2.099 175	0.054 625 49	0.115 962 59	1.598 951	3.493 864	10.283 300
6	2.690 439	0.031 867 53	0.079 661 53	1.570 592	3.707 107	12.961 910

As indicated in Eq. 9.9, the index of refraction is given by

$$n = \sqrt{\epsilon/\epsilon_0} = \sqrt{1 + \chi_e}. \tag{12.79}$$

This can be related to the above expression for the susceptibility to obtain

$$n^2 - 1 = 16\pi N_i a_0^3 \sum_j \frac{f_{ij} Ry^2}{E_{ij}^2 - E^2}. \tag{12.80}$$

Since we have assumed that $E \ll E_{ij}$, this can be expanded as

$$\frac{1}{E_{ij}^2 - E^2} = \frac{1}{E_{ij}^2} \left(1 - \frac{E^2}{E_{ij}^2}\right)^{-1} \approx \frac{1}{E_{ij}^2} \left(1 + \frac{E^2}{E_{ij}^2} + \frac{E^4}{E_{ij}^4} + \cdots\right), \tag{12.81}$$

which yields

$$n^2 - 1 = 16\pi N_i a_0^3 \left[\sum_j \frac{f_{ij} Ry^2}{E_{ij}^2} + \frac{E^2}{Ry^2} \sum_j \frac{f_{ij} Ry^4}{E_{ij}^4} + \frac{E^4}{Ry^4} \sum_j \frac{f_{ij} Ry^6}{E_{ij}^6} + \cdots\right]. \tag{12.82}$$

Using the definition of Eq. 12.71, this can be written

$$n^2 - 1 = 16\pi N_i a_0^3 \left[S(-2) + \left(\frac{E}{Ry}\right)^2 S(-4) + \left(\frac{E}{Ry}\right)^4 S(-6) + \cdots\right]. \tag{12.83}$$

A computation of these quantities $S(-k)$ has been made [9] for H, He, Ne, Ar, Kr, and Xe by Bell and Kingston and the values are given in Table 12.3.

13

Coherence and anisotropic excitation

A solitary ant, afield, cannot be considered to have much of anything on his mind; indeed, with only a few neurons strung together by fibers, he can't be imagined to have a mind at all, much less a thought. He is more like a ganglion on legs. Four ants together, or ten, encircling a dead moth on a path, begin to look more like an idea.

– Lewis Thomas [183]

When the intensity of the decay of a spectral line was written in the form

$$I_{ik}(t) = N_i(t)A_{ik}, \tag{13.1}$$

this was done with an implicit assumption that the nature of the excitation was such that the decay could be described by an instantaneous population of the eigenstates of the free atom. This will be true only if the eigenstates of the excitation Hamiltonian are the same as the eigenstates of the decay Hamiltonian, or if the turnoff of the excitation process is sufficiently slow and adiabatic so that the phase information from the excitation is randomized. If these conditions are not satisfied, then the decay eigenstates will consist of coherent admixtures of the excitation eigenstates (and vice versa), and the oscillating dipole moments in different decay transitions will have phases that are coherently related to each other. If the various transitions that share coherences are not spectrally resolved from one another then interference effects will result. In such cases the time dependence of the emitted radiation cannot be prescribed simply by the populations (which correspond to squared amplitudes), but must take into account the coherences. A convenient means for accomplishing this is the density matrix formulation.

13.1 Density matrix representation

13.1.1 Formulation

Let us assume two separate perturbations: an excitation perturbation, and a decay perturbation. We shall consider the expectation values of an arbitrary operator O in both representations. We write the excitation Hamiltonian as

$$[H_0 + V_{\text{excite}}]\Psi_n = -\frac{\hbar}{\mathrm{i}}\frac{\partial \Psi_n}{\partial t}, \tag{13.2}$$

and the expectation value of O is given by

$$\langle O \rangle = \sum_n g_n \langle \Psi_n | O | \Psi_n \rangle, \tag{13.3}$$

where g_n is the statistical weight of the level n. We write the decay Hamiltonian as

$$[H_0 + V_{\text{decay}}]\Phi_n = -\frac{\hbar}{i}\frac{\partial \Phi_n}{\partial t}. \tag{13.4}$$

Since decay occurs, we assume that the energy eigenstate is complex, and of the form $\epsilon_j - i\Gamma_j/2$, so the wave function can be written

$$\Phi_n = \phi_n \exp[-i(\epsilon_j - i\Gamma_j/2)t/\hbar]. \tag{13.5}$$

We can expand the excitation eigenstates on the decay eigenstates

$$\Psi_n = \sum_j a_{nj}(0)\exp[-i(\varepsilon_j - i\Gamma_j/2)t/\hbar]\phi_j \equiv \sum_j a_{nj}(t)\phi_j, \tag{13.6}$$

and the expectation value of O can be written

$$\langle O \rangle = \sum_n g_n \left\langle \sum_j a_{nj}(t)\phi_j | O | \sum_k a_{nk}(t)\phi_k \right\rangle. \tag{13.7}$$

Exchanging the orders of summation,

$$\langle O \rangle = \sum_j \sum_k \left[\sum_n g_n a_{nj}^*(t)a_{nk}(t) \right] \langle \phi_j | O | \phi_k \rangle. \tag{13.8}$$

The bracketed expression is the density matrix

$$\rho_{kj} \equiv \sum_n g_n a_{nj}^*(t)a_{nk}(t). \tag{13.9}$$

The matrix element of O in the decay representation can be written as

$$O_{jk} \equiv \langle \phi_j | O | \phi_k \rangle, \tag{13.10}$$

so the expectation value of the operator in the decay representation can be written

$$\langle O \rangle = \sum_j \sum_k \rho_{kj} O_{jk} = \text{Trace}\{\rho(t)0\}, \tag{13.11}$$

which is the trace (sum of the diagonal elements) of the product of the density matrix and the operator matrix. The time dependence of the density matrix can be displayed explicitly

$$\rho_{kj}(t) = \rho_{kj}(0)\exp[-i(\varepsilon_k - \varepsilon_j)t/\hbar - (\Gamma_k + \Gamma_j)t/2\hbar]. \tag{13.12}$$

Examples of the application of this formalism are given below.

13.1.2 Example: expectation values for a P state with coherently excited magnetic sublevels

We can compute the expectation of angular momentum operators in an arbitrary representation through the expression

$$\langle L_k \rangle = \text{Trace} \{ \rho \mathsf{L}_k \} \tag{13.13}$$

using the expectation values in their eigen representation and the density matrix of the desired representation

$$(L_k)_{mm'} = \langle \ell m \, | L_k | \, \ell m' \rangle. \tag{13.14}$$

The formulation of the matrix elements for the components L_k in the representation $|\ell m\rangle$ is thoroughly discussed in most quantum mechanics textbooks (e.g., Refs. [30, 155]). The total magnitude and z-component are given by the diagonal relationships

$$L^2 |\ell m\rangle = \ell(\ell + 1)\hbar^2 |\ell m\rangle \tag{13.15}$$

$$L_z |\ell m\rangle = m\hbar \, |\ell m\rangle. \tag{13.16}$$

The x- and y-components can be obtained from the raising and lowering operators $L_\pm = L_x \pm iL_y$, which have the property

$$L_\pm |\ell m\rangle = \sqrt{(\ell \mp m)(\ell \pm m + 1)} \, \hbar \, |\ell \, m \pm 1\rangle. \tag{13.17}$$

The other Cartesian components are then obtained from $L_x = (L_+ + L_-)/2$ and $L_y = (L_+ - L_-)/2i$.

Labeling the rows and columns by $m = +1, 0, -1$ and $m' = +1, 0, -1$ (abbreviated by $+, 0, -$), these matrices are given by

$$L^2 = \begin{pmatrix} 2 & 0 & 0 \\ 0 & 2 & 0 \\ 0 & 0 & 2 \end{pmatrix} \hbar^2; \qquad L_z = \begin{pmatrix} 1 & 0 & 0 \\ 0 & 0 & 0 \\ 0 & 0 & -1 \end{pmatrix} \hbar;$$

$$L_x = \begin{pmatrix} 0 & 1 & 0 \\ 1 & 0 & 1 \\ 0 & 1 & 0 \end{pmatrix} \frac{\hbar}{\sqrt{2}}; \qquad L_y = \begin{pmatrix} 0 & -i & 0 \\ i & 0 & -i \\ 0 & i & 0 \end{pmatrix} \frac{\hbar}{\sqrt{2}}.$$

Forming the desired matrix products and taking the traces

$$\langle L_x \rangle = \text{Trace} \begin{pmatrix} \rho_{++} & \rho_{+0} & \rho_{+-} \\ \rho_{0+} & \rho_{00} & \rho_{0-} \\ \rho_{-+} & \rho_{-0} & \rho_{--} \end{pmatrix} \begin{pmatrix} 0 & 1 & 0 \\ 1 & 0 & 1 \\ 0 & 1 & 0 \end{pmatrix} \frac{\hbar}{\sqrt{2}}$$

$$= (\rho_{+0} + \rho_{0+} + \rho_{0-} + \rho_{-0})\hbar/\sqrt{2}$$

$$\langle L_y \rangle = \text{Trace} \begin{pmatrix} \rho_{++} & \rho_{+0} & \rho_{+-} \\ \rho_{0+} & \rho_{00} & \rho_{0-} \\ \rho_{-+} & \rho_{-0} & \rho_{--} \end{pmatrix} \begin{pmatrix} 0 & -i & 0 \\ i & 0 & -i \\ 0 & i & 0 \end{pmatrix} \frac{\hbar}{\sqrt{2}}$$

$$= (\rho_{+0} - \rho_{0+} + \rho_{0-} - \rho_{-0}) \, i \, \hbar/\sqrt{2}$$

$$\langle L_z \rangle = \text{Trace} \begin{pmatrix} \rho_{++} & \rho_{+0} & \rho_{+-} \\ \rho_{0+} & \rho_{00} & \rho_{0-} \\ \rho_{-+} & \rho_{-0} & \rho_{--} \end{pmatrix} \begin{pmatrix} 1 & 0 & 0 \\ 0 & 0 & 0 \\ 0 & 0 & -1 \end{pmatrix} \hbar$$

$$= (\rho_{++} - \rho_{--}) \hbar$$

$$\langle L_x^2 \rangle = \text{Trace} \begin{pmatrix} \rho_{++} & \rho_{+0} & \rho_{+-} \\ \rho_{0+} & \rho_{00} & \rho_{0-} \\ \rho_{-+} & \rho_{-0} & \rho_{--} \end{pmatrix} \begin{pmatrix} 1 & 0 & 1 \\ 0 & 2 & 0 \\ 1 & 0 & 1 \end{pmatrix} \frac{\hbar^2}{2}$$

$$= (\rho_{++} + \rho_{+-} + 2\rho_{00} + \rho_{-+} + \rho_{--}) \hbar^2/2$$

$$\langle L_y^2 \rangle = \text{Trace} \begin{pmatrix} \rho_{++} & \rho_{+0} & \rho_{+-} \\ \rho_{0+} & \rho_{00} & \rho_{0-} \\ \rho_{-+} & \rho_{-0} & \rho_{--} \end{pmatrix} \begin{pmatrix} 1 & 0 & -1 \\ 0 & 2 & 0 \\ -1 & 0 & 1 \end{pmatrix} \frac{\hbar^2}{2}$$

$$= (\rho_{++} - \rho_{+-} + 2\rho_{00} - \rho_{-+} + \rho_{--}) \hbar^2/2$$

$$\langle L_z^2 \rangle = \text{Trace} \begin{pmatrix} \rho_{++} & \rho_{+0} & \rho_{+-} \\ \rho_{0+} & \rho_{00} & \rho_{0-} \\ \rho_{-+} & \rho_{-0} & \rho_{--} \end{pmatrix} \begin{pmatrix} 1 & 0 & 0 \\ 0 & 0 & 0 \\ 0 & 0 & 1 \end{pmatrix} \hbar^2$$

$$= (\rho_{++} + \rho_{--}) \hbar^2$$

$$\langle L^2 \rangle = \text{Trace} \begin{pmatrix} \rho_{++} & \rho_{+0} & \rho_{+-} \\ \rho_{0+} & \rho_{00} & \rho_{0-} \\ \rho_{-+} & \rho_{-0} & \rho_{--} \end{pmatrix} \begin{pmatrix} 1 & 0 & 0 \\ 0 & 1 & 0 \\ 0 & 0 & 1 \end{pmatrix} 2\hbar^2$$

$$= (\rho_{++} + \rho_{00} + \rho_{--}) 2\hbar^2$$

$$\langle L_x L_y \rangle = \text{Trace} \begin{pmatrix} \rho_{++} & \rho_{+0} & \rho_{+-} \\ \rho_{0+} & \rho_{00} & \rho_{0-} \\ \rho_{-+} & \rho_{-0} & \rho_{--} \end{pmatrix} \begin{pmatrix} i & 0 & -i \\ 0 & 0 & 0 \\ i & 0 & -i \end{pmatrix} \frac{\hbar^2}{2}$$

$$= (\rho_{++} + \rho_{+-} - \rho_{-+} - \rho_{--}) i\hbar^2/2$$

$$\langle L_x L_z \rangle = \text{Trace} \begin{pmatrix} \rho_{++} & \rho_{+0} & \rho_{+-} \\ \rho_{0+} & \rho_{00} & \rho_{0-} \\ \rho_{-+} & \rho_{-0} & \rho_{--} \end{pmatrix} \begin{pmatrix} 0 & 0 & 0 \\ 1 & 0 & -1 \\ 0 & 0 & 0 \end{pmatrix} \frac{\hbar^2}{\sqrt{2}}$$

$$= (\rho_{+0} - \rho_{-0}) \hbar^2/\sqrt{2}$$

$$\langle L_y L_z \rangle = \text{Trace} \begin{pmatrix} \rho_{++} & \rho_{+0} & \rho_{+-} \\ \rho_{0+} & \rho_{00} & \rho_{0-} \\ \rho_{-+} & \rho_{-0} & \rho_{--} \end{pmatrix} \begin{pmatrix} 0 & 0 & 0 \\ i & 0 & i \\ 0 & 0 & 0 \end{pmatrix} \frac{\hbar^2}{\sqrt{2}}$$

$$= (\rho_{+0} + \rho_{-0}) i\hbar^2/\sqrt{2}.$$

By invoking symmetries in the Hamiltonian (invariance under rotation, reflection, inversion, hermiticity, etc.) it is often possible [94] to substantially reduce the number of independent density matrix elements.

13.2 Stokes parameters

The density matrix formulation permits a full description of the excitation process in the decay representation, conveying the phase information in the coherent amplitudes of the matrix. This is in contrast to a description in terms of level populations, which involves only the incoherent squares of the diagonal elements of the density matrix. However, to incorporate this information into the description of the decay process, it is necessary to make a comprehensive formulation of the emitted radiation. The radiation will contain its own anisotropies, in response to the excitation anisotropies of the atomic system.

The task of fully characterizing the state of the emitted radiation is efficiently accomplished by the specification of the Stokes parameters [179]. These quantities comprise a four-vector that can be manipulated using a matrix algebra known as Mueller calculus [189]. Through these operators it is possible to trace the effect on the light produced by its passage through various optical analyzing devices.

To illustrate these methods, consider an electromagnetic wave represented by the vector components

$$
\begin{aligned}
E_x &= \text{Real}\left\{ E_x^{(0)} e^{i\,(kz - \omega t + \varphi_x)} \right\} \\
E_y &= \text{Real}\left\{ E_y^{(0)} e^{i\,(kz - \omega t + \varphi_y)} \right\}.
\end{aligned}
\tag{13.18}
$$

This has been written in complex notation to facilitate linear mathematical manipulation, and the need to extract the real part will be implicitly assumed in the notation of this section. In terms of these quantities, the Stokes parameters are defined as

$$
\begin{aligned}
I &\equiv E_x^* E_x + E_y^* E_y \\
Q &\equiv E_x^* E_x - E_y^* E_y \\
U &\equiv E_x^* E_y + E_y^* E_x \\
V &\equiv -\mathrm{i}(E_x^* E_y - E_y^* E_x)\,,
\end{aligned}
\tag{13.19}
$$

where the asterisk denotes the complex conjugation. Here I measures the intensity, Q/I measures the linear polarization, V/I measures the circular polarization, and U/Q measures the angle of the plane of polarization. The effect of the insertion of various optical analyzing devices can be specified by a matrix transformation of the form

$$
\begin{pmatrix} I' \\ Q' \\ U' \\ V' \end{pmatrix}
=
\begin{pmatrix}
T_{11} & T_{12} & T_{13} & T_{14} \\
T_{21} & T_{22} & T_{23} & T_{24} \\
T_{31} & T_{32} & T_{33} & T_{34} \\
T_{41} & T_{42} & T_{43} & T_{44}
\end{pmatrix}
\begin{pmatrix} I \\ Q \\ U \\ V \end{pmatrix}.
\tag{13.20}
$$

The usefulness of this formulation can be seen by considering the effect of some specific optical elements.

13.2.1 Linear polarizer

If the light is viewed though a linear polarizer that is oriented at an angle α to the x-axis, the transmitted radiation will be linearly polarized along α, and can be reprojected along the directions x and y to obtain

$$E'_x = (E_x \cos\alpha + E_y \sin\alpha)\cos\alpha$$
$$E'_y = (E_x \cos\alpha + E_y \sin\alpha)\sin\alpha. \tag{13.21}$$

Inserting these into Eqs. 13.19, the resultant transformed Stokes parameters are given by

$$I' = E'^*_x E'_x + E'^*_y E'_y$$
$$= [E^*_x E_x \cos^2\alpha + E^*_y E_y \sin^2\alpha + (E^*_x E_y + E^*_y E_x)\sin\alpha\cos\alpha] \times (\cos^2\alpha + \sin^2\alpha)$$
$$Q' = E'^*_x E'_x - E'^*_y E'_y$$
$$= [E^*_x E_x \cos^2\alpha + E^*_y E_y \sin^2\alpha + (E^*_x E_y + E^*_y E_x)\sin\alpha\cos\alpha] \times (\cos^2\alpha - \sin^2\alpha)$$
$$U' = E'^*_x E'_y + E'^*_y E'_x$$
$$= [E^*_x E_x \cos^2\alpha + E^*_y E_y \sin^2\alpha + (E^*_x E_y + E^*_y E_x)\sin\alpha\cos\alpha] \times 2\sin\alpha\cos\alpha$$
$$V' = -i(E'^*_x E'_y - E'^*_y E'_x) = 0. \tag{13.22}$$

Using the double-angle formulae

$$2\cos^2\alpha = 1 + \cos 2\alpha$$
$$2\sin^2\alpha = 1 - \cos 2\alpha$$
$$2\sin\alpha\cos\alpha = \sin 2\alpha \tag{13.23}$$

these expressions become

$$I' = \frac{1}{2}[(E^*_x E_x + E^*_y E_y) + (E^*_x E_x - E^*_y E_y)\cos 2\alpha + (E^*_x E_y + E^*_y E_x)\sin 2\alpha]$$

$$= \frac{1}{2}(I + Q\cos 2\alpha + U\sin 2\alpha)$$

$$Q' = \frac{1}{2}[(E^*_x E_x + E^*_y E_y) + (E^*_x E_x - E^*_y E_y)\cos 2\alpha + (E^*_x E_y + E^*_y E_x)\sin 2\alpha]\cos 2\alpha$$

$$= \frac{1}{2}(I\cos 2\alpha + Q\cos^2 2\alpha + U\sin 2\alpha\cos 2\alpha)$$

$$U' = \frac{1}{2}[(E^*_x E_x + E^*_y E_y) + (E^*_x E_x - E^*_y E_y)\cos 2\alpha + (E^*_x E_y + E^*_y E_x)\sin 2\alpha]\sin 2\alpha$$

$$= \frac{1}{2}(I\sin 2\alpha + Q\sin 2\alpha\cos 2\alpha + U\sin^2 2\alpha)$$

$$V' = 0. \tag{13.24}$$

Expressed as the transformation of Eq. 13.20, the corresponding matrix is

$$T_P(\alpha) = \frac{1}{2} \begin{pmatrix} 1 & \cos 2\alpha & \sin 2\alpha & 0 \\ \cos 2\alpha & \cos^2 2\alpha & \sin 2\alpha \cos 2\alpha & 0 \\ \sin 2\alpha & \sin 2\alpha \cos 2\alpha & \sin^2 2\alpha & 0 \\ 0 & 0 & 0 & 0 \end{pmatrix} \tag{13.25}$$

which has the special cases

$$T_P(0°) = \frac{1}{2} \begin{pmatrix} 1 & 1 & 0 & 0 \\ 1 & 1 & 0 & 0 \\ 0 & 0 & 0 & 0 \\ 0 & 0 & 0 & 0 \end{pmatrix} \tag{13.26}$$

$$T_P(90°) = \frac{1}{2} \begin{pmatrix} 1 & -1 & 0 & 0 \\ -1 & 1 & 0 & 0 \\ 0 & 0 & 0 & 0 \\ 0 & 0 & 0 & 0 \end{pmatrix} \tag{13.27}$$

$$T_P(\pm 45°) = \frac{1}{2} \begin{pmatrix} 1 & 0 & \pm 1 & 0 \\ 0 & 0 & 0 & 0 \\ \pm 1 & 0 & 1 & 0 \\ 0 & 0 & 0 & 0 \end{pmatrix}. \tag{13.28}$$

13.2.2 Retarding plate

If the light is viewed through a birefringent crystal that retards the phase of the x-component relative to that of the y-component by an angle ϕ, the transmitted radiation can be written

$$\begin{aligned} E'_x &= E_x \\ E'_y &= E_y e^{i\phi}. \end{aligned} \tag{13.29}$$

Using the Euler relationship

$$e^{\pm i\phi} = \cos \phi \pm i \sin \phi \tag{13.30}$$

the Stokes parameters become

$$\begin{aligned} I' &= E'^*_x E'_x + E'^*_y E'_y = E^*_x E_x + E^*_y E_y = I \\ Q' &= E'^*_x E'_x - E'^*_y E'_y - E^*_x E_x \quad E^*_y E_y = Q \\ U' &= E'^*_x E'_y + E'^*_y E'_x = E^*_x E_y e^{i\phi} + E^*_y E_x e^{-i\phi} \\ &= (E^*_x E_y + E^*_y E_x) \cos \phi + i (E^*_x E_y - E^*_y E_x) \sin \phi \\ &= U \cos \phi - V \sin \phi \\ V' &= -i(E'^*_x E'_y - E'^*_y E'_x) = -i(E^*_x E_y e^{i\phi} + E^*_y E_x e^{-i\phi}) \\ &= -i(E^*_x E_y - E^*_y E_x) \cos \phi + (E^*_x E_y + E^*_y E_x) \sin \phi \\ &= V \cos \phi + U \sin \phi, \end{aligned} \tag{13.31}$$

and the transformation can be written

$$T_R(\phi) = \begin{pmatrix} 1 & 0 & 0 & 0 \\ 0 & 1 & 0 & 0 \\ 0 & 0 & \cos\phi & -\sin\phi \\ 0 & 0 & \sin\phi & \cos\phi \end{pmatrix}, \tag{13.32}$$

which has the special cases of a quarter-wave plate

$$T_R(\pm 90^\circ) = \begin{pmatrix} 1 & 0 & 0 & 0 \\ 0 & 1 & 0 & 0 \\ 0 & 0 & 0 & \mp 1 \\ 0 & 0 & \pm 1 & 0 \end{pmatrix}. \tag{13.33}$$

13.2.3 Combined optical elements

Linear polarizers, retarding plates, and other optical elements can be combined by matrix addition and multiplication. A few examples will be presented below.

Linear polarization analyzer

The matrix describing the difference between measurements made with a linear polarizer oriented first along the x-axis and then along the y-axis is

$$T_P(0^\circ) - T_P(90^\circ) = \frac{1}{2}\begin{pmatrix} 1 & 1 & 0 & 0 \\ 1 & 1 & 0 & 0 \\ 0 & 0 & 0 & 0 \\ 0 & 0 & 0 & 0 \end{pmatrix} - \frac{1}{2}\begin{pmatrix} 1 & -1 & 0 & 0 \\ -1 & 1 & 0 & 0 \\ 0 & 0 & 0 & 0 \\ 0 & 0 & 0 & 0 \end{pmatrix} = \begin{pmatrix} 0 & 1 & 0 & 0 \\ 1 & 0 & 0 & 0 \\ 0 & 0 & 0 & 0 \\ 0 & 0 & 0 & 0 \end{pmatrix}. \tag{13.34}$$

If this operation is performed on an arbitrarily polarized light source

$$\begin{pmatrix} 0 & 1 & 0 & 0 \\ 1 & 0 & 0 & 0 \\ 0 & 0 & 0 & 0 \\ 0 & 0 & 0 & 0 \end{pmatrix}\begin{pmatrix} I \\ Q \\ U \\ V \end{pmatrix} = \begin{pmatrix} Q \\ I \\ 0 \\ 0 \end{pmatrix} \tag{13.35}$$

it yields a total intensity that is proportional to the linear polarization of the source.

Circular polarizer

A circular polarizer can be constructed by first inserting a 45° linear polarizer to produce E_x- and E_y-components that are in phase, and then inserting a quarter-wave plate to advance the phase of, e.g., the y-component by 90° relative to the x-component. This yields a matrix

$$T_R(90°)T_P(45°) = \frac{1}{2}\begin{pmatrix} 1 & 0 & 0 & 0 \\ 0 & 1 & 0 & 0 \\ 0 & 0 & 0 & -1 \\ 0 & 0 & 1 & 0 \end{pmatrix}\begin{pmatrix} 1 & 0 & 1 & 0 \\ 0 & 0 & 0 & 0 \\ 1 & 0 & 1 & 0 \\ 0 & 0 & 0 & 0 \end{pmatrix} = \frac{1}{2}\begin{pmatrix} 1 & 0 & 1 & 0 \\ 0 & 0 & 0 & 0 \\ 0 & 0 & 0 & 0 \\ 1 & 0 & 1 & 0 \end{pmatrix}. \quad (13.36)$$

When this acts on unpolarized light it yields

$$\frac{1}{2}\begin{pmatrix} 1 & 0 & 1 & 0 \\ 0 & 0 & 0 & 0 \\ 0 & 0 & 0 & 0 \\ 1 & 0 & 1 & 0 \end{pmatrix}\begin{pmatrix} I \\ 0 \\ 0 \\ 0 \end{pmatrix} = \frac{1}{2}\begin{pmatrix} I \\ 0 \\ 0 \\ I \end{pmatrix}. \quad (13.37)$$

The presence of a fourth component indicates circularly polarized light.

Circular polarization analyzer

If the ordering of the polarizer and the quarter-wave plate is reversed, the device becomes an analyzer. By first inserting the quarter-wave plate and then using the polarizer to study the result, the combined elements can be used to determine the state of polarization of the source. This yields the matrix

$$T_P(45°)T_R(90°) = \frac{1}{2}\begin{pmatrix} 1 & 0 & 1 & 0 \\ 0 & 0 & 0 & 0 \\ 1 & 0 & 1 & 0 \\ 0 & 0 & 0 & 0 \end{pmatrix}\begin{pmatrix} 1 & 0 & 0 & 0 \\ 0 & 1 & 0 & 0 \\ 0 & 0 & 0 & -1 \\ 0 & 0 & 1 & 0 \end{pmatrix} = \frac{1}{2}\begin{pmatrix} 1 & 0 & 0 & -1 \\ 0 & 0 & 0 & 0 \\ 1 & 0 & 0 & -1 \\ 0 & 0 & 0 & 0 \end{pmatrix}. \quad (13.38)$$

Using this device, the circular polarization of the source can be determined. This is done by combining the results of three measurements: one with the retarder removed and the polarizer set at 0°; a second with the retarder removed and the polarizer set at 90°; and a third with the retarder in place and the polarizer set at 45°. If the results of the first two measurements are added and their sum subtracted from twice the result of the third measurement, the matrix algebra yields

$$T_P(0°) + T_P(90°) - 2T_P(45°)T_R(90°) =$$

$$\frac{1}{2}\begin{pmatrix} 1 & 1 & 0 & 0 \\ 1 & 1 & 0 & 0 \\ 0 & 0 & 0 & 0 \\ 0 & 0 & 0 & 0 \end{pmatrix} + \frac{1}{2}\begin{pmatrix} 1 & -1 & 0 & 0 \\ -1 & 1 & 0 & 0 \\ 0 & 0 & 0 & 0 \\ 0 & 0 & 0 & 0 \end{pmatrix} - \begin{pmatrix} 1 & 0 & 0 & -1 \\ 0 & 0 & 0 & 0 \\ 1 & 0 & 0 & -1 \\ 0 & 0 & 0 & 0 \end{pmatrix} = \begin{pmatrix} 0 & 0 & 0 & 1 \\ 0 & 1 & 0 & 0 \\ -1 & 0 & 0 & 1 \\ 0 & 0 & 0 & 0 \end{pmatrix}. \quad (13.39)$$

Operation of this matrix on a light source of arbitrary polarization yields

$$\begin{pmatrix} 0 & 0 & 0 & 1 \\ 0 & 1 & 0 & 0 \\ -1 & 0 & 0 & 1 \\ 0 & 0 & 0 & 0 \end{pmatrix}\begin{pmatrix} I \\ Q \\ U \\ V \end{pmatrix} = \frac{1}{2}\begin{pmatrix} V \\ Q \\ -I+V \\ 0 \end{pmatrix} \quad (13.40)$$

and the total intensity transmitted is proportional to the degree of circular polarization.

Table 13.1. Stokes parameters for the
He I 1s2s $^1S_0$–1s2p $^1P_1^o$ transition with a
130-keV beam incident on an exciter foil tilted at $\alpha = 30°$.
(Quoted uncertainties are given in parentheses.)

$\vartheta(°)$	Q/I	U/I	V/I
90	0.123(29)	−0.042(25)	0.114(68)
53	0.087(15)	−0.069(31)	0.093(29)

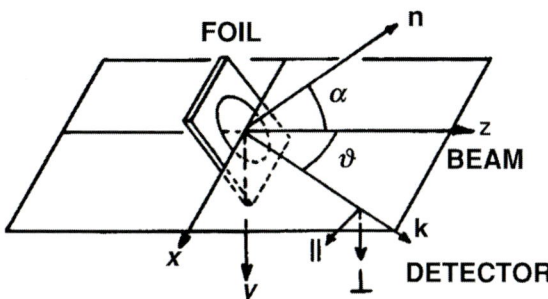

Fig. 13.1. Geometry for the beam–tilted-foil measurement of elliptic polarization. (After Ref. [12]).

Many other types of optical devices can be represented and predictively manipulated by matrix multiplication using this powerful technique.

13.3 Application to a measurement of elliptic polarization

The use of density matrix and Stokes parameter methods can now be applied to the interpretation of specific experimental data. An instructive and conceptually clear example can be drawn from the field of beam–foil spectroscopy. In a measurement [12] made by passing a fast (130-keV) beam of helium ions through a thin foil that was tilted relative to the beam, it was observed that elliptically polarized light is emitted in the He I transition 1s2s $^1S_0$−1s3p $^1P_1^o$ at 5016 Å. As labeled in Fig. 13.1, the beam travels along the z-axis, the foil is tilted about the x-axis at an angle $\alpha = 30°$ to the beam, and the light is viewed in the x–z plane at angles $\vartheta = 90°$ and $53°$ to the beam.

The Stokes parameters were measured using a quarter-wave plate and a linear polarizer, and the results are presented in Table 13.1.

To obtain the density matrix from these data, we must first recognize the symmetries. Reflection symmetry about the y–z plane leads to the relationships

$$\rho_{++} = \rho_{--}; \quad \rho_{+-} = \rho_{-+}; \quad \rho_{0+} = \rho_{0-}; \quad \rho_{+0} = \rho_{-0}. \tag{13.41}$$

Hermiticity requires that

$$\rho_{0+} = \rho_{+0}^*. \tag{13.42}$$

The normalization of the total population yields

$$\rho_{++} + \rho_{00} + \rho_{--} = 1. \tag{13.43}$$

With all of these factors taken into account, only four independent quantities remain: ρ_{++}, ρ_{+-}, Real$\{\rho_{0+}\}$, and Imag$\{\rho_{0+}\}$. The density matrix thus becomes

$$\rho = \begin{pmatrix} \rho_{++} & \rho_{0-}^* & \rho_{+-} \\ \rho_{0-} & 1 - 2\rho_{++} & \rho_{0-} \\ \rho_{+-} & \rho_{0-}^* & \rho_{++} \end{pmatrix}. \tag{13.44}$$

In terms of these four independent amplitudes, the Stokes parameters can be written (to within an overall constant I_0 that involves detection efficiencies, solid angle viewed, etc.)

$$I = I_0[\sin^2 \vartheta - \rho_{++}(1 - 3\cos^2 \vartheta) + \rho_{+-}(1 - \cos^2 \vartheta)]$$
$$Q = I_0[\sin^2 \vartheta - \rho_{++}(3 - 3\cos^2 \vartheta) + \rho_{+-}(1 + \cos^2 \vartheta)]$$
$$U = I_0[2\sqrt{2}\,\mathrm{Imag}\{\rho_{0+}\}\sin \vartheta)]$$
$$V = I_0[2\sqrt{2}\,\mathrm{Real}\{\rho_{0+}\}\sin \vartheta)]. \tag{13.45}$$

If viewed at 90°,

$$\frac{Q}{I} = \frac{1 - 3\rho_{++} - \rho_{+-}}{1 - \rho_{++} + \rho_{+-}} = 0.123$$

$$\frac{U}{I} = \frac{2\sqrt{2}\,\mathrm{Imag}\{\rho_{0+}\}}{1 - \rho_{++} + \rho_{+-}} = -0.042$$

$$\frac{V}{I} = \frac{2\sqrt{2}\,\mathrm{Real}\{\rho_{0+}\}}{1 - \rho_{++} + \rho_{+-}} = 0.114, \tag{13.46}$$

whereas, when viewed at the "magic angle" $\cos(54.73°)=1/\sqrt{3}$,

$$\frac{Q}{I} = \frac{1 - 3\rho_{++} - 2\rho_{+-}}{1 + \rho_{+-}} = 0.087$$

$$\frac{U}{I} = \frac{2\sqrt{3}\,\mathrm{Imag}\{\rho_{0+}\}}{1 + \rho_{+-}} = -0.069$$

$$\frac{V}{I} = -\frac{2\sqrt{3}\,\mathrm{Real}\{\rho_{0+}\}}{1 + \rho_{+-}} = 0.093. \tag{13.47}$$

Solving these simultaneously (correcting the equations for the fact that the data in Table 13.1 were taken at 53° rather than 54.73°) yields

$$\rho_{++} = 0.309$$
$$\rho_{+-} = -0.008$$
$$\rho_{0-} = -0.027 - 0.015i = -0.031\exp(-0.507i). \tag{13.48}$$

Recalling the various symmetry equivalences from Eqs. 13.41–13.43, the density matrix is given by

$$
\rho = \begin{pmatrix} 0.309 & -0.031e^{-0.507i} & -0.008 \\ 0.031e^{+0.507i} & 0.382 & -0.031e^{+0.507i} \\ -0.008 & -0.031e^{-0.507i} & 0.309 \end{pmatrix}. \tag{13.49}
$$

Inserting these values into the relationships obtained for the orbital angular momentum operators in Section 13.1.2,

$$
\begin{aligned}
\langle L_y \rangle &= \langle L_z \rangle = \langle L_x L_y \rangle = \langle L_x L_z \rangle = 0 \\
\langle L_x \rangle &= 2\sqrt{2}\,\text{Real}\{\rho_{0-}\}\hbar \\
&= 2\sqrt{2}(-0.027)\hbar = -0.076\hbar \\
\langle L_x^2 \rangle &= (1 - \rho_{++} + \rho_{+-})\hbar^2 \\
&= (1 - 0.309 - 0.008)\hbar^2 = 0.683\hbar^2 \\
\langle L_y^2 \rangle &= (1 - \rho_{++} + \rho_{+-})\hbar^2 \\
&= (1 - 0.309 + 0.008)\hbar^2 = 0.699\hbar^2 \\
\langle L_z^2 \rangle &= 2(\rho_{++})\hbar^2 \\
&= 2(0.309)\hbar^2 = 0.618\hbar^2 \\
\langle L_y L_z \rangle &= \sqrt{2}(\rho_{0-}^*)i\,\hbar \\
&= (0.044i)e^{0.507i}\hbar^2. \tag{13.50}
\end{aligned}
$$

Some interesting conclusions concerning the expectation values of the angular momentum can be drawn from these results. Notice that

$$
\frac{\langle L_x \rangle}{\sqrt{\langle L_x^2 \rangle}} \equiv \frac{\text{CW} - \text{CCW}}{\text{CW} + \text{CCW}} = -0.092 \tag{13.51}
$$

so we can say, classically, that 20% more atoms are orbiting counterclockwise (CCW) than clockwise (CW) (about the x-axis in Fig. 13.1) when exiting the foil. Since

$$
\langle L_x^2 \rangle - \langle L_y^2 \rangle = -0.016\hbar^2 \tag{13.52}
$$

there is very little vector redistribution of the angular momentum, and the anisotropy is primarily a pseudovector redistribution (i.e., CW vs CCW). Since

$$
2\langle L_x^2 \rangle - \langle L^2 \rangle = -0.643\hbar^2 \tag{13.53}
$$

it is not a pure state along the x-axis. If that were the case this quantity would vanish, since $2\ell^2 - \ell(\ell + 1) = 0$ for an $\ell = 1$ state. Finally, since

$$
\frac{\langle L_y L_z \rangle}{\sqrt{\langle L_y^2 \rangle \langle L_z^2 \rangle}} = 0.067e^{-0.507i} \tag{13.54}
$$

there is a significant amount of correlation between the y- and z-motions (again, classically interpretable as a preponderance of counterclockwise orbits about the x-axis in Fig. 13.1).

13.4 Alignment and orientation

The various dynamical symmetries of the density matrix can be further elucidated through a decomposition into irreducible tensor operators, in terms of which the elements become state multipoles. In this representation a spherical tensor $\rho_q^{(k)}$ can be written as

$$\rho_q^{(k)} = \sum_{mm'} (-1)^{l'+m'} \sqrt{2k+1} \begin{pmatrix} l' & l & k \\ m' & -m & q \end{pmatrix} \rho_{mm'}, \tag{13.55}$$

where the quantity enclosed in parenthesis is a Wigner 3-j symbol, which is an alternative form of a Clebsch–Gordan coefficient. A similar spherical tensor transformation for $k = 1$ was utilized in the previous section when L_x, L_y, L_z were written in terms of L_+, L_-, L_z. These two formulations each provide separate insights into the relationships among the amplitudes. The representation $\rho_{mm'}$ shows the coherences among the specific quantum states, whereas the representation $\rho_q^{(k)}$ categorizes them according to multipolarities. In the example of the previous section, the independent density matrices have the following equivalences

$$\rho_0^{(2)} = \sqrt{30}\,(\rho_{++} - 1/3)$$
$$\rho_2^{(2)} = \sqrt{5}\,\rho_{+-}$$
$$i\rho_1^{(2)} = \sqrt{10}\,\mathrm{Imag}\{\rho_{0+}\}$$
$$\rho_1^{(1)} = -\sqrt{6}\,\mathrm{Real}\{\rho_{0+}\}. \tag{13.56}$$

Note that the first three involve spherical tensors of rank 2 whereas the fourth involves a spherical tensor of rank 1. Anisotropic excitation of rank 1 is referred to as "orientation" (in nuclear physics parlance, "polarization"), and the amplitudes depend not only on the magnitude, but also the sign of the quantum number m. Anisotropic excitation of rank 2 is referred to as "alignment," and the amplitudes depend on m^2, but do not depend on the sign of m. If the excitation involves alignment but not orientation, radiation will contain linear polarization, but not circular or elliptic polarization. The spherical tensor formulation can thus provide a useful method for identifying symmetries in systems of greater complexity than the singlet-P example considered here.

13.4.1 Cascade transfer of alignment and orientation in decay-curve measurements

The instantaneous excitation of impulsively excited atomic levels contains two separable contributions. One is the remnant of the initial excitation, and a second results from the continuous repopulation by cascades from higher states. In general these two contributions are incoherent, and can be considered individually although they are superimposed in the measurement. The remnant of the initial excitation decays with a single exponential meanlife that is common to all magnetic substates of the level, whereas the cascade contribution involves a multiexponential sum for each cascade, which includes its own meanlife and the meanlives of all indirect cascades that feed through it.

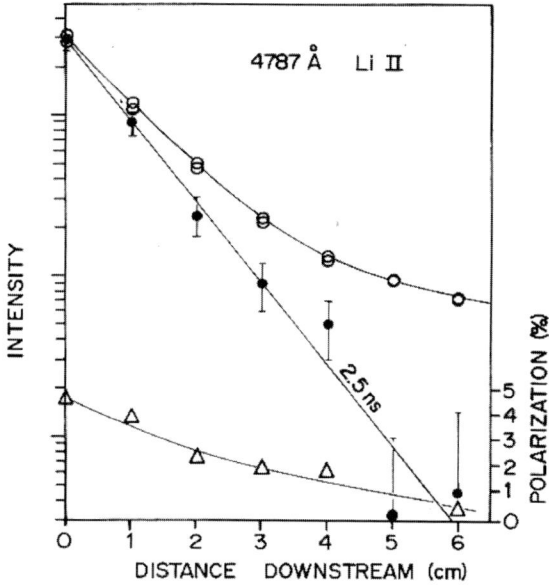

Fig. 13.2. Lifetime measurement using the decay of linear polarization. (From Ref. [11].)

If the primary level is anisotropically excited, the alignment and orientation in the remnant of the initial population should remain constant in time, since all magnetic substates decay with the same lifetime. If there is alignment and orientation in the population of the cascade levels, the degree to which that anisotropy is transferred to the primary levels is not complete, and depends on the quantum numbers of the decay process. For example, the transfer of alignment can be unity for an yrast chain (f–d–p etc.), but in general is less than unity. Moreover, it has been observed that the states of high n and high ℓ often exhibit less alignment and less orientation than do low-lying states. If there is little transfer of alignment and orientation, then it is sometimes possible to discriminate against the cascade repopulation component by studying the time rate of decay not of the total intensity, but rather of the linear or circular polarization of the sample.

13.4.2 Cascade-free lifetime measurements utilizing anisotropies

A beam–foil measurement [11] of the lifetime of the 2s4d $^1D_2$ level in Li II that utilized the alignment of the initial excitation is shown in Fig. 13.2. The light from the 2s3p $^1P_1$–2s4d $^1D_2$ transition was viewed at a right angle to the beam in light polarized parallel and perpendicular to the plane of the beam and photon. The instrumental polarization was minimized through the use of a Hanle depolarizer [115] (combined wedges of quartz and fused silica that scramble the polarization prior to input to the optical detection system) and corrected for. The open circles denote the two decay curves measured in polarized light, the closed circles represent their difference. The triangles show the change of polarization of the total radiation as cascade repopulation "washes out" the polarization due to the initial population of the primary level. Although the polarization was small, reasonable accuracies were achieved by increasing the data collection time with increasing time-since-excitation, so as to maintain approximately constant statistical accuracies over the decay curve.

Similar results were obtained using the decay of orientation in decay-curve measurements of the transitions at 3433 Å in O VI [123]. The radiation arises from a blend of the 6h–7i, 6g–7h and 6f–7g transitions, and showed over 20% circular polarization in beam–foil excitation of a 4-MeV oxygen beam incident on a foil tilted at 40°. Studies of the normalized difference between decay curves measured in right and left circularly polarized light indicated substantial reduction of cascade effects.

13.5 Quantum-beat spectroscopy

One of the most striking expositions of the hidden periodicities of quantum mechanics is provided by the phenomenon of zero-field quantum beats. This can exhibit itself when the particles in an accelerator beam are impulsively excited upon their sudden emergence from a thin foil, and the radiation they subsequently emit displays an oscillating intensity.

A conceptual description of this process is possible. The excitation eigenstates inside the solid foil are expected to differ from those of the free atom. The high-speed passage of an atom through a region of dense charge inside the foil produces motional magnetic fields of sufficient strength to decouple the $\mathbf{J}$ vectors of the electrons and cause the $\mathbf{L}$ and $\mathbf{S}$ vectors to precess independently. The emergence of the atom from the foil is very sudden, so the recoupling of the angular momenta occurs within a very short time interval. Moreover, the direction of the beam relative to the surface of the foil provides a preferred axis, so it is expected that there will be spatial anisotropies in the various excitation amplitudes. These will result in anisotropies and coherences in the density matrix elements of the decay representation.

As an example, consider the excitation and decay of the 2p levels in hydrogen. Within the foil it is likely that these decouple into states of good $L = 1$ and $S = \frac{1}{2}$ which precess independently because of their differing g-factors. Upon emergence from the foil, they recouple to form states of good $J = \frac{1}{2}$ and $J = \frac{3}{2}$, with mixed amplitudes (in current parlance, they are "entangled") due to their shared phases of the L- and S-states from which they were both formed. In a quantum mechanical explanation, the excitation eigenstates can reach the ground state of the free hydrogen atom by either of two paths (like a two-slit experiment), provided by the 2p $^2P_{1/2}$ and 2p $^2P_{3/2}$ decay eigenstates. This produces an interference frequency corresponding to the fine structure energy divided by $\hbar$.

A classical vector model description is also possible, and involves picturing the decoupled $\mathbf{L}$ and $\mathbf{S}$ of an ensemble of atoms simultaneously recoupling to form two coherent populations, one with $J = \frac{1}{2}$ and one with $J = \frac{3}{2}$. For each population, $\mathbf{L}$ and $\mathbf{S}$ precess about $\mathbf{J}$, emitting a dipole pattern of radiation that sweeps through space at its own characteristic frequency. Owing to their simultaneous formation, the radiation from the two coherent populations interferes at the difference frequency, again the fine structure energy divided by $\hbar$. Although there is no time variation in the total radiation emitted into 4π steradians, the spatial rotation of the interference pattern produces quantum beats when viewed within a limited solid angle.

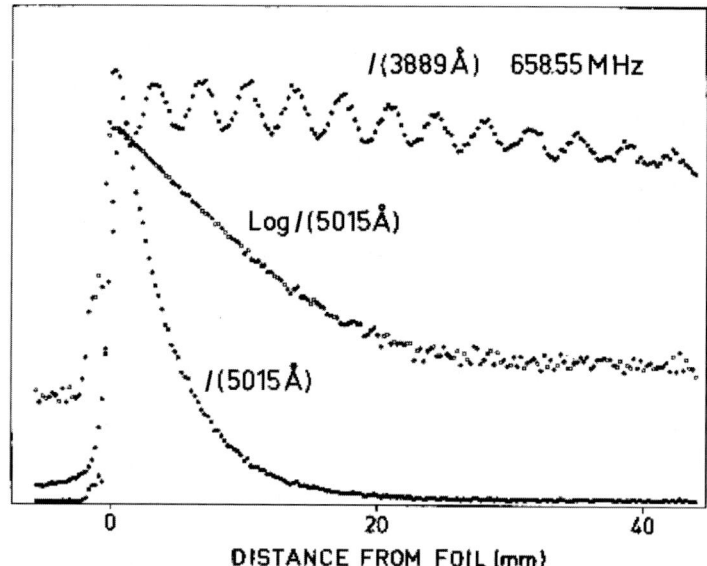

Fig. 13.3. Simultaneous measurements of decay curves of the 1s3p $^1P_1^o$ and ^3P_J levels. The quantum-beat pattern of the unresolved triplet decay provides a time calibration for the measurement of the lifetime of the singlet. (From Ref. [4].)

An example [4] of a quantum-beat pattern is shown in Fig. 13.3. Here, zero-field beats are superimposed on the unresolved decay curves of the 1s2s $^3S_1$–1s3p $^3P_{0,1,2}$ multiplet at 3889 Å. The beats occur because the 1s3p $^3P_{0,1,2}^o$ levels are coherently excited due to the sudden excitation. (The fine structure spacing of these levels is small compared to the energy spread introduced by the uncertainty principle because of the rapid emergence of the atom from the foil.) The beat frequency observed corresponds to the $^3P_1^o$–$^3P_2^o$ splitting, which is 685.55 ± 0.15 MHz.

Also displayed in Fig. 13.3 are the linear and logarithmic plots of the decay curve of the 1s2s $^1S_0$–1s3p $^1P_1^o$ transition at 5015 Å. In this experiment [4] the 5015 and 3889 Å decay curves were measured simultaneously, using two separate spectrometers that viewed the interaction chamber at right angles to each other and to the beam. In this manner, the lifetime of the singlet level could be placed on a time scale obtained from the quantum-beat frequency of the triplet levels.

An extension of this two-spectrometer method [5] is shown in Fig. 13.4. Again the 1s3p ($^3P_1$–$^3P_2$) beat frequency of the 3889 Å line is used as a time calibration, but here it is in conjunction with a measurement of the beat frequency of the unresolved and coherently excited 1s2p ^3P–1s4d ^3D multiplet at 4471 Å. The 1s3p ($^3P_1$–$^3P_2$) beat frequency of 658.55 ± 0.15 MHz provides the time calibration for measurement of the 1s4d ($^3D_1$–$^3D_2$) beat frequency, which yields 553.0 ± 0.7 MHz. The common time base of the simultaneously measured zero-field quantum beats thus permits high-precision specification of these relative frequencies.

Quantum beats can also be produced by the imposition of external fields, which lead to eigenstates that differ between the excitation and decay processes. An example was already

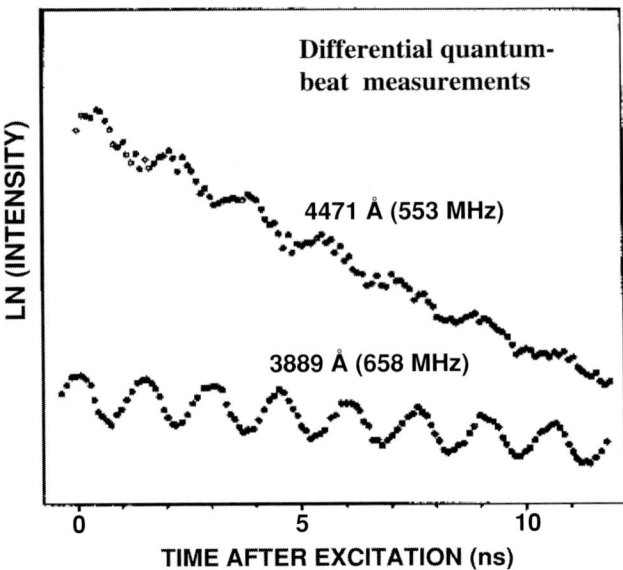

Fig. 13.4. Simultaneously measured zero-field quantum-beat patterns 1s2p ^{3}P–1s4d (^{3}D$_1$–^{3}D$_2$) and the 1s2s ^{3}S–1s3p (^{3}P$_1$–^{3}P$_2$) decay at 3889 Å in He I. The common time base permits high-precision specification of the relative frequencies. (From Ref. [5].)

given in Section 10.4 with magnetic fields, where the time-integrated case of the Hanle effect was considered. If the time evolution of this process is followed, the imposition of an external magnetic field mixes the magnetic substates of the angular momentum vectors perpendicular to the field, and produces coherences. In this case there is a clear classical analogue in the Larmor precession. If there is anisotropic excitation within the plane of precession, then there will be a rotation of the radiation pattern, and hence oscillations in the radiation emitted into a restricted solid angle. Again, as in the case of zero-field quantum beats, there are no oscillations in the total emitted intensity into all 4π steradians, but rather a rotation of the radiation pattern that can be used to specify, for example, the magnetic g-factors of the states.

In the cases of the zero-field quantum beats and the weak magnetic field beats, the states that were mixed by the excitation possess the same lifetime, so there are no temporal beats in the total emitted radiation. An alternative situation exists in the case of an impressed electric field, known as Stark beats. The electric field induces an electric dipole moment, which mixes states of opposite parity. States of differing parity usually have quite different lifetimes. Thus, because of the entanglements between dissimilar lifetimes, real temporal oscillations in the total emitted radiation occur.

For light sources that possess coherent, anisotropic excitation, quantum-beat spectroscopy can be used very effectively to measure fine and hyperfine structure splitting and magnetic g-factors. In cases where the energy splittings and g-factors are known, the method can be used to calibrate time scales or beam energies.

13.6 **Level crossing and optical double resonance spectroscopy**

If a light source of anisotropic atomic excitation is available (for example, through selective excitation by a laser beam with preferred directions of its propagation and optical polarization vectors), the emitted light pattern will also exhibit a preferential pattern in its angular distribution and polarization. It is possible to alter that pattern through the imposition of a second perturbation that places the atom in a state that is an admixture of the eigenstates of the primary excitation mechanism.

One example is the level crossing method, in which an external field is impressed that draws together two levels that differ in magnetic quantum number. For example, a magnetic field can slow a counterclockwise orbit and speed a clockwise orbit until their energies coincide, at which point they form a coherent admixture. This admixture alters the angular distribution and polarization of the radiation pattern, signaling the precise value of the impressed field at which the level crossing occurs.

A similar effect occurs in the optical double resonance method, in which a second source of excitation (for example a second laser) is used to induce transitions (and thereby mixing) between two levels. This also changes the radiation pattern, and permits the precise determination of the frequency of the second source of radiation at which the transitions occur.

Since the radiation pattern of E1 emission is characterized by the proverb "under the candle it is darkest," these methods correspond to looking under the candle, and observing a sudden flash of light for a specific value of the coherence-producing perturbation.

References

1. A. J. Ångström, 'Optiska undersökningar,' Kongliga Svenska Vetenskaps Akademiens Handlingar 1852, 229–323 (1852)
2. G. O. Abell, *Exploration of the Universe* (New York, Holt, Rinehart and Winston, 1975)
3. E. Arimondo, M. Inguscio and P. Violino, 'Hyperfine structure in the alkali atoms,' *Rev. Mod. Phys.* **49**, 31–75 (1977)
4. G. Astner, L. J. Curtis, L. Liljeby, S. Mannervik and I. Martinson, 'A high precision beam–foil meanlife measurement of the 1s3p $^1P_1$ level in He I,' *Z. Physik A* **279**, 1–6 (1976)
5. G. Astner, L. J. Curtis, L. Liljeby, S. Mannervik and I. Martinson, 'Measurements of the $n\,^3D_1$– $n\,^3D_2$ fine-structure separations in ^{4}He I by the beam–foil quantum-beat method,' *J. Phys. B* **9**, L345–8 (1976)
6. J. J. Balmer, 'Notiz über die Spectrallinien des Wasserstoffs' *Verhandlungen der Naturforschenden Gessellshaft in Basel* **7**, 548–60, 750–2 (1885)
7. L. Banet, 'Evolution of the Balmer series,' *Am. J. Phys.* **34**, 496–503 (1966)
8. D. R. Bates and A. Damgaard, 'The calculation of the absolute strengths of spectral lines,' *Phil Trans Roy. Soc. A* **242**, 101–22 (1949)
9. R. J. Bell and A. E. Kingston, 'The van der Waals interaction of two or three atoms,' *Proc. Phys. Soc.* **88**, 901–7 (1966)
10. H. J. Bernstein and A. V. Phillips, 'Fiber bundles and quantum theory,' *Scientific American* **245**, 123–37 (1981)
11. H. G. Berry, L. J. Curtis and J. L. Subtil, 'Cascade-induced alignment in changes in intensity-decay curves,' *J. Opt. Soc. Am.* **62**, 771–3 (1972)
12. H. G. Berry, L. J. Curtis, D. G. Ellis and R. M. Schectman, 'Anisotropy in the beam–foil light source,' *Phys. Rev. Lett.* **32**, 751–4 (1974)
13. I. B. Bersuker, 'Effect of core electrons on the transition of optical electrons,' *Opt. Spektrosk.* **3**, 97–103 (1957)
14. H. A. Bethe and E. E. Salpeter, *Quantum Mechanics of One- and Two-Electron Atoms* (New York, Plenum Publ. 1977)
15. K. Bockasten, 'A study of C IV: Term values, series formula, and Stark effect,' *Arkiv för Fysik* **10**, 567–84 (1956)
16. N. Bohr, 'On the constitution of atoms and molecules,' *Phil. Mag.* **26**, 1, 476, 857 (1913); 'On the quantum theory of radiation and the structure of the atom,' *ibid.* 394 (1915)
17. I. S. Bowen, Unpublished, cited by D. Huff and W. V. Houston in 'The appearance of "forbidden lines" in spectra,' *Phys. Rev.* **36**, 842–6 (1930)
18. G. Breit, 'Does the electron have an intrinsic magnetic moment?' *Phys. Rev.* **72**, 984 (1947)

19. L. Brillouin, 'Remarques sur la mécanique ondulatoire.' *Le Journal de Physique et le Radium* **7**, 353–68 (1926)

20. R. Bunsen and G. Kirchhoff, 'Chemical analysis by spectrum observations,' *Phil. Mag.* **20** 89–109 (1860)

21. A. Burgess and M. J. Seaton, 'A general formula for the calculation of atomic photoionization cross sections,' *Mon. Not. Roy. Astron. Soc.* **120**, 121–51 (1960) [tables preprinted in *Rev. Mod. Phys.* **30**, 992–3 (1958)]

22. J. Callaway, R. W. LaBahn, R. T. Pu and W. M. Duxler, 'Extended polarization potential: application to atomic scattering,' *Phys. Rev.* **168**, 12–21 (1968)

23. H. B. G. Casimir and D. Polder, 'The influence of retardation on London–van der Waals forces,' *Phys. Rev.* **73**, 360–72 (1948)

24. A. J. Cannon and E. C. Pickering, 'The Henry Draper Catalogue,' The Observatory, 1918–1924 (Annals of the Astronomical Observatory of Harvard College 91–99, Cambridge MA)

25. S. Chandrasekhar, 'On the continuous absorption coefficient of the negative hydrogen ion,' *Astrophys. J.* **102**, 223–31 (1945)

26. M. Chowdhury, 'Selection rules for processes involving photon–molecule interaction: a symmetry-conservation-based approach bypassing transition matrix elements,' *J. Chem. Education* **73**, 743–6 (1996)

27. G. M. Clemence, 'The relativity effect in planetary motions,' *Rev. Mod. Phys.* **19**, 361–4 (1947)

28. D. R. Cok and S. R. Lundeen, 'Atomic-beam measurements of helium F–G, G–H, and H–I intervals,' *Phys. Rev. A* **23**, 2488–95 (1981); **24**, 3283(E) (1981)

29. E. U. Condon and G. H. Shortley, *The Theory of Atomic Spectra* (Cambridge, Cambridge University Press, 1935)

30. R. D. Cowan, *The Theory of Atomic Structure and Spectra* (University California Press, Berkeley, 1981)

31. L. J. Curtis, 'A diagrammatic mnemonic for calculation of cascading level populations,' *Am. J. Phys.* **36**, 1123–5 (1968)

32. L. J. Curtis, 'Predictions of the early meanlife theory of Wilhelm Wien,' *J. Opt. Soc. Am.* **63**, 105–7 (1972)

33. L. J. Curtis, 'Convolution of a driven excitation into the decay curve of an arbitrarily cascaded and blended level,' *J. Opt. Soc. Am.* **64**, 495–7 (1974)

34. L. J. Curtis, 'Lifetime measurements,' in *Beam–Foil Spectroscopy*, S. Bashkin, ed. (Berlin, Springer, 1976) pp. 63–109

35. L. J. Curtis, 'On the αZ expansion of the Dirac energy of a one-electron atom,' *J. Phys. B* **10**, L641–5 (1977)

36. L. J. Curtis, 'Concept of the exponential law prior to 1900,' *Am. J. Phys.* **46**, 896–906 (1978)

37. L. J. Curtis, 'Use of hydrogenic transition probabilities for non-penetrating Rydberg states with core polarization,' *J. Phys. B* **12**, 509–13 (1978)

38. L. J. Curtis, 'Dipole polarisabilities for single valence electron ions,' *Phys. Scr.* **21**, 162–4 (1980)

39. L. J. Curtis, 'Cancellations in atomic dipole moments in the Cu isoelectronic sequence,' *J. Opt. Soc. Am.* **71**, 566–8 (1981)

40. L. J. Curtis, 'Theoretical estimates of the quadrupole polarizability and dynamical polarizability in single valence electron ions' *Phys. Rev. A* **23**, 362–5 (1981)

41. L. J. Curtis, 'A semiclassical formula for the term energy of a many-electron atom,' *J. Phys. B* **14**, 1373–86 (1981)

42. L. J. Curtis, 'Lifetime measurements in highly ionized atoms,' *Phys. Scr.* **T8**, 77–83 (1984)

43. L. J. Curtis, 'Explicit functional approximation for the Z dependence of self-energy radiative corrections in hydrogenlike ions,' *J. Phys. B* **18**, L651–6 (1985); erratum **19**, 1699 (1986)

44. L. J. Curtis, 'Semiempirical methods for systematization of high-precision atomic data,' *Comments At. Mol. Phys.* **16**, 1–19 (1985)

45. L. J. Curtis, 'Isoelectronic studies of the $5s^2\ {}^1S_0$–$5s5p\ {}^{1,3}P_J$ intervals in the Cd sequence,' *J. Opt. Soc. Am.* **3**, 177–82 (1986); erratum **5**, 2399 (1988)

46. L. J. Curtis, 'Bengt Edlén's Handbuch der Physik article – 26 years later,' *Phys. Scr.* **35**, 805–10 (1987).

47. L. J. Curtis, 'Semiempirical formulations of spectroscopic data and calculations,' *Nucl. Instr. Meth. in Phys. Res.* **B31**, 146–52 (1988)

48. L. J. Curtis, 'Semiclassical specification of singlet–triplet mixing angles, oscillator strengths, and g-factors in the nsn'p, nsn'p^5, np^2, and np^4 configurations,' *Phys. Rev. A* **40**, 6958–68 (1989)

49. L. J. Curtis, 'Semiempirical confrontations between theory and experiment in highly ionised complex atoms,' *Phys. Scr.* **39**, 447–57 (1989)

50. L. J. Curtis, 'Classical mnemonic approach for obtaining hydrogenic expectation values of r^p,' *Phys. Rev. A* **43**, 568–9 (1991)

51. L. J. Curtis, 'Isoelectronic smoothing of line strengths in intermediate coupling,' *Phys. Scr.* **43**, 137–43 (1991)

52. L. J. Curtis, 'Use of intermediate coupling relationships to test measured branching fraction data,' *J. Phys. B* **31**, L769–74 (1998)

53. L. J. Curtis, 'Intermediate coupling branching fractions for UV transitions in ions of the Si and Ge sequences,' *J. Phys. B* **33**, L259–63 (2000)

54. L. J. Curtis, 'Branching fractions and transition probabilities for Ga II, In II and Tl II from measured lifetime and energy level data,' *Phys. Scr.* **62**, 31–5 (2000)

55. L. J. Curtis, 'Branching fractions for $5s^25p^2$–$5s^25p6s$ supermultiplet in the Sn isoelectronic sequence,' *Phys. Scr.* **63**, 104–7 (2001)

56. L. J. Curtis and W. H. Smith, 'Radiative-lifetime and absolute-oscillator-strength studies for some resonance transition in Si I, II, and III,' *Phys. Rev. A* **9**, 1537–42 (1974)

57. L. J. Curtis and P. Erman, 'Distortion effects in the measurement of long optical lifetimes,' *J. Opt. Soc. Am.* **67**, 1218–30 (1977)

58. L. J. Curtis and D. G. Ellis, 'A formula for cancellation disappearances of atomic oscillator strengths,' *J. Phys. B* **11**, L543–6 (1978)

59. L. J. Curtis and P. S. Ramanujam, 'Semiclassical formulation of term energies and electrostatic intervals in He I,' *Phys. Rev. A* **25**, 3090–6 (1982)

60. L. J. Curtis and P. S. Ramanujam, 'Ground-state fine structures for the B and F isoelectronic sequences using the extended regular doublet law,' *Phys. Rev. A* **26**, 3672–5 (1982)

61. L. J. Curtis and P. S. Ramanujam, 'Isoelectronic predictions for ground-state ^{2}P fine structures using the extended regular doublet law,' *Phys. Scr.* **27**, 417–21 (1983)

62. L. J. Curtis and R. R. Silbar, 'Self-consistent core potentials for complex atoms: a semiclassical approach,' *J. Phys. B* **17**, 4087–101 (1984)

63. L. J. Curtis and D. G. Ellis, 'Predictive systematization of the $2s^2$–$2s2p$ resonance and intercombination transitions in the Be isoelectronic sequence,' *J. Phys. B* **29**, 645–54 (1996)

64. L. J. Curtis, R. M. Schectman, J. L. Kohl, D. A. Chojnacki, and D. R. Shoffstall, 'New cascade analysis techniques for determining spontaneous atomic transition probabilities,' *Nucl. Instr. Meth.* **90**, 207–16 (1970)

65. L. J. Curtis, J. Bromander and H. G. Berry, 'A meanlife measurement of the 3d ^{2}D resonance doublet in Si II by a technique which exactly accounts for cascading,' *Phys. Lett.* **34A**, 169–70 (1971)

66. L. J. Curtis, J. Reader, S. Goldsmith, B. Denne and E. Hinnov, '4s^{2}4p ^{2}P intervals in the Ga isoelectronic sequence from Rb^{6+} to In^{18+},' *Phys. Rev. A* **29**, 2248–50 (1984)

67. L. J. Curtis, R. R. Haar and M. Kummer, 'An expectation value formulation of the perturbed Kepler problem,' *Am. J. Phys.* **55**, 627–31 (1987)

68. L. J. Curtis, Z. B. Rudzikas and D. G. Ellis, 'Empirical determination of intermediate-coupling amplitudes and transition rates from spectroscopic data,' *Phys. Rev. A* **44**, 776–9 (1991)

69. L. J. Curtis, D. G. Ellis and I. Martinson, 'Data-based predictions of line strengths in alkali-metal-like isoelectronic sequences,' *Phys. Rev. A* **51**, 251–6 (1995)

70. L. J. Curtis, S. T. Maniak, R. W. Ghrist, R. E. Irving, D. G. Ellis, E. Träbert, J. Granzow, P. Bengtsson and L. Engström, 'Measurements and data-based predictions for $\Delta n = 1$ resonance and intercombination transitions in the Be and Ne sequences,' *Phys. Rev. A* **51**, 4575–82 (1995)

71. L. J. Curtis, D. G. Ellis, R. Matulioniene and T. Brage, 'Relativistic empirical specification of transition probabilities from measured lifetime and energy level data,' *Phys. Scr.* **56**, 240–4 (1997)

72. L. J. Curtis, R. T. Deck and D. G. Ellis, 'Limitations on the precision of atomic meanlife measurements,' *Phys. Lett. A* **230**, 330–5 (1997)

73. L. J. Curtis, R. Matulioniene, D. G. Ellis and C. F. Fischer, 'A predictive data-based exposition of 5s5p ^{1,3}P$_1$ lifetimes in the Cd isoelectronic sequence,' *Phys. Rev. A* **62**, 052513:1–7 (2000)

74. L. J. Curtis, R. E. Irving, M. Henderson, R. Matulioniene, C. F. Fischer and E. H. Pinnington, 'Measurements and predictions of the 6s6p ^{1,3}P$_1$ lifetimes in the Hg isoelectronic sequence,' *Phys. Rev. A* **63**, 042502:1–7 (2001)

75. A. Dalgarno and J. T. Lewis, 'The representation of long range forces by series expansions,' *Proc. Roy. Soc. (London)* **A233**, 57–64 (1955)

76. A. Dalgarno and J. T. Lewis, 'The exact calculation of long-range forces between atoms by perturbation theory,' *Proc. Roy. Soc. (London)* **A233**, 70–4 (1955)

77. A. Dalgarno and A. E. Kingston, 'The refractive indices and Verdet constants of the inert gases,' *Proc. Roy. Soc. (London)* **A259**, 424–9 (1960)

78. C. G. Darwin, 'The wave equations of the electron,' *Proc. Roy. Soc. (London)* **A118**, 654–80 (1928)

79. J. T. Davies and J. M. Vaughan, 'A new tabulation of the Voigt profile,' *Astrophys. J.* **137**, 1302–5 (1963)

80. R. T. Deck and J. D. Walker, 'The connection between spin and statistics,' *Phys. Scr.* **63**, 7–14 (2001)

81. J. B. Delos, S. K. Knudson and D. W. Noid, 'Highly excited states of a hydrogen atom in a strong magnetic field,' *Phys. Rev. A* **28**, 7–21 (1983)

82. P. A. M. Dirac, 'The evolution of the physicist's picture of nature,' *Scientific American* **208**, 45–54 (May 1963)

83. C. Eckart, 'The application of group theory to the quantum dynamics of monatomic atoms,' *Rev. Mod. Phys.* **2**, 305–80 (1930)

84. B. Edlén, 'Wellenlängen und Termsysteme zu den Atomspektren der Elemente Lithium, Beryllium, Bor, Kohlenstoff, Stickstoff und Sauerstoff,' *Nova Acta Regiae Societatis Scientiarum Upsaliensis, ser. 4* (1934), **9**, no. 6, 1–153

85. B. Edlén, 'S I-ähnliche Spektren der Elemente Titan bis Eisen, Ti VII, V VIII, Cr IX, Mn X und Fe XI,' *Z. Physik* **104**, 188–93 (1936); 'Zur Kenntnis der Cl I-äntlichen Spektrum Cl I, A II, K III, Ca IV, Ti VI, V VII, Cr VIII, Mn IX, Fe X und Co XI,' *Z. Physik* **104**, 407–16 (1937)

86. B. Edlén, 'An attempt to identify the emission lines in the spectrum of the solar corona,' *Arkiv för Matematik, Astronomi och Fysik* **28B**, 1–4 (*1941*); 'Die Deutung der Emissionslinien im Spektrum der Sonnenkorona,' *Z. Astrophys.* **22**, 30–64 (1942)

87. B. Edlén, 'Atomic Spectra,' *Handbuch der Physik* **27** (Berlin, Springer, 1964) pp. 80–220

88. B. Edlén, 'The transitions 3s–3p and 3p–3d, and the ionization energy of the Na I iso-electronic sequence,' *Phys. Scr.* **17**, 565–74 (1976)

89. B. Edlén, personal communication to the author (1976).

90. A. Einstein, 'Zur Elektrodynamik bewegter Körper,' *Ann. d. Physik* **17**, 891 (1905)

91. A. Einstein, 'Zum Quantensatz von Sommerfeld und Epstein,' *Verhand. Deut. Phys. Ges.* **19**, 82 (1917) [English translation by C. Jaffé, Joint Institute for Laboratory Astrophysics (JILA) Report 116, University Colorado, 1980]

92. A. Einstein, Zur Quantentheorie der Strahlung,' *Physikalische Zeitshcrift* **18** 121–8 (1917) [English translation by H. A. Boorse and L. Motz, *The World of the Atom* (Basic Books, NY, 1966) pp. 890–901]

93. A. M. Ellis, 'Spectroscopic selection rules: the role of photon states,' *J. Chem. Education* **76**, 1291–4 (1999)

94. D. G. Ellis, 'Density-operator description of foil-excited atomic beams: zero-field quantum beats in $L–S$ coupling without cascades,' *J. Opt. Soc. Am.* **63**, 1232–5 (1973)

95. A. Ernst and J.-P. Hsu, 'First proposal of the universal speed of light by Voigt in 1887,' *Chinese J. Phys.* **39**, 211–30 (2001); see also 'The dawn of Lorentz and Poincaré Invariance (1887–1905) – First proposal of the universal speed of light by Voigt in 1887,' in Chapter 1 of *Lorentz and Poincaré Invariance – 100 years of Relativity*, J.-P. Hsu and Y.-A. Zhang, eds. (River Edge NJ, World Scientific 2001) pp. 3–24

96. E. Fermi, *Nuclear Physics* (Chicago IL, University of Chicago Press, 1950)

97. R. P. Feynman, *The Feynman Lectures on Physics*, (Reading MA, Addison-Wesley, 1963) pp. 1–2

98. L. L. Foldy and S. A. Wouthuysen, 'On the Dirac theory of spin $\frac{1}{2}$ particles and its non-relativistic limit,' *Phys. Rev.* **78**, 29–36 (1950)

99. J. von Fraunhofer, 'Bestimmung des Brechungs- und Farbenzerstreuungs-Vermögens verschiedener Glasarten, in Bezug auf die Vervollkommung achromatischer Fernröhre,' Denkschriften der Königlichen Akademie der Wissenschaften zu München **5**, 193–226 (1814)

100. H. Friedrich and J. Trost, 'Phase loss in WKB waves due to reflection by a potential,' *Phys. Rev. Lett.* **76**, 4869–73 (1996)

101. T. F. Gallagher, *Rydberg Atoms*, (Cambridge, Cambridge University Press, 1994)

102. J. D. Garcia, 'Quantum solutions and classical limits for strong Coulomb fields,' *Phys. Rev. A* **34**, 4396–8 (1986)

103. R. Garstang, 'Hyperfine structure and intercombination line intensities in the spectra of magnesium, zinc, cadmium, and mercury,' *J. Opt. Soc. Am.* **52**, 845–51 (1962)

104. W. R. S. Garton and F. S. Tomkins, 'Diamagnetic Zeeman effect and magnetic configuration mixing in long spectral series of Ba I,' *Astrophys. J.* **158**, 839–45 (1969)

105. C. G. Gillispie, *Pierre-Simon Laplace 1749–1827: a Life in the Exact Sciences* (Princeton, Princeton University Press, 1997)

106. S. L. Glashow and A. H. Rosenfeld, 'Eightfold-way assignments for Y_1^* (1660) and other baryons,' *Phys. Rev. Lett.* **10**, 192–6 (1963) [footnote 3]

107. H. Goldstein, *Classical Mechanics* (Cambridge MA, Addison-Wesley, 1980)

108. I. S. Gradsteyn and I. M. Ryzhic, *Table of Integrals, Series, and Products* (New York, Academic Press, 1965), formulae 3.661-3,4.

109. L. C. Green, P. R. Rush and C. D. Chandler, 'Oscillator strengths and matrix elements for the electric dipole moment for hydrogen,' *Astrophys. J. Suppl.* **3**, 37–50 (1957)

110. G. Greenstein, *Portraits of Discovery: Profiles in Scientific Genius* (New York, John Wiley, 1998).

111. W. Grotrian, 'Zur Frage der Deutung der Linien im Spektrum der Sonnenkorona,' *Naturwiss.* **27**, 214 (1939)

112. J. R. Grover, 'Shell-model calculations of the lowest-energy nuclear excited states of very high angular momentum,' *Phys. Rev.* **157**, 832 (1967)

113. R. R. Haar and L. J. Curtis, 'The Thomas precession gives $g_e - 1$, not $g_e/2$,' *Am. J. Phys.* **55**, 1044–5 (1987)

114. S. Hameed, A. Herzenberg and M. G. James, 'Core polarization corrections to oscillator strengths in the alkali atoms,' *J. Phys. B* **1**, 822–30 (1968)

115. W. Hanle, 'Messung des Polarisationsgrades von Spektralinien,' *Z. Instrumentenkunde* **51**, 488–90 (1931)

116. J. E. Hardis, L. J. Curtis, P. S. Ramanujam, A. E. Livingston, and R. L. Brooks, 'Measurement of the transition probability of the $2s^2$ $^1S_0$–$2s3p$ $^3P_1^o$ intercombination line in Ne VII.' *Phys. Rev. A* **27**, 257–61 (1983)

117. D. R. Hartree, 'The wave-mechanics of an atom with a non-Coulomb central field. Part I. Theory and methods,' *Proc. Cambridge Phil. Soc.* **24**, 89–132 (1928)

118. S. Hawking, *A Brief History of Time*, (New York, Bantam Books, 1988)

119. M. Henderson, L. J. Curtis, R. Matulioniene, D. G. Ellis, and C. E. Theodosiou, 'Lifetime measurements in Tl III and the determination of the ground-state dipole polarizabilities for Au I – Bi V, *Phys. Rev. A* **56**, 1872–8 (1997)

120. J. Herschel, 'Quetelet on Probabilities,' *Edinburgh Reviews* **92**, 1–57 (1850)

121. K.-N. Huang, 'Energy-level scheme and transition probabilities in P-like ions,' *At. Data Nucl. Data Tables* **30**, 313–421 (1984)

122. K.-N. Huang, 'Energy-level scheme and transition probabilities in Si-like ions,' *At. Data Nucl. Data Tables* **32**, 503–645 (1985)

123. S. Huldt, L. J. Curtis, B. Denne, L. Engström, K. Ishii, and I. Martinson, 'Observation of strong orientation effects for levels in highly ionized oxygen,' *Phys. Lett.* **66A**, 103–5 (1978)

124. F. Hund, *Linienspektren und Periodisches System der Elemente* (Berlin, Julius Springer, 1927) p. 124

125. J. D. Jackson, *Classical Electrodynamics* (New York, John Wiley, 1998)

126. J. H. Jeans, *The Mysterious Universe, by Sir James Jeans* (Cambridge, England, The University Press, 1930)

127. J. E. Jones (later J. E. Lennard-Jones), 'On the determination of molecular fields – II. From the equation of state of a gas,' *Proc. Roy. Soc. (London)* **A106**, 463–77 (1924)

128. W. R. Johnson, D. Kolb and K.-N. Huang, 'Electric-dipole, quadrupole and magnetic dipole susceptibilities and shielding factors for closed-shell ions of the He, Ne, Ar, Ni(Cu^+), Kr, Pb, and Xe isoelectronic sequences,' *At. Data Nucl. Data Tables* **28**, 333–40 (1983)

129. J. B. Keller, 'Corrected Bohr–Sommerfeld quantum conditions for nonseparable systems,' *Annals of Phys.* **4**, 180–8 (1958)

130. E. J. Kelsey and L. Spruch, 'Retardation effects on high Rydberg states: a retarded R^{-5} polarization potential,' *Phys. Rev. A* **18**, 15–28 (1978); 'Retardation effects and the vanishing as $R \sim \infty$ of the nonadiabatic R^{-6} interaction of the core and a high Rydberg electron,' *loc. cit.* 1055–6; 'Vacuum fluctuation and retardation effects on long-range potentials,' *loc. cit.* 845–52

131. W. H. King, *Isotope Shifts in Atomic Spectra* (New York, Plenum Press, 1984)

132. G. W. King and J. H. VanVleck, 'Relative intensities of singlet-singlet and singlet-triplet transitions,' *Phys. Rev.* **56**, 464–5 (1939)

133. A. I. Kitaigorodskii, Lecture, Amsterdam, (1975)

134. C. J. Kleinman, Y. Hahn and L. Spruch, 'Dominant nonadiabatic contribution to the long-range electron–atom interaction,' *Phys. Rev.* **165**, 53–62 (1968)

135. T. C. Koopmans, 'Über die Zuordaung von Wellenfunktionen und Eigenwerten zu den einzelnen Elektronen eines Atoms,' *Physica* **1**, 104–13 (1934)

136. H. Kopfermann, *Nuclear Moments* (New York, Academic Press, 1953)

137. W. Kuhn, 'On the total strength of the absorption lines starting from the same state,' *Z. Physik* **33**, 408 (1925)

138. A. Landé, 'Über den anomalen Zeemaneffekt (Teil I),' *Z. Physik* **5**, 231–41 (1921)

139. A. Landé, 'Termstruktur und Zeemaneffekt der Multipletts,' *Z. Physik* **15**, 189–205 (1923); 'Termstruktur und Zeemaneffekt der Multipletts, Zweite Mitteilung,' *Z. Physik* **19**, 112–23 (1923)

140. R. E. Langer, 'On the connection formulas and the solutions of the wave equations,' *Phys. Rev.* **51**, 669–76 (1937)

141. O. Laporte, 'Die Struktur des Eisenspektrums. Teil 1, *Z. Physik* **23**, 135–75 (1924)

142. J. Larmor, 'On the theory of the magnetic influence on spectra; and the radiation from moving ions,' *Phil. Mag.* **44**, 512 (1897)

143. Z.-S. Li, J. Norin, A. Persson, C.-G. Wahlström. S. Svanberg, P. S. Doidge and E. Biémont, 'Radiative properties of neutral germanium obtained from excited-state lifetime and branching-ratio measurements and comparison with theoretical calculations,' *Phys. Rev. A* **60**, 198 (1999)

144. F. London, 'Zur Theorie und Systematik der Molekularkräfte,' *Z. Physik* **63**, 245–79 (1930)

145. H. A. Lorentz, 'La théorie électromagétique de Maxwell et son application aux corps mouvants,' *Archives Néelandaises des Sciences Exactes* **25**, 363–552 (1892)

146. G. Ludwig, 'Einfluss der Polarisation des inneren Elektrons im Felde des äusseren auf die Terme des Spektrums eines Zwei-Elektronensystems (insbesondere He),' *Helv. Phys. Acta* **7**, 273–84 (1934)

147. S. T. Maniak and L. J. Curtis, 'A comment on labeling conventions in isoelectronic sequences,' *Phys. Rev. A* **42**, 1821–3 (1990)

148. W. C. Martin and W. L. Wiese, 'Atomic Spectroscopy' *Atomic, Molecular, & Optical Physics Handbook*, G. W. F. Drake, ed. (Woodbury NY, AIP Press, 1996) pp. 135–153

149. V. P. Maslov, 'Théorie des perturbations et methods asymptotiques,' (Denod, Gauthier-Villars, Paris, 1972)

150. R. D. Mattuck, *A Guide to Feynman Diagrams in the Many-Body Problem*, Second Edition (New York, McGraw-Hill, 1976) p. 1

151. W. McGucken, *Nineteenth-Century Spectroscopy: Development of the Understanding of Spectra, 1802–1897*, (Baltimore, Johns Hopkins University Press, 1969)

152. J. D. McGervey, 'Reduced mass and the orbit radius,' *Am. J. Phys.* **53**, 909 (1985)

153. W. N. Mei, 'Comment on phase space integration method for bound states' *Am. J. Phys.* **66**, 541–2 (1998)

154. T. Melvill, 'Observations on light and colours,' 1752 [The paper was read before the Edinburgh Philosophical Society on 3 January and 7 February in 1752. Melvill died in December 1753 and the paper was published posthumously in *Physical and Literary Essays*, vol. 2 (Edinburgh, G. Hamilton and J. Balfour, 1756) pp. 12–90. The paper has been reprinted in *J. Roy. Astron. Soc. Canada* **8**, 231 (1914)]

155. E. Merzbacher, *Quantum Mechanics* (New York, Wiley, 1961) p. 474

156. A. A. Michelson, 'Recent advances in spectroscopy' in *Nobel Lectures in Physics (1901–1921)* (Singapore, World Scientific Publ., 1998) pp. 171–2

157. S. Mrozowski, 'Über die Hyperfeinstruktur der verbotenen Quecksilberline 2655,8 Å (6 $^3P_0$– 6 $^1P_0$),' *Z. Physik* **108** 204–11 (1938)

158. C. J. Overbeck, cited in *Am. J. Phys.* **46**, 323 (1976)

159. W. K. H. Panofsky and M. Phillips, *Classical Electricity and Magnetism* (Reading MA, Addison-Wesley, 1955) pp. 334–5

160. S. Pasternack, 'On the mean value of r^s for Keplerian systems,' *Proc. Natl. Acad. Sci.* **23**, 91–4 (1939)

161. S. Pasternack, 'Transition probabilities of forbidden lines,' *Astrophys. J.* **97**, 129 (1940)

162. W. Pauli, Jr., 'Über den Zusammenhang des Abschlusses Elektronen-gruppen in Atom mit der Komplexstruktur der Felder,' *Z. Physik* **31**, 765 (1925)

163. W. Pauli, Jr., 'Zur Quantenmechanik des magnetischen Elektrons,' *Z. Physik* **43**, 601 (1927)

164. C. H. Payne, 'Stellar atmospheres,' *The Observatory*, Harvard College Observatory, Monograph no. 1 (Ph.D. Thesis, 1925)

165. E. Poincaré, *Science et Méthode* (Paris, Flammarion, 1908) [*Science and Method* (New York, Dover, 1952) translation]

166. L. A. J. Quetelet, *Sur l'Homme et le Développment de ses Facultés, Essai d'une Physique Social* (Paris, Bachelier, 1835)

167. G. Racah, 'Theory of atomic spectra,' *Phys. Rev.* **62**, 438 (1942); **63**, 367 (1943); **76**, 1352 (1949)

168. J. Reader, V. Kaufman, J. Sugar, J. O. Ekberg, U. Feldman, C. M. Brown, J. F. Seely and W. L. Rowan, '3s–3p, 3p–3d, and 3d–3f transitions of sodiumlike ions,' *J. Opt. Soc. Am.* B **4**, 1821–8 (1987)

169. F. Reiche and W. Thomas, 'On the number of dispersion electrons which are related to a stationary state,' *Z. Physik* **34**, 510 (1925)

170. W. Ritz, 'On a new law of series spectra,' *Astrophys. J.* **28**, 237–43 (1908)

171. H. N. Russell and F. A. Saunders, 'New regularities in the spectra of the alkaline earths,' *Astrophys. J.* **61**, 38 (1925)

172. H. N. Russell, A. G. Shenstone and L. A. Turner, 'Report on notation for atomic spectra,' *Phys. Rev.* **33**, 900–6 (1929)

173. J. R. Rydberg, 'Recherches sur la constitution des spectres d'émission des élements chimiques,' *Kongliga Svenska Vetenskaps-Akademiens Handlinger* **23**, 6 (1888)

174. M. N. Saha, 'Ionisation in the solar chromosphere,' *Phil. Mag.* **40**, 472 (1920); 'On a physical theory of stellar spectra,' *Proc. Roy. Soc. (London)* **A99**, 135 (1921)

175. E. H. Saloman and Y.-K. Kim, 'Energy levels and transition probabilities in the ground-state configuration of sulfur-like ions,' *At. Data Nucl. Data Tables* **41**, 339 (1988)

176. G. H. Shortley, 'The computation of quadrupole and magnetic-dipole transition probabilities,' *Phys. Rev.* **57**, 225–34 (1940)

177. J. C. Slater, 'The theory of complex spectra,' *Phys. Rev.* **34**, 1293–332 (1929)

178. I. I. Sobelman, *Atomic Structure and Radiative Transitions*, (Berlin, Springer–Verlag, 1979)

179. G. G. Stokes, 'On the composition and resolution of streams of polarized light from different sources,' *Trans. Cambridge Phil. Soc.* **9**, 399–416 (1852)

180. J. Sugar and A. Musgrove, 'Energy levels of krypton, Kr I through Kr XXXVI,' *J. Chem. Phys. Ref. Data* **20**, 859–915 (1991)

181. C. E. Theodosiou, L. J. Curtis and C. A. Nicolaides, 'Determination of dipole polarizabilities for Mg^+ and Ca^+ ions from precision lifetime measurements and transition-moment cancellations,' *Phys. Rev. A* **52**, 3677–80 (1995)

182. L. W. Thomas, 'The motion of the spinning electron,' *Nature* **117**, 514 (1926); 'The kinematics of an electron with an axis,' *Phil. Mag.* **3**, 1–22 (1927)

183. L. Thomas, *The Lives of a Cell* (New York, Viking Press, 1974) p. 12

184. P. A. Tipler, *Foundations of Modern Physics* (New York, Worth Publishers, 1969)

185. G. E. Uhlenbeck and S. A. Goudsmit, 'Ersetzung der Hypothese vom unmechanischen Zwang durch eine Forderung bezüglich des inneren Verhaltens jedes einzelnen Elektrons,' *Naturwiss.* **13**, 953–4 (1925)

186. J. D. van der Waals, Ph.D. Thesis (Leiden, 1873) [*Physical Memoirs* **1**, 333 (1890)]

187. W. Voigt, 'Über das Doppler'sche Princip,' *Nachrichten Ges. Göttingen*, 10. Maerz, No. 2 (1887) pp. 41–51

188. W. Voigt, 'Über das Gesetz der Intensitätsverteilung innerhalb der Linien eines Gasspektrums,' *Münch. Ber.* **1912**, 603 (1912)

189. M. J. Walker, 'Matrix calculus and the Stokes parameters of polarized radiation,' *Am. J. Phys.* **22**, 170–4 (1954) and references therein.

190. I. Waller, 'Der Starkeffekt zweiter Ordnung bei Wasserstoff und die Rydbergkorrektion der Spektra von He und Li^+,' *Z. Physik* **38**, 635 (1926)

191. V. Weisskopf and E. Wigner, 'Berechnung der naturlichen Linienbreite auf Grund der Diracschen Lichttheorie,' *Z. Physik* **63**, 54 (1930)

192. H. E. White, *Introduction to Atomic Spectra* (New York, McGraw-Hill, 1934) p. 16

193. E. P. Wigner, 'Einige Folgerungen aus der Schrödingerschen Theorie für die Termstrukturen,' *Z. Physik* **43**, 624–52 (1927)

194. H. C. Wolfe, 'Multiplet splitting and intensities of intercombination lines: part I,' *Phys. Rev.* **41**, 443–58 (1932)

195. W. H. Wollaston, 'A method of examining refractive and dispersive powers by prismatic reflection,' *Phil. Trans. Roy. Soc.* **92**, 365–80 (1802)

196. D. R. Wood and K. L. Andrew, 'Arc spectrum of lead,' *J. Opt. Soc. Am.* **58**, 818–29 (1968)

197. D. R. Wood, K. L. Andrew, A. Giacchetti and R. D. Cowan, 'Zeeman effect and intensity anomalies in Pb I,' *J. Opt. Soc. Am.* **58**, 830–6 (1968)

198. G. K. Woodgate, *Elementary Atomic Structure* (London, McGraw-Hill, 1970)

199. W. Yei, A. Sieradzan and M. D. Havey, 'Delayed-detection of atomic Na 3p $^2P_{3/2}$ hyperfine structure using polarization quantum-beat spectroscopy,' *Phys. Rev. A* **48**, 1909–15 (1993)

200. P. Zeeman, 'On the influence of magnetism on the nature of light emitted by a substance,' *Phil. Mag.* **43**, 226–39 (1897)

201. M. L. Zimmerman, J. C. Castro and D. Kleppner, 'Diamagnetic structure of Na Rydberg states,' *Phys. Rev. Lett.* **40**, 1083–6 (1978)

Index

Italics are used to denote figures and tables.